POWER GENERATION CHOICES:
COSTS, RISKS AND EXTERNALITIES

Proceedings of an International Symposium
Washington, (USA), 23-24 September 1993

Organised by
the OECD Nuclear Energy Agency
and the Oak Ridge National Laboratory

In co-operation with
American Nuclear Society
Canadian Nuclear Association
Central Research Institute of Electric Power Industry
European Nuclear Society
International Atomic Energy Agency

PUBLISHER'S NOTE

The following texts have been left in their original form to permit faster distribution at lower cost.

NUCLEAR ENERGY AGENCY
ORGANISATION FOR ECONOMIC CO-OPERATION AND DEVELOPMENT

ORGANISATION FOR ECONOMIC CO-OPERATION AND DEVELOPMENT

Pursuant to Article 1 of the Convention signed in Paris on 14th December 1960, and which came into force on 30th September 1961, the Organisation for Economic Co-operation and Development (OECD) shall promote policies designed:

— to achieve the highest sustainable economic growth and employment and a rising standard of living in Member countries, while maintaining financial stability, and thus to contribute to the development of the world economy;

— to contribute to sound economic expansion in Member as well as non-member countries in the process of economic development; and

— to contribute to the expansion of world trade on a multilateral, non-discriminatory basis in accordance with international obligations.

The original Member countries of the OECD are Austria, Belgium, Canada, Denmark, France, Germany, Greece, Iceland, Ireland, Italy, Luxembourg, the Netherlands, Norway, Portugal, Spain, Sweden, Switzerland, Turkey, the United Kingdom and the United States. The following countries became Members subsequently through accession at the dates indicated hereafter: Japan (28th April 1964), Finland (28th January 1969), Australia (7th June 1971), New Zealand (29th May 1973) and Mexico (18th May 1994). The Commission of the European Communities takes part in the work of the OECD (Article 13 of the OECD Convention).

NUCLEAR ENERGY AGENCY

The OECD Nuclear Energy Agency (NEA) was established on 1st February 1958 under the name of the OEEC European Nuclear Energy Agency. It received its present designation on 20th April 1972, when Japan became its first non-European full Member. NEA membership today consists of all European Member countries of OECD as well as Australia, Canada, Japan, Republic of Korea, Mexico and United States. The Commission of the European Communities takes part in the work of the Agency.

The primary objective of NEA is to promote co-operation among the governments of its participating countries in furthering the development of nuclear power as a safe, environmentally acceptable and economic energy source.

This is achieved by:

— *encouraging harmonization of national regulatory policies and practices, with particular reference to the safety of nuclear installations, protection of man against ionising radiation and preservation of the environment, radioactive waste management, and nuclear third party liability and insurance;*

— *assessing the contribution of nuclear power to the overall energy supply by keeping under review the technical and economic aspects of nuclear power growth and forecasting demand and supply for the different phases of the nuclear fuel cycle;*

— *developing exchanges of scientific and technical information particularly through participation in common services;*

— *setting up international research and development programmes and joint undertakings.*

In these and related tasks, NEA works in close collaboration with the International Atomic Energy Agency in Vienna, with which it has concluded a Co-operation Agreement, as well as with other international organisations in the nuclear field.

ORGANISATION DE COOPÉRATION ET DE DÉVELOPPEMENT ÉCONOMIQUES

En vertu de l'article 1ᵉʳ de la Convention signée le 14 décembre 1960, à Paris, et entrée en vigueur le 30 septembre 1961, l'Organisation de Coopération et de Développement Économiques (OCDE) a pour objectif de promouvoir des politiques visant :

— à réaliser la plus forte expansion de l'économie et de l'emploi et une progression du niveau de vie dans les pays Membres, tout en maintenant la stabilité financière, et à contribuer ainsi au développement de l'économie mondiale ;

— à contribuer à une saine expansion économique dans les pays Membres, ainsi que les pays non membres, en voie de développement économique ;

— à contribuer à l'expansion du commerce mondial sur une base multilatérale et non discriminatoire conformément aux obligations internationales.

Les pays Membres originaires de l'OCDE sont : l'Allemagne, l'Autriche, la Belgique, le Canada, le Danemark, l'Espagne, les États-Unis, la France, la Grèce, l'Irlande, l'Islande, l'Italie, le Luxembourg, la Norvège, les Pays-Bas, le Portugal, le Royaume-Uni, la Suède, la Suisse et la Turquie. Les pays suivants sont ultérieurement devenus Membres par adhésion aux dates indiquées ci-après : le Japon (28 avril 1964), la Finlande (28 janvier 1969), l'Australie (7 juin 1971), la Nouvelle-Zélande (29 mai 1973) et le Mexique (18 mai 1994). La Commission des Communautés européennes participe aux travaux de l'OCDE (article 13 de la Convention de l'OCDE).

L'AGENCE DE L'OCDE POUR L'ÉNERGIE NUCLÉAIRE

L'Agence de l'OCDE pour l'Énergie Nucléaire (AEN) a été créée le 1ᵉʳ février 1958 sous le nom d'Agence Européenne pour l'Énergie Nucléaire de l'OECE. Elle a pris sa dénomination actuelle le 20 avril 1972, lorsque le Japon est devenu son premier pays Membre de plein exercice non européen. L'Agence groupe aujourd'hui tous les pays Membres européens de l'OCDE, ainsi que l'Australie, le Canada, la République de Corée, les États-Unis, le Japon, et le Mexique. La Commission des Communautés européennes participe à ses travaux.

L'AEN a pour principal objectif de promouvoir la coopération entre les gouvernements de ses pays participants pour le développement de l'énergie nucléaire en tant que source d'énergie sûre, acceptable du point de vue de l'environnement, et économique.

Pour atteindre cet objectif, l'AEN :

— *encourage l'harmonisation des politiques et pratiques réglementaires notamment en ce qui concerne la sûreté des installations nucléaires, la protection de l'homme contre les rayonnements ionisants et la préservation de l'environnement, la gestion des déchets radioactifs, ainsi que la responsabilité civile et l'assurance en matière nucléaire ;*

— *évalue la contribution de l'électronucléaire aux approvisionnements en énergie, en examinant régulièrement les aspects économiques et techniques de la croissance de l'énergie nucléaire et en établissant des prévisions concernant l'offre et la demande de services pour les différentes phases du cycle du combustible nucléaire ;*

— *développe les échanges d'information scientifiques et techniques notamment par l'intermédiaire de services communs ;*

— *met sur pied des programmes internationaux de recherche et développement, et des entreprises communes.*

Pour ces activités, ainsi que pour d'autres travaux connexes, l'AEN collabore étroitement avec l'Agence Internationale de l'Énergie Atomique de Vienne, avec laquelle elle a conclu un Accord de coopération, ainsi qu'avec d'autres organisations internationales opérant dans le domaine nucléaire.

Foreword

Historically, choices between different means of producing electricity have rested largely on attempting to minimise the costs of electricity production. In recent years additional factors including so-called externalities, notably environmental and safety issues as well as macroeconomic and strategic considerations, have also become part of the decision-making process regarding technology choices.

The papers presented at the symposium examined the situation of nuclear power in different countries, the state of the art in economic analysis of nuclear power and other fuels and energy forms used in electricity generation from a total cost perspective within a pragmatic decision-making context. They also covered environmental, health, trade, security of supply, risks related to public acceptance, etc.) as well as the criteria (methods, assumptions, etc.) which can be used in economic comparisons of electricity generation.

The proceedings of this symposium provide a comprehensive examination of the total costs of nuclear power and other generating technologies. They include the papers presented as well as a few additional papers included in a poster session. The opinions, conclusions and recommendations are those of the authors only, and do not necessarily correspond to the views of any national authority or international organisation. The report is published under the responsibility of the Secretary-General.

Avant-propos

Traditionnellement, les modes de production d'électricité étaient choisis essentiellement dans le souci de réduire au minimum les coûts de la production d'électricité. Depuis quelques années, d'autres facteurs notamment externes (dits externalités) liés à l'environnement ou à la sûreté, ou encore à des considérations macro-économiques et stratégiques, ont également pesé dans les décisions concernant les choix technologiques.

Les communications présentées à l'occasion du symposium passent en revue la situation de l'électronucléaire dans différents pays et font le point de l'analyse économique de la production d'électricité à partir de l'énergie nucléaire et des autres combustibles et formes d'énergie, en considérant leurs coûts globaux dans la perspective d'une prise de décision pragmatique. Elles traitent en outre certains aspects comme l'environnement, la santé, les échanges, la sécurité des approvisionnements, les risques (qui conditionnent l'adhésion du public) et certains paramètres (liés aux méthodes, hypothèses, etc.) susceptibles d'être utilisés dans les comparaisons économiques concernant la production d'électricité.

On trouvera dans le compte rendu de ce symposium une étude très complète de l'ensemble des coûts de l'énergie nucléaire et des autres technologies de production. Outre les communications, quelques documents supplémentaires soumis par ailleurs sont inclus. Les opinions, conclusions et recommandations n'engagent que leurs auteurs et ne reflètent pas nécessairement celles de gouvernements nationaux ou d'organisations internationales. Le rapport est publié sous la responsabilité de Secrétaire général.

TABLE OF CONTENTS
TABLE DES MATIÈRES

Opening Remarks
K. Uematsu, Director General, OECD/NEA (France) 11

Session 1/Séance 1
Chairman/Président: J.E. Gray

Global Electricity Demand and Supply Trends
J.S. Foster, WEC (Canada) 17

Current and Future Costs of Power Generation Technologies
S. Yoda, K. Nagano, K. Yamaji, CRIEPI (Japon) 31

How Costs, Risks, and Externalities Affect Decisions
Regarding Technology Choice
S. Tierney, USDOE (oral presentation, paper not available)

Social Costing Research: Status and Prospects
R.W. Fri, Resources for the Future, (United States) 55

Discussion 63

Remarks: Global Energy Outlook and Externalities
J.E. Gray, Past Chairman, U.S. Energy Association 71
(After dinner presentation)

Session 2/Séance 2

Chairman/Président: J. Yasinsky

Power Generation Choices: A Canadian Perspective
E.A. Marriage, Ontario Hydro
L. Masson, Hydro-Quebec
G.E. Gunter, New Brunswick Power 77

**How Generation Choices are influenced by Costs, Risks
and Externalities: The Generation Planning Process
in Ontario, Canada**
E.A. Marriage, M.S. Rodgers, Ontario Hydro 95
(Additional paper, not presented orally)

**Power Generation Choices: An International Perspective
on Costs, Risks and Externalities**
R. Carle, G. Moynet, Electricité de France 109

**Generation Choices as Influenced by Costs, Risks
and Externalities - Germany**
L. Strauss, Bayernwerk Aktiengesellschaft 125

**Generation Choices as Influenced by Costs, Risks
and Externalities - Sweden**
C.-E. Nyquist, Vattenfall 139

**The Economics of Nuclear Power and Competing
Technologies in the U.K.**
B.L. Eyre, P.M.S. Jones, AEA Technology 159

Discussion 183

Session 3/Séance 3
Chairman/Président: R. Carle

Meeting/Managing the Demand for Electricity
E. Linn Draper, Jr., American Electric Power Co., Inc. 193

Sustainable Development and Choices of
Electricity Generating Technologies
Y. Akiyama, The Kansai Electric Power Co., Inc.
(Japan) 205

Electric Power for a Sustainable Development -
The Italian Scenario
S. Barabaschi, ANSALDO 231

Low Environmental Impact: Swiss Power Generation
K. Küffer, NOK 251

Discussion 273

Session 4/Séance 4
Chairman/Président: R. Shelton

Outlook for Costs by Energy Source
L.J. Williams, J. Fortune, G. Booras,
EPRI (United States) 287

Outlook for Risks by Energy Sources
J. Grawe, VDEW, & University of Stuttgart (Germany) 305

The Evaluation of External Costs from
Energy Sources - the EC-US Fuel Cycle Study
A. Krupnick, Resources for the Future
A. Markandya, Metroeconomica & Harvard Institute for
    International Development
R. Lee, Oak Ridge National Laboratory
P. Valette, Commission of the European Communities 331

**Perspective on Energy Security and Other Non-Environmental
Externalities in Electricity Generation**
D.R. Bohi, Resources for the Future (United States) 351

**Prospects for Internalisation of Externalities -
Where Do We Stand - What is Ahead ?**
P.M.S. Jones (United Kingdom) 369

Discussion 389

*
*   *

Closing Remarks
P. Girouard, OECD/NEA (France) 395

LIST OF PARTICIPANTS/LISTE DES PARTICIPANTS 401

OPENING REMARKS

by

Dr. Kunihiko UEMATSU
Director-General
OECD Nuclear Energy Agency
Paris, France

Ladies and Gentlemen,

It is with great pleasure that I open this International Symposium on Power Generation Choices: An International Perspective on Costs, Risks and Externalities. This Symposium is organised by the OECD Nuclear Energy Agency and by the Oak Ridge National Laboratory, and is supported by five international organisations in the energy field, the American Nuclear Society, the Canadian Nuclear Association, the Central Research Institute of the Electric Power Industry, the European Nuclear Society and the International Atomic Energy Agency.

The speakers that you will hear have very broad experience in the energy field, and are in an excellent position to put costs, risks and externalities of all forms of energy into their proper perspective.

You will get a very broad view from many countries. We live in a global context, not only geographical, but scientific and social. The world is not big enough for one country not to impact others through its use of eneergy. Technology is traded between countries. Social concerns originating in one country are often expressed a short time later by other countries. There is a need to integrate all aspects among countries and among the various ways of viewing problems to allow more rational and easier decision making. There has been a fragmentation of approaches and difficulties in co-ordinating policies. One needs only to cite carbon and energy taxes to underline the point. It is my hope that the presentations here will foster mutual understanding, and that the goal of a common approach could be envisioned.

At a recent OECD/IEA (International Energy Agency) Ministerial level meeting, energy ministers were confronted with the forecast of a 50% increase in world energy consumption in the next 20 years. That carried with it severe environmental implications. There is a need to address such implications - from global, regional and national standpoints - in a cost-effective way. This is one of the important issues that this Symposium will be discussing.

The mix of fuels used for electricity generation has to take account of energy security, environment, safety and costs. The Energy Ministers recognised the need for undistorted energy prices and free and open trade. It is difficult to achieve this goal when numerous other political considerations, and even methods of evaluation of costs, risks and externalities deform the decision making process. An international consensus on methodologies has not yet been achieved, and certainly not been accepted by decision makers.

Environmental issues have dominated the headlines in recent years. The UNCED (United Nations Conference on Environment and Development) Conference in Rio has been a major source of impetus. As has the IPCC (Intergovernmental Panel on Climate Change). A tremendous amount of work on improving environmental impacts had been initiated even before this. Various international organisations, including the OECD and the CEC, as well as national bodies and many others have been doing the basic research work. The larger question, on how to use the results, remains. How does one choose the technology which is the most cost effective as well as environmentally benign? How can one provide an overall ranking of the different power generation systems? Many attempts at comparisons have been made, all with limited success. How can one evaluate the impacts of an industry complying with regulatory limits? Are the costs, as presently calculated, "correct", or should externalities be factored in. If the externalities are to be factored in, how does one evaluate them? The environmental-economic approach of estimating impacts and merging the information with cost data has been endorsed by OECD Environment Ministers, whose communiqué says that " ... sound economic analysis of costs and benefits and their distribution, coupled with scientific assessment of relative risks, is the optimal basis for setting priorities, environmental goals and choices ...".

If I may turn to my specific area of expertise, nuclear power, I will briefly deal in turn with the costs, risks and externalities. Our study, done jointly with the IEA, on electricity generating costs has analysed the cost of different fuels in several countries. In this study, out of 15 cases submitted, nuclear power is the preferred choice, based on cost, for 13 countries, using a 5% discount rate. At a 10% discount rate, five countries find that nuclear is cheaper, with three additional countries finding no difference between coal and nuclear. Essentially, at the higher discount rate, gas, coal and nuclear come out as equal. Nuclear should make a substantial contribution to the overall energy supply mix. The importance of that contribution is recognised by various governments. However, apart from France and Japan, nuclear power is either stagnant or retreating. Obviously other forces are at work. Public opinion is one large factor.

There are legitimate public concerns about safety and waste management. Our own work indicates that the technology exists to resolve those problems. The future of nuclear power would therefore seem to be primarily a political and public acceptability question. In the long term, I believe that comparative economics will be the determining factor, but the economics is determined by more than just the cost to utilities. There is the factor of risk and the cost borne by others - the externalities.

Too often when one thinks of risk in the context of power generation, one thinks of the possibility of a nuclear accident. One does not have in one's mind a picture of oil rigs on fire or of gas explosions. This is a problem of perception. The best data that we have been able to gather suggest that the number of deaths per unit of energy produced is several orders of magnitude lower in the OECD countries in the case of nuclear power compared to the alternatives. But when one deals with different types of environmental risks one has to deal with societal values, and this is very difficult to tackle. We all live with risk, and all view it differently.

There is another risk to consider - the financial risk - that risk has certainly slowed the progression of nuclear power. Many things contribute to financial risk, but I will mention only two: the time needed to recoup the high capital cost, and the regulatory and political uncertainty. Designers have worked very actively to reduce the cost of nuclear power plants and to reduce the construction time. Others have worked to stabilise the regulatory climate with arrangements such as one-step licensing. Time will tell if the results meet expectations.

We can feel that we are on secure ground when talking about externalities relating to nuclear power. From the onset of nuclear programmes, costs such as decommissioning had to be factored into the total cost. Rules such as ALARA (As Low As Reasonably Achievable) have forced implementation of health and environmental measures whenever economically justified. The use of nuclear power does not generate any greenhouse gases or acid gases. A recent study by an Expert Group concluded that the only externalities of significance not included in the cost to utilities were: first - certain infrastructures in the case of new nuclear programmes, and second - government guarantees in the case of third party liability agreements. Even in the case of the latter, the distortion in electricity price would be of the order of one or two percent.

Nuclear can and should stand on its own economic merits - presuming that the rules are well established and are applied evenly to all energy technologies.

It is my hope that all forms of energy will come to be considered as "normal" and "ordinary", and that adequate means can be discovered to ease comparisions, not only from the point of view of utilities, but aso from the governmental perspective. There is always a tendency to take a narrow tactical view, looking only at direct costs and immediate problems. To take up a theme presented at the last World Energy Congress, "Quality of life depends to a great extent on the availability of energy.". All forms of energy used wisely and appropriately contribute to that quality of life. Let us hope that a strategic view will be taken by all and that techniques for supporting that strategy will be developed.

I thank you all for coming, and I particularly wish to thank the distinguished team of speakers as well as the organisers who have, I am sure, laid the basis for a successful symposium. I wish you all an interesting, instructive and profitable two days.

WORLD ENERGY OUTLOOK

(from 1990 to 2010)

1 - World energy consumption, 50% increase.

2 - Oil consumption, 40% increase; mostly in non-OECD countries.

3 - Demand for coal, 45% increase.

4 - Natural gas consumption, 66% increase.

5 - Share of fossil fuels in world energy consumption will remain at 90%.

6 - CO_2 emissions by energy production, 50% increase.

Chairman/Président
J.E. Gray, Past Chairman, U.S. Energy Association

Global Electricity Demand and Supply Trends
J.S. Foster, Honorary President, WEC (Canada)

Current and Future Costs of Power Generation Technologies
S. Yoda, K. Nagano, K. Yamaji, CRIEPI (Japon)

**How Costs, Risks, and Externalities Affect Decisions
Regarding Technology Choice**
S. Tierney, Assistant Secretary of Energy, USDOE
(oral presentation, paper not available)

Social Costing Research: Status and Prospects
R.W. Fri, President, Resources for the Future (United States)

Discussion

Remarks: Global Energy Outlook and Externalities
J.E. Gray, Past Chairman, U.S. Energy Association
(After dinner presentation)

GLOBAL ELECTRICITY DEMAND AND SUPPLY TRENDS

Dr. John S. Foster
Hon. President, World Energy Council

Abstract

During the two decades from 1970 to1990 electricity use increased substantially in all parts of the world, doubling in industrialized regions and quadrupling in developingregions. Electricity growth exceeded that of the economies as a whole and electricity assumed an increasing share of final energy demand. Generating units doubledin size. Measures were taken to reduce noxious emissions and R & D laid the groundwork for more efficient and more environmentallycompatible power generation in the future.

Résumé

EVOLUTION DE LA DEMANDE ET DE L'OFFRE MONDIALES D'ELECTRICITE

Entre 1970 et 1990, la consommation d'électricité a considérablement augmenté dans toutes les régions du monde, doublant dans les régions industrialisées et quadruplant dans celles en développement. La croissance de la consommation d'électricité a été supérieure à celle des économies dans leur ensemble, et l'électricité occupe désormais une place de plus en plus importante dans la demande finale d'énergie. La puissance des unités de production a aujourd'hui doublé. Des mesures ont été prises pour réduire les émissions nocives et les travaux de R-D ont permis de jeter les bases d'une production plus efficace et plus respectueuse de l'environnement pour l'avenir.

GLOBAL ELECTRICITY DEMAND AND SUPPLY TRENDS

Electricity is the most convenient form in which energy can be transmitted, distributed and made available for conversion to the form of energy desired at the point of use for a large and increasing number of applications. As a consequence there has been a steadily growing demand for it, on the global scale, since its commercial supply was introduced a little over a hundred years ago.

This morning I propose to talk about some of the trends in that global demand and supply over the most recent two decades - 1970 to 1990. The trends I have selected are those related to general growth, geographical distrbution, sources of supply, electricity use in relation to regional economies and to total energy use, and improvements in supply processes.

The main source of the data is the *National Energy Data Reports* of the World Energy Council which are published at each triennial Congress. In addition some data have been taken from national publications and others have been drawn from the Enclopaedia Britannica's *Books of the Year* which incorporate data adapted from the United Nations' *Energy Statistics Yearbook.*

GROWTH IN ELECTRICITY USE

WORLD ELECTRICAL GENERATION
TERAWATTHOURS per YEAR

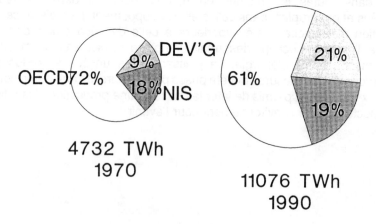

4732 TWh
1970

11076 TWh
1990

FIGURE 1

As indicated by the sizes of the pie diagrams in Figure 1 and by the Terawatthour values beneath them, the worldwide use of electrical energy increased 2.3 times between 1970 and 1990. This is an average rate of growth of 4% per year. During the period, OECD electricity use doubled but the OECD share of world electricity shrank from 72% to 61% as the use in the developing countries multiplied 5 times.

The integrated data in the pie charts give only the broadest impression of the change in electricity use between 1970 and 1990. There were of course quite different trends in different world regions. The next 3 figures illustrate this. The main purpose of this set of figures is to show patterns of trends for large blocs rather than particular trends for individual regions, although some of the latter have a special interest.

ELECTRICAL GENERATION
WEALTHIER ECONOMIES

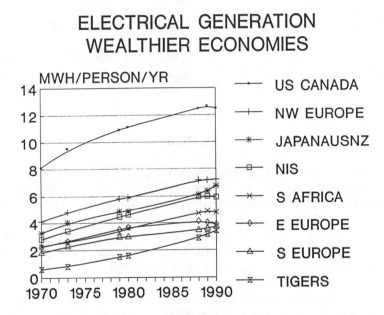

FIGURE 2

Figure 2 shows the trends in electrical generation per capita in the wealthier countries: the US and Canada; northwesternEurope; Japan, Australia & New Zealand; countries of the former Soviet Union; South Africa; Eastern Europe; Portugal, Spain, Italy & Greece; and the Young Tigers - Singapore, Hong Kong, Taiwan, and Korea. The main points of interest in this graph are that the trend for Canada and the US is of the same character as those for the rest of group - apparently saturating - and, further, that the slope for US-Canada over the period is about the same as that for the other regions. I will come back to the latter point later.

ELECTRICAL GENERATION
WEALTHY ECONOMIES

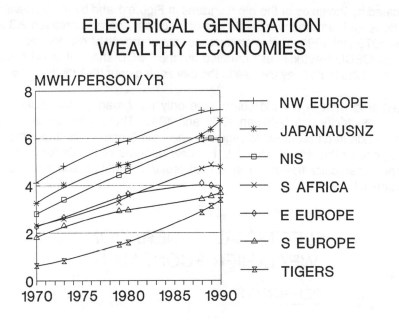

FIGURE 3

Figure 3 is the same as 2 except that the curve for Canada and the US has been deleted to improve the view of the curves for the other regions. The Young Tigers are a special case, beginning in 1970 with per capita electricity use at the level of that in the poorer regions, and ending close to the lower level in the wealthier regions. For the other regions, however, the trend in rate of growth is generally negative, although the JANZ (Japan, Australia, New Zealand) group and southern Europe do show a small uptick at the end of the period.

Although the main purpose of Figures 2 and 3 is to illustrate the apparently saturating trend in electricity use in the richer economies, there are one or two specific remarks I might make.

Political groupings, such as the OECD or the EEC, are frequently used as regions for presentation of energy statistics. Latitude and other local factors, however, are, not surprisingly, often more important. than political associations in this regard. The difference in per capita electricity use between southern Europe and northwestern Europe, the rest of the EEC, illustrates this.

One striking point brought out in this figure is the relatively high per capita electricity use in South Africa - higher than that in Eastern and southern Europe. This is for the total population of 30 million.

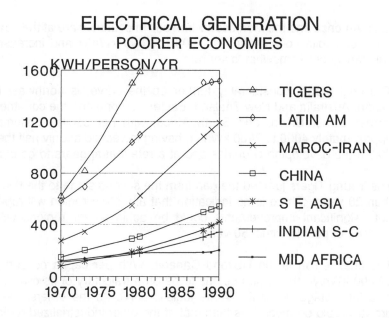

ELECTRICAL GENERATION
POORER ECONOMIES

FIGURE 4

The third graph in this set, Figure 4, shows the per capita electricity use in the poorer regions of the world. The scale of this graph is 5 times that of Figure 3. The regions covered are Latin America; the band of lands from Morocco across the north of Africa, and through the Middle East to Arabia and Iran; China; Malaysia, Indonesia, Thailand & the Philippines; Pakistan, India & Sri Lanka; and the lands of tropical Africa, from the Sudan to South Africa. Also shown on this graph are the Young Tigers again, to provide a bridge from the earlier graphs to this one. On this graph they start with Latin America and soar out of the picture by 1980.

The different character of the trends in electricity use between this graph and the earlier ones is very apparent. In generall the curves are exponential, with a positive change in the rate of growth. As for the wealthier regions, however, there are a couple of digressions from the general pattern. The growth in Latin America has moderated in the past few years; and growth of electricity use stalled in the 80's in tropical Africa.

The overall impression of this set of three graphs is that we are looking at snapshots of the progress of various world regions along a common logistic or S-curve.

China, the Indian sub-continent, Southeast Asia, and tropical Africa are in the earliest stage, with per capita consumption in the range of 300 to 500 kWh/yr, having doubled in each of the last 2 decades.

Latin America and the Morocco-to-Iran belt of countries are at the next stage, with per capita consumption at 1200 to 1500 kWh/yr and increasing at a substantial but somewhat lower rate.

The next group, considerably further up the curve, is northwest Europe; Japan, Australia and New Zealand; Eastern Europe and the countries of the former Soviet Union; and South Africa, with per capita consumption at approximately 4000 to 7000 kWh/yr, having increased at only half the rate of that in the developing countries and at a rate that appears to be declining.

The Young Tigers jumped the gap from the second stage to the third in less than 20 years. There is no indication that any other region will repeat that. but significant improvements could be achieved for regions with large masses of people within 50 years.

The final region is the US and Canada. With per capita consumption at 12,000 kWh/yr, the region seems to be on a trajectory of its own. If it were at a later stage along the same logistic curve as the others, the rate of growth would be much less than that of the other industrialized regions, but it is about the same. It is reasonable to expect that this difference between the US and Canada and the other industrialized regions will disappear but it looks as though it will require several decades for this to happen.

SOURCES OF ELECTRICITY

Of as much interest as where the electricity is being used, is the matter of where it comes from . What are the sources being used to provide this energy?

There is a significant difference between the developing and industrialized countries in this respect, too. The next pair of figures bring this out.

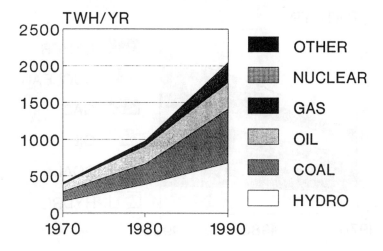

ELECTRICAL GENERATION SOURCES
DEVELOPING COUNTRIES

FIGURE 5

Figure 5 shows the sources of electricity in the Developing Countries - the countries that made up the regions that appeared in Figure 4, less the Young Tigers.

Growth in overall electricity use, as we have already discussed, was exponential - doubling each decade.

The sources of nearly all the electricity were hydro power and the traditional thermal sources - coal, oil and gas. Hydro power and the use of oil maintained a more or less linear growth during the past two decades. The exponential growth was provided by coal and gas. More than half of the coal use was in China and the expansion in gas use was mainly in the Morocco-to-Iran belt.

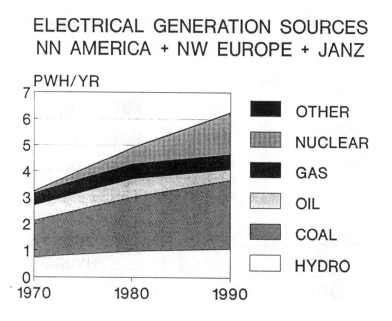

ELECTRICAL GENERATION SOURCES
NN AMERICA + NW EUROPE + JANZ

FIGURE 6

In the industrialized countries nuclear power had a major impact. Growth in overall electricity use was linear or even a little less.

Hydro power and the use of coal matched this linear growth. The use of gas remained more or less constant and the use of oil declined, particularly in the eighties as a consequence of measures taken in the seventies in reaction to the price upsets and uncertainties in the Middle East oil-producing areas during that decade.

Nuclear power increased exponentially at a dramatic rate - increasing many fold during the seventies and tripling in the eighties. By so doing it not only replaced the decline in the supply of electricity from oil but provided sufficient additional electricity to maintain the nearly linear growth in overall electricity supply. In twenty years the supply from this source had grown from practically nothing to equal that from hydro power, which had been built up over a century.

Of course, as this audience is only too aware, this is only part of the history of nuclear power during the past two decades. There are various levels of supply. So far we have been looking only at the level of the source of the electricity. Of course we are equally interested in the source of the source - the provision of the power plants, themselves. At this second level of supply the story of nuclear power was quite different over the past two decades.

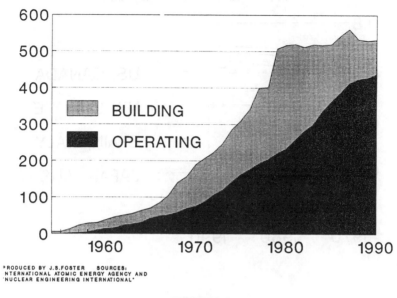

WORLD NUCLEAR-ELECTRIC UNITS

PRODUCED BY J.S.FOSTER   SOURCES:
INTERNATIONAL ATOMIC ENERGY AGENCY AND
'NUCLEAR ENGINEERING INTERNATIONAL'

FIGURE 7

Figure 7 shows the numbers of nuclear units in operation and under construction over a period that includes the past two decades. The sum is, of course, the total number of units committed. Its pattern is remarkable. If it hadn't happened, I wouldn't have believed it could happen - that, with the great body of diverse statistics underlying the pattern, it could take the form of an exponential curve lopped off sharply in mid-flight.

At the World Energy Congress in Munich in 1980, I made some remarks about the effect on other sources of energy if the hiatus in net nuclear power plant commitments, then 3 years old, were, for the sake of argument, to persist for 10 years. I really didn't believe it would endure so long. In the event, it has endured nearly twice that long, as new capacity, committed during a period of high growth, outran the actual subsequent demand.

ELECTRICITY AND THE ECONOMY

Another interesting trend is the relationship of electricity to the economy as a whole. The next pair of figures show the course of this relationship for OECD regions and for some developing regions over the past two decades. The economies are represented by the GDP data for each of the regions, expressed in constant monetary units (1985 US dollars).

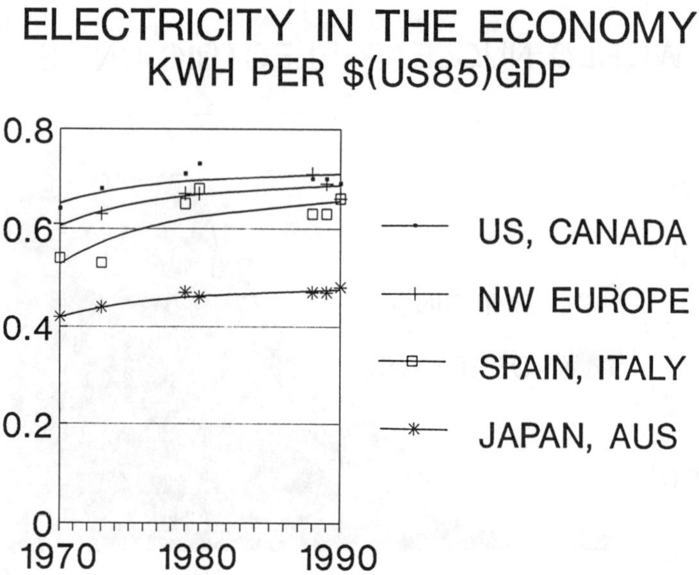

ELECTRICITY IN THE ECONOMY
KWH PER $(US85)GDP

FIGURE 8

Figure 8 shows the kilowatthours used per dollar of GDP in the United States and Canada, northwestern Europe, Spain and Italy, and Japan and Australia. Some countries have been omitted because of the lack of data.

The data for the first three regions are reasonably close together, at about 0.6 kWh per $GDP, increasing by about 10% over the period but at a declining rate. Japanese data largely determine the characteristics of the fourth curve. Although the shape is similar to the other three, and the increase over the period is also about 10%, the magnitude of the parameter is only about ¾ that in the other regions, or about 4½ kWh/$GDP.

We are very conscious that processes are becoming more efficient and we might be excused for thinking that electricity use per dollar of GDP would decline with time. This is true for any particular process; however, there are several other factors at work. As development proceeds electricity displaces other energy forms for many processes. Thermomechanical pulping for paper production, induction melting of steel, heat pump and microwave drying and heating are just some examples. In addition, lifestyles change in concert with these industrial developments and more electricity is used in residential and commercial applications. As a consequence electricity use in the economy as a whole has increased per dollar of GDP.

Several factors may explain the lower level of electricity use per dollar of GDP in Japan as compared with that in the other OECD countries. Greater efficiency may be a partial explanation. However, a lower ratio of primary to secondary industries is undoubtedly a factor as is, in all probability, a difference in lifestyle.

ELECTRICITY IN THE ECONOMY
KWH per $(US85)GDP

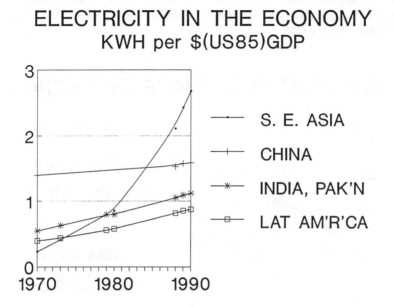

FIGURE 9

Figure 9 shows similar data for some of the developing regions of the world. Marked differences between these data and those for the industrialized countries are apparent. The levels of electricity use per $GDP are two or three times as high as in the OECD countries; the rate of increase in the levels is generally much higher and the rates are increasing, although these things are not so pronounced in China.

The probable reason for these differences is that the developing countries have large rural populations and corresponding large agrarian economies, while the growth in electricity use is largely in the rapidly expanding urban and industrialized areas. As a consequence the increase in electricity use is outstripping the growth in agrarian output and particularly in commercial agrarian output, especially when its value is converted to hard currency.

ELECTRICITY'S SHARE OF FINAL ENERGY DEMAND

I have alluded to the displacement of other forms of energy in industrial processes by electricity. The same thing has been happening in other parts of the economy - in residential heating for example. In addition there has been generally faster growth in the use of electricity than in the use of other energy forms in their established applications. The combination of substitution and generally higher growth in electricity use has resulted in electricity providing an increasingly large share of total energy use.

ELECTRICITY'S SHARE OF ENERGY

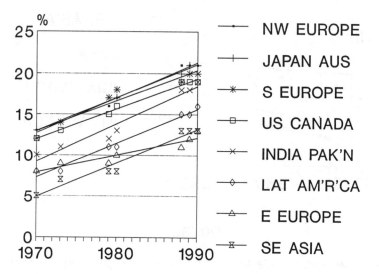

FIGURE 10

Figure 10 shows the increase in electricity's share of final energy demand in most of the regions of the world.

It must be noted that the final energy demand is the final energy derived from commercial energy supplies only. Available data do not permit inclusion of non-commercial sources on a consistent basis. Non-commercial supplies, however, generally insignificant in industrialized countries, are very important in most developing regions. Consequently, whereas the figure shows electricity's share of total energy use for the OECD regions and probably Eastern Europe, it overstates electricity's share of total energy by up to factor of 2 for the developing regions.

The newly independent countries of the former Soviet Union and tropical African countries are not included because of a lack of data. The trend line for the Young Tigers, also not shown, is very similar to that for India and Pakistan.

The growth of electricity's role in the delivery of energy over the past two decades is truly remarkable. Almost everywhere electricity's share of final commercial energy demand grew by an absolute 8 %. The only exception in the figure is Eastern Europe where the share increased by an absolute 5%. In the OECD countries the share grew from 12% to 20%, a 70% increase. In developing regions the share more than doubled.

SUPPLY TECHNOLOGIES

There is one further trend that requires mention. That is the changes in supply technologies. The seventies and eighties were a period of application and increase in the size of generating units rather than a time of notable advances in performance.

In the field of hydro-electricity, better metallurgy, machine design and manufacture, and understanding of soil mechanics, hydrology, dam construction and reservoir and dam behaviour led to better developments. The main evident phenomenom in the field, however, was the commissioning of several large installations, splendidly exemplified by Itaipu.

There had been steady advances in steam cycle parameters until the sixties. During the seventies and eighties, however, improvements were largely in control and measures to enhance reliability. Top cycle temperatures were already fully exploiting the potential in economically available materials, so there was little advance in generation efficiency, at least in the industrialized countries, where levels of 35% were typical for new plants. Again the main evident phenomenon was the general trend toward larger units. In the developing countries efficiencies did improve as unit sizes and cycle conditions followed those of an earlier stage of development in the industrialized countries. In China, for instance, efficiency of thermal power generation increased steadily from 29% to 31% during the eighties.

There was one trend, in relation to thermal power, particularly coal-fired plants, however,that was and remains extremely important. That is the introduction of flue gas desulphurization equipment and catalytic nitrogen convertors and the enactment of legislation and conventions that require the general use of such treatments in the future.

In nuclear power, too, plant cycles that had evolved in the fifties and sixties were employed in the plants that provided the energy in the seventies and eighties. Here, also, the most evident change was in unit size, as units grew in size from 500MW to 1300MW.

There were a few developments outside the mainstream. Improved geothermal plants were built; new designs of wind-generators wer installed; and some electricity was derived from a few solar-thermal installations and several small demonstrationor special purpose photovoltaic arrays. The total generation from all these sources was, of course, miniscule in relation to world electricity use.

There were activities off stage, however, that promise further improvements in performance in the future.

The development of robust, efficient gas turbines, derived from experience in the aircraft propulsion field, has opened up the practcal possibilty of highly efficient combined cycles. Several prototype plants using such cyles are already in operation or under construction.These include cycles using pressurized fluidized bed boilers, integrated gasification combined cycles, and simple gas-fired combined cycles. The first two coal-fired cycles offer efficiencies of about 45% and the gas-fired cycle, about 55%. Increasing application of these new systems will lead to increasingly better efficiency in power generation in the future while maintaining high emission quality.

In relation to nuclear power, experience was gained and progress made in various aspects of highly efficient nuclear fuel cycles employing both fast and thermal neutron fields. The effects of this on power generation efficiency is rather longer range than that of the changes in hydrocarbon cycles.

CONCLUDING REMARKS

There was substantial growth in electricity use throughout the world from 1970 to 1990, doubling in the industrialized countries and quadrupling in the developing countries. In the developing countries electricity use was still rising exponentially. The increase in use of electricity in the industrialized countries, however, is decelerating. It is quite likely that the use of electricity in these countries is approaching a sort of plateau that will probably endure until electricity begins to be used on a significant scale for transportation. When this happens will depend on the relative costs of electricity-based vs coal-based propulsion after the decline in availability of the natural fluid hydrocarbons, a few decades from now.

During the 1970-90 period, electricity's share of final commercial energy demand increased substantially, exceeding a level of 20% in the most industrialized regions. Its use also grew faster than the economy in all world regions, although, again, the increase seemed to be levelling off in the most industrialized countries.

From all this, the most important prospects for the next century seem to be that electricity use will continue to grow fastest in the developing countries; that electricity will continue to provide an increasing share of final energy; that nuclear power will resume supplying an increasing share of that electricity; and that major progress can be expected in electricity generation efficiency and in the reduction of adverse effects on the environment.

CURRENT AND FUTURE COSTS OF POWER GENERATION TECHNOLOGIES

Susumu Yoda, Koji Nagano and Kenji Yamaji
Central Research Institute of Electric Power Industry
Tokyo, Japan

Abstract

The objective of this paper is to identify the issues for further elaboration of power generation cost evaluations and to explore trends of power generation technologies in the future. While environmental externalities and social costs for energy security are partly internalised in current electricity prices, range of cost components and time horizon for cost evaluation should be enlarged to deal with emerging issues such as global environmental protection. Under the present circumstances, nuclear power is one of the most economical and stable sources of electricity in most of the industrialised countries. However, attention should also be paid to new trends in technology innovation, such as new economies associated with decentralised power sources and integration of energy systems.

COÛTS ACTUELS ET FUTURS DES TECHNOLOGIES DE PRODUCTION D'ÉLECTRICITÉ

Résumé

L'objet de la présente communication est de recenser les questions qu'il importe de résoudre pour évaluer les coûts de la production d'énergie et étudier les perspectives d'évolution des technologies de production. Si les coûts externes d'environnement et les coûts sociaux associés à la sécurité énergétique sont partiellement pris en compte dans le prix actuel de l'électricité, il convient d'élargir la gamme des facteurs de coûts et l'horizon fixé pour l'évaluation des coûts, de façon à tenir compte de problèmes nouveaux tels que la protection de l'environnement de notre planète. Actuellement, l'énergie nucléaire est l'une des sources de production d'électricité les plus économiques et les plus stables, dans la plupart des pays industrialisés. Il importe toutefois de prêter attention au progrès de l'innovation technologique, notamment aux nouvelles économies associées aux sources d'énergie décentralisées et à l'intégration des systèmes énergétiques.

1. Introduction

Electricity changed human history.   In 1879, Thomas Edison invented an incandescent light bulb.   More importantly, he set the stage for the growth of the electric utility industry in 1882 by supplying electricity to some 100 buildings in New York for his new light bulbs.   The genius of Mr. Edison combined two economies of scale both on the supply side and demand side of electricity. Combining a central power station with a network of customers was a great breakthrough.   The invention of power grid, or "Edison System", realized economies of scale in generation technologies as well as economies involved in a network; namely, levelized demand profile, reliability of supply, and standardization of equipment.   Since every energy resource can be converted to electricity, we have been able to choose the least cost resource or the best combination of resources in each region for electricity generation; thus, together with technological progresses, the cost of electricity has been reduced steadily, or at least kept stable within a reasonable range.   Scale economies realized by the Edison System and this economic advantage in selecting diversified energy resources created a strong positive feedback loop between the reduction in the cost of electricity and the size of electricity supply, which enabled the accelerated expansion of electric utility industry.

However, recently, the positive feedback loop seems not to be functioning as well.   Many industrialized countries have entered a matured stage of development; central power stations are already large enough in terms of scale economies; and, supply conditions of energy resources are no longer so stable as they once were.   Moreover, the threat of global environmental disruptions poses a philosophical question regarding the concept of economic development itself.   These problems are reflection of a more deeply rooted human challenge. Humankind on the earth is now being forced to choose a future under the three conflicting requirements; economic growth, energy and other natural resources and food, and preservation of the environment.   This challenge may be thought of as a "TRILEMMA".   The electric utility industry must also cope with the trilemma.   We must explore a possibility of new economies in electricity supply system which is more compatible to the situation of trilemma and leading to a sustainable development of mankind.

The key issues to be addressed in this context are as follows: First, the factors missing in conventional cost evaluation, such as environmental externalities and the cost for energy securities, should be identified.   The methodology of cost evaluation should be enlarged to deal with new economies for sustainable development.   Secondly, whether the scale economies still exist or not should be carefully evaluated both for individual technologies and for electricity supply system as a whole.

2. Issues in Evaluating Electric Power Generation Costs

2.1. Range of Cost Components

Figure 1 summarizes the coverage of components most commonly used to evaluate generation costs. In general, evaluations are made up of three major items; capital costs, operation and maintenance costs, and fuel costs. It should be noted that in the case of fossil fired generation technologies, de-sulphur and de-nitrogen treatments not only require capital investments but also decrease overall generation efficiency. This issue will be discussed further in Section 3.2 of this paper.

Fuel cost calculations require careful attention to projecting not only the price paid for the fuels and their related services (including enrichment and fabrication of nuclear fuel) but also currency exchange rates. This is particularly true in Japan, since only a small portion of coal is produced domestically and nearly all other fuel requirements are imported including uranium.

Figure 1   Basic Components of Generation Costs

Capital Cost   +   Operation and   + Fuel Cost   =   Total Cost
                   Maintenance Cost

– Construction   – Salaries   – Fuel Purchase
– Interest   – Materials   (incl. Nuclear fuel
– Auxiliary   . – Supplies   cycle services)
  Equipment   – Transportation
  (e.g. de–SOx,   – Waste Treatment and
  de–NOx)   Disposal
– Dismantling

(Note) Arranged from OECD/NEA(1983).

2.2. Integrated System Costs versus Costs of Individual Technologies

Generation cost may be expressed and evaluated by several different approaches. Table I summarizes the purposes and corresponding methodologies of evaluation.

To judge which generation technology is the most economical, each candidate technology should be evaluated independently. Such costs for individual technologies can be valuated either as 'first-year cost' or as 'levelized cost' per unit of generated electricity (kWh). The former, which is the sum of the capital costs, operation and maintenance costs, and fuel costs incurred in the first year of the operation of the plant under consideration, is the most simple index for technology assessment. In addition, the first-year cost approach is quite compatible with the accounting system in a corporation, and thus of practical importance for utility management. The latter, which is the total costs discounted to a reference year divided by the discounted total electricity generated over the life-time of the plant. Since all related costs to the plant, incurred before starting the operation such as interests during construction, as well as incurred after the operation such as dismantling and waste management, are included in the levelized cost, it provides the most comprehensive index for technology comparison.

On the other hand, 'system cost' evaluates the cost of the whole generation system. This is particularly useful to assess the impact of additional installation of a new plant to the generation system. The system cost in such a case may be influenced not only by the technology chosen for the new plant but also by the existing generation mix, particularly the load profile and dispatch.

Table I Basic Approaches for Calculating Costs of Electricity Generation

METHODS	BASIS	REMARKS
First–Year Costs	Costs based on a share of capital plus operating and fuel costs in the first year of the plant operation divided by the output in-the year.	Essentially an accounting approach. Various ways of calculating capital charge. Nominal or real term.
Levelized Costs	Costs over the lifetime of the plant, including capital, operating, fuel costs, are discounted to a present value, and then divided by the total discounted output.	Reasonable basis for technology comparisons. Nominal or Real term.
System Costs	Effect of a new station on total costs of operating a national grid system is simulated based on assumed demand, investment and load profile.	Basis for investment decision. Results depend on the grid system itself.

(Note) Arranged from OECD/NEA(1983).

2.3. Externalities

The externalities related to electricity generation are categorized by two different aspects. First, they are classified by their causes as either 1) environmental, or 2) non-environmental externalities. The environmental externalities caused by electricity generation include influences such as acid deposition by sulphur oxide and nitrogen oxide, global warming by carbon dioxide, impacts of thermal effluent, and submergence of lands by dam construction. The non-environmental factors include user costs for consumptions of exhaustible resources, and security costs for supply of energy, particularly oil.

On the other hand, externalities are distinguished between; 1) those already dealt with, and 2) the remaining as not internalized yet. Some of the externalities are already internalized by financial or institutional schemes, such as abatements costs of polluting agents, taxes charged on resource consumptions, subsidies for R&D, etc.

As discussed in Section 3.2, the actual status of internalization of those external costs may vary by country, at different degrees in various schemes. It is foreseen that further internalization of remaining externalities will expose significant influences on cost structure of electricity generation, and thus choices of generation sources; therefore, we should pay careful attention to the consistency with the existing arrangements of external costs.

2.4. Life Cycle Analysis

In conventional cost analyses, the costs incurred directly related to power generation are taken into account. As long as the direct costs, such as purchased prices of construction materials and fuel, appropriately reflect the related indirect costs, these direct cost calculations are considered to represent properly the total cost of generation technology.

The idea of life cycle analysis has gained prominence, particularly along the context of environmental externalities. For example, it has been recognized that all the emitted carbon should be taken into account when we evaluate the global warming impact of electricity generation, both emissions directly connected with generation and indirect ones including productions of cement and other materials for plant construction, transportation of those materials as well as fuel, and the waste treatment and disposal. The extension of cost analysis to incorporate all the indirect aspects will be also useful and important, particularly for evaluating possible outcome of remaining external effects in detail, as well as assessing schemes of their internalization.

3. Current Status of Generation Costs

3.1. Economic Comparison among Generation Technologies

To focus on the status quo of generation technologies and their economics in particular, it may be helpful to take Japan's situation as an example. There are two sources of generation cost evaluation published yearly in Japan; by the Agency of Natural Resources and Energy (ANRE) of the Ministry of International Trade and Industry (MITI), and by The Institute of Energy Economics (IEE).

Figure 2 shows the trend of cost comparisons reported by ANRE every year. The fossil fired generation technologies based on coal, LNG (liquefied natural gas) and oil, and nuclear power generation technology are compared. The underlying assumptions are; 1) the costs are given assuming a new commission of a plant of each technology at the year of evaluation, 2) the 'first-year cost' evaluation is applied; however, the other methodology of 'levelized cost' has been used alternatively since 1985, and 3) every plant is assigned to base-load operation at the uniform utilization factor of 70%.

Figure 2   Trend of the Generation Costs by ANRE

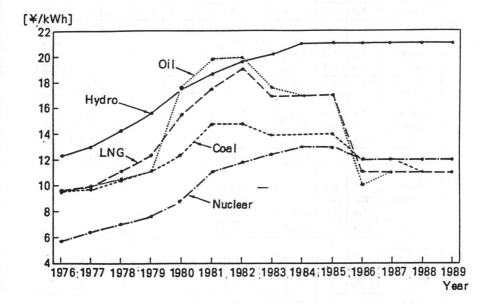

Table II, in turn, represents the trend of evaluated costs of generation by IEE. They employed the levelized cost calculation, except for the earlier evaluations in 1982 and 1983 by the first-year cost method.   The assumptions and conditions of the evaluation are similar to those by ANRE.

Table II Generation Costs Evaluated by IEE

[Yen/kWh]

	1982	1983	1984	1985	1986	1987	1988	1989	1990	1991	1992
Nuclear											
Capital	5.5	7.5	6.3	6.3	6.3	6.3	6.2	6.3	6.5	6.4	6.40
Operation	3.0	2.0	2.4	2.2	2.2	2.2	2.2	2.2	2.4	2.3	2.33
Fuel	3.5	3.2	3.0	2.3	2.1	1.8	1.7	1.7	1.7	1.8	1.48
Total	12.0	12.7	11.6	10.7	10.6	10.2	10.1	10.3	10.8	10.6	10.21
Coal											
Capital	3.3	5.9	5.1	5.4	5.4	5.4	4.9	4.9	5.0	5.1	5.39
Operation	1.5	1.9	2.3	2.2	2.2	2.2	2.1	2.1	2.2	2.2	2.36
Fuel	8.5	5.3	5.9	5.5	3.5	2.7	3.7	4.1	3.5	3.5	3.23
Total	14.3	13.1	13.3	13.1	11.1	10.3	10.7	11.1	10.6	10.8	10.98
Oil											
Capital	2.5	3.4	3.1	3.1	3.1	3.4	3.5	3.4	3.5	3.9	3.83
Operation	1.0	0.9	1.1	1.0	1.1	1.1	1.2	1.2	1.3	1.4	1.36
Fuel	15.1	12.8	13.5	11.3	6.2	6.7	6.2	6.9	6.6	6.7	5.78
Total	18.6	17.1	17.8	15.4	10.4	11.2	10.9	11.4	11.4	12.0	10.97
LNG											
Capital	3.7	5.0	4.4	4.3	4.3	4.4	3.9	4.1	4.1	4.1	4.15
Operation	2.0	1.2	1.4	1.3	1.4	1.3	1.2	1.3	1.3	1.3	1.36
Fuel	13.5	10.5	10.8	8.7	5.9	5.4	5.6	5.9	5.5	5.2	4.74
Total	19.2	16.7	16.8	4.3	11.6	11.1	10.8	11.3	10.9	10.6	10.25
LNG-CC(*)											
Capital	–	–	–	–	–	–	–	–	4.3	4.5	–
Operation	–	–	–	–	–	–	–	–	1.4	1.5	–
Fuel	–	–	–	–	–	–	–	–	5.0	4.7	–
Total	–	–	–	–	–	–	–	–	10.8	10.7	–

(Note)   Sources : Yuasa (1992) for 1982–1991, and Seki (1992) for 1992. First-year cost method for 1982 and 1983; levelized cost method thereafter.

(*) LNG fired combined cycle power plant.

When the first-year cost method is used as shown in Figure 2, nuclear power became slightly more costly than the fossil fired generations after the collapse of fossil fuel prices in the late 1980s. In the levelized cost evaluations in both by ANRE and by IEE, however, nuclear has always kept stable as the most economical source of electricity. The other technologies based on fossil fuels experienced considerable fluctuations of generation costs after the oil crises.

As for the international perspective of generation cost comparison, OECD/NEA (1986) provides an extensive study of comparative cost evaluation between nuclear and coal, which was updated from OECD/NEA (1983) study. The levelized cost calculation was also employed in these analyses. Table III summarizes their results, in terms of the ratio of assessed generation costs of coal to nuclear power. Direct comparisons between countries are not possible nor appropriate as underlying data assumptions vary significantly. The table III reveals that nuclear generation is superior to coal fired generation in almost all the industrialized countries, although coal might compete with nuclear in those specific regions where cheap coal resources are available. The changes between the two analyses are not very significant, and the most of them are explained by the difference of assumed prices of coal or currency exchange rates. These results also explain the reason why nuclear has been recognized as an important power source in many industrialized countries.

TABLE III Cost Ratio of Coal Versus Nuclear

	OECD/NEA(1983)	OECD/NEA(1986)
Belgium	1.39	1.62
Finland	–	1.33
France	1.75	1.80
West Germany	1.64	1.41 / 2.02
Italy	1.57 —	1.40
The Netherlands	1.29	1.31
Norway	1.42	1.20 / 1.38
Spain	–	1.19
Sweden	1.33	–
United Kingdom	1.43	1.40 / 1.71
Canada/		
Central	1.42	1.44
Eastern Coal	1.12	1.18
Western Coal	0.72	0.66
Japan	1.51	1.37
United States/		
Central	1.01	0.83 / 1.11
Eastern	–	1.08
Rocky Mountain	–	0.77

3.2. Treatments of Externalities in Generation Cost Structures

To examine ways and degrees of influence of externalities on generation costs, here we take Japan's current arrangements as an example. While some externalities are included in the cost structure in a form of costs of pollutant abatements, taxation is also used as a way to reflect the social costs of energy consumptions.

The energy taxes related to electricity generation in Japan are, as summarized in Table IV, the tariffs on imported crude oil and heavy fuel etc., the domestic oil consumption tax, and the electric power development promotion tax.

The revenue from the oil tariff, at 315 Japanese yen per kl of crude oil and 2,520 yen per kl of heavy oil, etc., is utilized as the special account for coal within the national budget, such as restructuring of coal mining industries and regional developments of coal producing areas, etc. The fund for the account was 104.3 billion yen for fiscal 1992.

The revenue from the oil tax, at 2,040 yen per kl of oil and 720 yen per ton of LNG, is once put into the general account of the national budget, but then transferred to the special account of oil and its alternative energy. In the total fund of 535.4 billion yen in fiscal 1992, the largest portion of the fund is used for the oil and LPG reserves, occupying 349.1 billion yen, 65% of the account. The remaining portion is divided into two categories of usage: 1) the expenditures related to oil such as development and exploration of resources and restructuring of industries, with 156.4 billion yen, and 2) development of alternatives like solar energy, liquefaction and other advanced technologies of coal utilization, with the fund of 39.9 billion yen in fiscal 1992, respectively. In the latter, coal related expenditures amount 29.5 billion yen.

The revenue from the electric power development promotion tax, at 0.445 yen per kWh sold by utilities, is put in the special account for electric power development promotion, and divided to the two sub-accounts; the fund corresponding to 0.16 yen per kWh in the revenue is set in the account for electric power siting, used as rebates and subsidies for regions where central power stations are sited, mainly nuclear. The rest corresponding to 0.285 yen per kWh goes to the account for electric power diversification, such as resource developments of hydro and geothermal, and technology R&D of renewables, coal utilization, fuel cell, superconductivity, and advanced nuclear technologies. Nuclear related expenditures have the largest share occupying about 76% of the special account.

All these arrangements are designed to use their revenues for specific purposes. Yokobori (1992) justified these systems as they have been based upon the recognition of these social costs, such as for preventing environmental damages, for security of supply and for local developments. This is true, as long as the schemes reflect these social costs correctly. In case when the circumstances would have changed, however, these systems should be carefully adjusted.

Table IV   Energy Taxes and the Related Accounts of the National Budget
in Fiscal Year 1992 in Japan

A. The Tariffs on Imported Crude Oil and Heavy Fuel etc.

Crude Oil	315 yen/kl =>	The Special Account for Coal	104.3 billion
Gasoline	1,430	Restructuring of Coal	
Kerosene	580	Industry	24.8
Diesel Oil	1,290	Developments of Coal⁻	
Heavy Oil	2,520	Producing Regions	11.3
		Accidents of Coal Mines	48.3
		Labor for Coal Mines	16.2
		Others	3.8

B. The Oil Consumption Tax (*)

Oil	2,040 yen/kl =>	The Special Account for Oil and its Alternatives	
LPG	670		535.4 billion
LNG	720	Measures for Oil	495.6
		Resource Developments	101.9
		Reserves-of Oil and LPG	349.1
		Restructuring of Oil Industry	40.7
		Others	3.1
		Measures for Alternatives	39.9
		Supply Security	1.3
		Commercialization	8.4
		Technology Developments	28.2
		Others	2.0

C. The Electric Power Development Promotion Tax

0.445 yen/kWh =>   The Special Account for Electric Power
Development Promotion   412.4 billion

The Account for Electric Power Siting	191.8	The Account for Electric Power Diversification	220.6
Electric Power Siting Promotion Funds	82.7	Nuclear	26.9
		Coal Utilization	15.7
Nuclear Safety and Other Research Contracts	70.2	Hydro	4.8
		Geothermal	12.6
Special Grants for Nuclear Plant Siting and Electricity Export	29.0	Solar	11.8
		Fuel Cell	6.6
		Superconductivity	6.2
Hydro Power	6.1	R&D by Science and Technology Agency	106.0
Others	3.8	Others	30.0

41

(Note)   – Source: MITI (ed.), Sekiyu-Shiryou (1992).
         – All the amounts are in Japanese Yen.
         – As for the Special Account for Electric Power Development
         Promotion, the amounts of sources and use do not match due to carry
         overs from previous years.
         (*) Excluding gasoline tax, diesel oil tax and jet fuel tax. The revenue
         from these taxes is used for transportation infrastructure.

Besides these national taxes, there are several local taxes charged to the utilities, such as the nuclear fuel tax at 7% of the value of nuclear fuels fed into reactors. In fiscal year 1990, Japanese electric utilities paid 6.3% of sales for those energy taxes and public imposts including the other public charges, such as water supply and traffic tolls. This percentage is large when compared with 0.9% of the average for the Japanese industry as a whole. The corporate tax payment of the electric utilities in 1990 was 2.3% of sales. Thus, the total tax burden ratio of the utilities was as high as 8.6%, which is sharply contrasted with 2.2% of the whole industry sector average.

In addition to the taxation to deal with externalities, the measures of desulphurization and denitrification have been introduced in the 1970s. As most of the fossil fired power plants are equipped with those measures in Japan, the externalities of sulphur and nitrogen emissions are dealt with as abatement costs. CRIEPI analyzed the latest status of these technologies attached to coal and LNG fired power stations (Hondo and Uchiyama (1993)). Taking into account both the direct expenditures and efficiency losses incurred by those equipments, it is evaluated that de-SOx and de-NOx treatments in coal fired plants occupy 24% of the generation cost per kWh, and de-NOx in LNG power plants occupies 4% of the generation cost, as shown in Figure 3.

Figure 3   Pollutant Abatement Costs in Coal and LNG Fired Stations in Japan

(A) Coal Fired Technology   (B) LNG Fired Technology

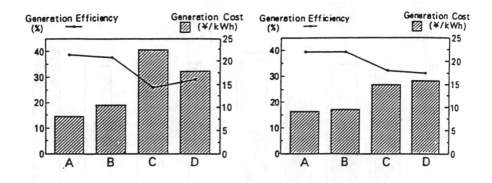

(Note) Case A :   Without Abatements,
              B :   Current status in Japan with both de-SOx and de-NOx for
                    Coal and de-NOx for LNG,
              C :   Current status plus amine absorption for CO_2 abatement,
              D :   Current status plus pure O_2 burning and PSA adsorption for
                    CO_2 abatement.

Modifying the generation cost comparison by IEE (Seki (1992)), effects of these externality treatments are illustrated in Figure 4, where the national tax related arrangements are re-rationed in proportion to their usage. As shown in Figure 4, several externality factors are already dealt with by these schemes to a considerable amount and in a wide variety.

Figure 4 Generation Cost Structures and Externalities

(a) Externalities as Energy Taxes   (b) Abatement Costs of SOx and NOx

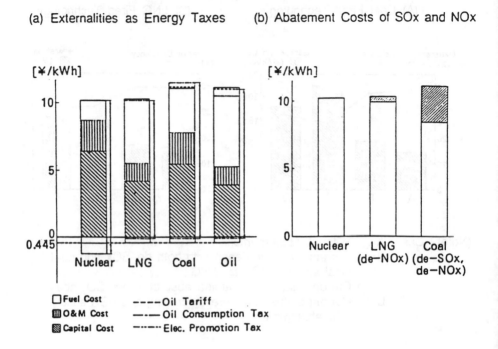

(Note)   – The revenue of the electric power development tax was divided into that used for nuclear and that for others, and then we calculated imaginary tax rates so as to raise the corresponding revenue from nuclear and the other electricity generations respectively.

– The revenue of the oil tariff charged on oil is applied for the special account for coal. We calculated an imaginary tax rate for coal so as to raise the corresponding revenue.

– A roughly equal fund to the revenue of the oil tax on oil is used for oil, such as strategic reserves. Only a negligible portion of the fund is used for LNG. We calculated an imaginary tax rate on coal so as to raise the corresponding revenue of the oil tax used for coal.

3.3. Scale Economies of Generation Technologies

Scale economy of production process is defined as a phenomenon where the production cost per unit of produced goods becomes smaller as the production capacity increases. This characteristic is expressed by 'scale exponent' defined as;

$$C = k \times P^{a-1}$$

where, C : unit cost of production at the capacity P,
        k : constant,
        a : scale exponent.

The scale exponent takes its value in a range between 0 to 1. The smaller the scale exponent, the larger the benefit of scale expansion.

Recently, however, questions have been raised whether the scale economies of electricity generation technologies still exists for the further capacity increment. Let's take nuclear generation technology as an example. IAEA (1984) made a survey about the scale economies of small and medium sized nuclear power plants. The reported scale exponents of nuclear reactors from related literature are in the range of 0.4–0.8. There seems to have been a consensus that nuclear reactor technologies appreciate substantial scale economy effects.

In the case of Japan, the overnight construction costs per unit of generation capacity of the nuclear power reactors currently under operation have increased gradually. This phenomenon is explained by the tendency to put higher priorities on the multi-barrier safety features on reactor designs, which led to not only increase of capital investments but also prolonged lead-time. According to a statistical analysis on the construction costs of nuclear reactors in Japan (Tajino (1988)), the scale economies of nuclear power plants still exist, although the cost reductions gained are not sufficient to offset the tendency of cost increases.

Another way to explore scale economies is to analyze macroscopic cost functions of a utility as a whole, or generation section as a whole within a utility, applying econometric methodology. CRIEPI made an analysis for Japan's utilities' cost functions (Nakanishi and Itoh (1988)), and concluded that expansions of generation capacity have not led to any decrease of the average cost of generation, and thus the scale economy might have been lost recently. In the analysis for the whole utilities including transmission and distribution sections, however, revealed that their scale economies are shrinking but still exist.

Note that the definitions of scale economies are slightly different between these examples. Moreover, both methodologies require further improvements to obtain clear conclusions. Here we can say only that there is no evidence if the scale economies of generation technologies have lost. In the meantime, we should pay careful attention to the trends of innovation, particularly 'down-sizing' of generation technologies. This issue will be discussed in detail in Section 4.2.

3.4. Comparative Cost Evaluation of Generation Mixes

Among the methodologies explained in Section 2.2, the integrated system cost evaluation provides additional implications to the individual technology assessments. Integrated system cost evaluations often employ optimization analysis, where the capacity of each generation type is determined so that the total cost of generation system is minimum while meeting the assumed load profile. The method is thus useful for development planning of power plants.

Nanahara et al. (1989) introduced a notable extension of the methodology. Figure 5 shows the fundamental idea of 'robustness' of a supply system. A 'robust' system can be defined as a system that is not greatly be affected by changes of constraints or conditions, though it is not necessarily an optimal under the assumed circumstances. We introduce a robustness index as defined by the following equation:

$$R(x) = \max_i [f_i(x) - F_i]$$

where, x : a generation mix,
        i : represents a set of conditions,
        $f_i(x)$: the system cost of the generation mix x under the
                condition i,
        $F_i = \min f_i(x)$: the system cost of the optimal generation
                mix under the condition i.

The value in the parenthesis can be referred to the 'regret' value in system analyses. Thus, R(x) may be called the maximum regret. The optimal generation mix under the robustness principle should be x* which minimizes R(x).

Figure 5 Concept of Robustness

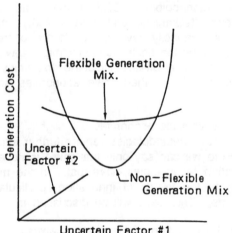

Figure 6 shows an example of evaluations in which both the cost minimization and minimization of R(x) are considered. There exists a trade-off relation between the cost and the robustness, and the generation mixes represented by

46

the points on the frontier curve between the optimal solutions, (I) with minimum cost, and (II) with minimum R(x), are all Pareto optimum solutions. The proposed methodology suggest that depending on the attitude towards risks involved in electricity system development, the optimal generation mix may change.

Figure 6   An Example of Trade-off between Cost and Robustness

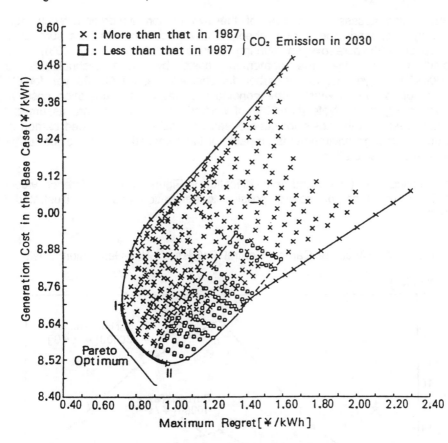

The methodology presented here is still under development. Although further improvements are required, such attempts to deal with uncertainties surrounding energy supply planning will be regarded as one of crucial importance.

4. Prospects of Power Costs

4.1. Effects of Carbon Dioxide Emission Control on Generation Costs

Among the remaining externalities, carbon emission control will have the largest impacts on generation costs. The externality with carbon emissions can be internalized either as a carbon surcharge or as carbon abatement costs. CRIEPI analyzed the macro-economic impacts of carbon tax in Japan (Nagata, Yamaji and Sakurai (1991)), and concluded that by imposing a tax at 4,000 yen per t-C (ton of carbon) starting 1990 and increasing the tax rate by 4,000 yen per t-C every year, Japan's carbon emission can be almost stabilized at the 1988 level up to 2005. The tax rate in 2005 is to reach as high as 64,000 yen per t-C.

CRIEPI also assessed the cost of the two carbon absorption technologies currently proposed; the amine absorption process and the pressure swing adsorption with pure-oxygen burning (Hondo and Uchiyama (1993)). The results revealed that the abatement costs by these technologies are 36,000-59,000 yen per ton of carbon for coal and 49,000-58,000 yen for LNG per ton of carbon, of which corresponding changes of the generation costs are shown in Figure 3. Note that 90% of emitted carbon dioxide are assumed to be absorbed, and the costs for deep ocean disposal of the absorbed carbon are included, although environmental impacts of the disposed carbon dioxide are not taken into consideration.

Again taking Seki (1992) and its modification in Figure 4(a) as reference values, the influence of carbon control policies to-generation costs are illustrated in Figure 7.

Figure 7   Effects of Carbon Control Policies to The Generation Costs

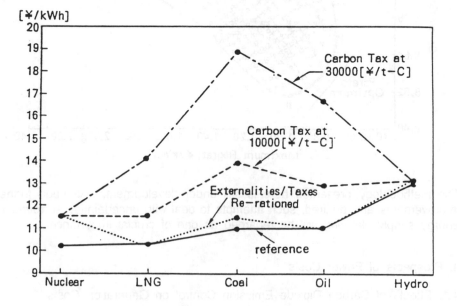

Any of those stringent policies of carbon control, such as the emission stabilization by taxation or absorbing all the carbon from flue gas, will force significant increase of generation costs, roughly 80% for LNG and more than doubling for coal and oil.

Meanwhile, there are many more options other than those in electricity sector for atmospheric carbon abatement, such as promotions of energy conservation and efficiency improvement or preservation of forest resources and reforestation. The carbon tax, if introduced, should not be viewed solely as a penalty to control carbon emission, but also as a means to raise funds for promoting these other measures.

Needless to say, the present tax systems vary broadly across countries. The influences of carbon abatement costs or carbon taxes may thus vary. It is of crucial importance to assure consistency of carbon abatement policies with the existent tax schemes and the performance of the macro economy.

4.2. Developments and Commercialization of Decentralized Power Sources

Industrialized countries have enjoyed scale economies in the development and management of electricity supply systems for decades. Recently, however, decentralized power sources are attracting more serious attentions and may play more important roles in electricity grids near future. This might be recognized as reflection of a new trend of generation technology innovation as 'down-sizing'. In particular, it has become more attractive for utilities to construct modular power stations which consist of several units of highly efficient LNG combined cycle plant at roughly 100–200 MWe, instead of stations with a single large unit. Although gas combined cycle plants remain as central stations, this direction of development will lead to a more general trend towards decentralized power sources.

As decentralized power sources, the following technologies are now considered promising; the ones based on fossil fuels, such as gas turbines and fuel cells, and those utilizing renewable energy sources, such as photovoltaic cell and wind power.

The merits of decentralized supplies can be attributed to several factors. First, these plants are installed as directly connected to the demand. This eliminates the burden of energy transmission and distribution, and enables utilization of both electricity and heat. Consequently, the plants can be designed to be optimal for specific cogeneration pattern at high efficiencies.

Secondly, most of the renewable energy will be introduced as decentralized generation plants. To meet the targets of carbon emission reductions, all the countries without exception must make efforts to raise their reliance upon renewable energy. The promotion of decentralized sources should be realized in a harmony with this tendency.

Moreover, decentralized power sources might be capable of improving the stability of the electricity network. Compared to a situation where a large-scale central station had an accident, decentralized units are easier to be backed up and allow more holonic coordination of their operations.

Finally, decentralized sources will be important as the first and the most practical access to electricity in developing countries. Although electricity is one of the most crucial necessities for the countries' future developments, it is difficult to implement large-scale grids with limited financial allowances and social infrastructures. Decentralized systems may serve as pioneers of electricity utilizations in rural regions, where in fact renewables might be the only available energy resources.

For the commercialization of decentralized sources, several schemes are available to promote. Surplus power beyond the local demand might be purchased by utilities at appropriate prices. The local or the central government might waive taxes or compensate some portion of investment costs for their installation. Such arrangements, though they should of course be carefully chosen upon implementation, will be required in near term in order to lower the initial barriers of their commercialization and to lead to cost reductions.

Uchiyama and Imamura (1992) made a simulation study of market penetration of photovoltaic cell and wind power in Japan. Figure 8 shows an example for photovoltaic, suggesting that such additional promotion measures are effective for stimulating initial introductions and consequent cost reductions by learning effects, which then enable further market penetration as a positive feedback.

Figure 8 A Simulation of Photovoltaic Cell Commercialization in Japan

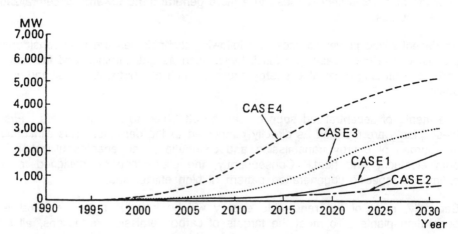

(Note) Case   1: Reference case, with the excess power purchased by utilities at 70% of the marginal cost of photovoltaic generator,

                    2: Without excess power purchase by utilities,

                    3: With the excess power purchased at the marginal costs,

                    4: Reference case plus subsidy of 25% of investments.

4.3. Costs of Nuclear Power in the Future

As discussed in Section 3.3, nuclear power development has benefitted from its scale economies. In the meantime, large-scale nuclear power stations have been installed at remote sites. Recently, to find sites of new installations has become more difficult, partly reflecting anti-nuclear movements in most of the industrialized countries. It should not be overlooked as well that remote sitings require high voltage, long distance transmission which is costly and has been receiving concerns over its environmental impacts.

There is little evidence yet whether the scale economies of nuclear power plants have been lost. Cost reductions should be pursued further as before in accordance with scale economies, through standardization and learning which enable smooth applications and provisions of licenses.

Meanwhile, a new approach to nuclear development might enroll possibilities to create flexible options to cope with new trends of generation technologies. As discussed in Section 4.2, it seems that the trend of down-sizing of power plants will prevail electricity supply systems in the future. To apply the notion to nuclear might have twofold importance. First, specially designed small scale reactors, perhaps enhanced with inherent safety features, may be constructed in urban areas, which not only reduce transmission requirements but also enable better utilization of energy as cogeneration supply of electricity and heat. Heat supplies for industrial production will be specially important in view of cascade heat utilization at improved efficiencies.

Not only for urban utilizations, advanced nuclear rector concepts may also be beneficial to developing countries that have too limited financial allowances to purchase expensive energy resources. As well as enhancements of inherent safety, such reactors should be designed to allow easy operation and maintenance, less frequent and easy fuel exchange and handling. Compact designs might have another attractiveness as they might enable factory fabrication of the whole plants, reducing much of on-site construction and enhancing high standards of quality assurances.

As observed in Chapter 3, nuclear power is the only non-fossil sources of electricity which can compete economically with fossil fired generation technologies at present. However, the circumstances surrounding nuclear energy in most of industrialized countries do not allow easy optimism to the future of nuclear power. As well as conventional efforts to improve the current technology, new imaginations will be needed to extend nuclear energy to wider applications. It would be important to seek ways for nuclear technology to adjust changing tides in the world energy markets.

5. Concluding Remarks

This paper has attempted to provide a review of conventional methodologies and results of power generation cost evaluations, with emphasis on the missing factors in such evaluations. In addition, we tried to explore new trends in innovation by focusing on the scale economies of power supply systems.

We need new economies in electricity supply for sustainable development of mankind. Cost of power generation should be evaluated in an enlarged framework in terms of scope, time and space. To meet such requirements, we have identified some future directions in power technology development, such as decentralized power sources and integration of energy systems.

Internalization of remaining externalities are recognized globally as an efficient way to reflect global and regional environmental concerns to societies, and hence as one of the largest issues to be addressed in the formation of cost or price of electricity. As every country has constructed a unique system of the externality treatments, in different ways and at different levels, serious attention should be paid to maintain consistency of the whole scheme upon actual implementation of a new internalization arrangement.

As an economically viable CO_2 free power technology, and as a major energy source to replace limited fossil fuel resources in the long-run, nuclear power keeps an unique position in the new economies for sustainable development. As well as conventional efforts of technology improvements based on its scale economies, new imaginations will be necessary for nuclear technology to adjust to these new trends of innovation.

REFERENCES

Hondo, H. and Y. Uchiyama (1993); Economic Analysis of Gas Emission Control of Fossil-fired Power Plant, Denryoku-keizai Kenkyuu (Economics and Electric Utilities) No.32, Central Research Institute of Electric Power Industry (in Japanese).

IAEA (1984); Scaling Factors of SMPR –Considerations for Comparative Evaluations of Small and Large Nuclear Power Plants-, International Atomic Energy Agency, Vienna.

Ministry of International Trade and Industry (ed.) (1992); Sekiyu-shiryou (Data-book of Oil) (in Japanese).

Nagata, Y., K. Yamaji and N. Sakurai (1991); CO2 Reduction by Carbon Taxation and Its Economic Impacts, Research Report No.Y91002, Central Research Institute of Electric Power Industry (in Japanese).

Nakanishi, Y. and N. Itoh (1988); Economies of Scale in Japanese Electric Utilities, Research Report No.Y87017, Central Research Institute of Electric Power Industry (in Japanese).

Nanahara, T., K. Takahashi, Y. Nonaka and F. Arakawa (1989); Approach to Evaluation of Flexibility of Generation Mix, Proceedings for IFAC/IFORS/IAEE International Symposium on Energy Systems, Management and Economics (ESME 89), Tokyo.

OECD/NEA (1983); The Cost of Generating Electricity in Nuclear and Coal Fired Power Stations, OECD, Paris.

OECD/NEA (1986); Projected Costs of Generating Electricity from Nuclear and Coal-fired Power Stations for Commissioning in 1995, OECD, Paris.

Seki, K. (1992); An Estimation of Generation Costs by Sources, Enerugi-keizai (Energy Economics) Vol.17, No.12, The Institute of Energy Economics (in Japanese).

Tajino, K. (1988); On the Scale Economies and Interests during Construction of Nuclear Power Plants, Enerugi-keizai (Energy Economics) Vol.14, No.6, The Institute of Energy Economics (in Japanese).

Uchiyama, Y. and E. Imamura (1992); Development of Market Potential Model to Analyze Dispersed Power Generation Systems -Prospect and Market Potential of Photovoltaic System and Wind Power System-, Denryoku-keizai Kenkyuu (Economics and Electric Utilities) No.31, Central Research Institute of Electric Power Industry (in Japanese).

Yokobori, K. (1992); Internalization of Externalities as a Means to Enhance Public Acceptability of Energy -Case of Japan-, the 15th Congress, World Energy Council, Madrid.

Yuasa, T. (1992); Prospects of Generation Costs by Sources -A Scenario Analysis-, Enerugi-Keizai (Energy Economics) Vol.18, No.11, The Institute of Energy Economics (in Japanese).

SOCIAL COSTING RESEARCH: STATUS AND PROSPECTS

Robert W. FRI
President, Resources for the Future

Abstract

Internalising the costs of environmental and other externalities in electricity prices will, in principle, allocate resources to power generation more efficiently than command-and-control regulation. Recent research has made progress toward developing methods for calculating these full social costs. This research has already proved useful, especially in guiding state-level experiments in the use of social costing. Although difficult methodological issues remain, future research also promises to help policy makers use a variety of policy instruments more precisely and effectively. For this to happen, however, there must be a close link between policy and research communities in defining the research agenda.

ETUDE DES COUTS SOCIAUX : SITUATION ACTUELLE ET PERSPECTIVES

Résumé

L'internalisation des coûts d'environnement et d'autres externalités dans les prix de l'électricité conduira, en principe, à une affectation plus rationnelle des ressources à la production d'électricité, que ne le permettrait une réglementation contraignante. Récemment, les travaux de recherche ont fait progresser la mise au point de méthodes de calcul de l'ensemble de ces coûts sociaux. Ces travaux se sont déjà révélés utiles, en particulier pour orienter la prise en compte expérimentale des coûts sociaux au niveau national. Bien que certains problèmes méthodologiques subsistent, les recherches futures devraient aider les décideurs à utiliser un large éventail de moyens d'action avec plus de précision et d'efficacité. Pour ce faire, une étroite coopération entre les pouvoirs publics et les chercheurs sera toutefois indispensable pour définir le calendrier des recherches.

By now, the desirability of "internalizing externalities" is an almost commonplace idea. I suppose that it is about time, since the concept of externalities has developed in the economic literature for some sixty years. It was not until about three years ago, however, that social and physical scientists began to make a comprehensive and systematic effort to estimate the actual value of externalities in real situations. One of the leading efforts to do so is being sponsored by the Department of Energy and is being conducted by Resources for the Future and the Oak Ridge National Laboratory.

The interest is quantifying externalities is closely related to the prevailing urge to rely on markets rather than traditional regulation to achieve a variety of policy ends. An externality is, roughly speaking, a cost that is not reflected in a market transaction. Pollution is the canonical example, for the use of the air and water has no price in ordinary markets, but degradation of these resources certainly has a cost. If one knew this external cost, then the market price could be adjusted to reflect it. By internalizing the externality in this way, or "getting the prices right" in the current jargon, markets can allocate resources efficiently to pollution control. As command-and-control regulation becomes increasingly less efficient and more cumbersome, the market alternative looks ever more attractive.

"Getting the prices right" is one of those straightforward sounding ideas that one is more likely to endorse the less one knows about it. For example, the DOE project has demonstrated that producing the numbers needed to quantify an externality requires not only difficult calculations, but also advances in both theory and methods. I do not intend to take up these research issues today, however.

What I will try to do today is to suggest how the policy landscape looks from the research perspective, and in particular to elaborate on two conclusions about the status of social costing research. First, I believe that the work to date already has suggested specific ways in which social costing can help policy. But, second, it appears to me that if future research is to be equally helpful to policy, it will need clearer guidance from policy makers.

To address these conclusions, I must draw extensively on work done by my colleagues at Resources for the Future, and especially on a paper authored by Dallas Burtraw and Alan J. Krupnick for the Office of Technology Assessment. I have shamelessly borrowed their thinking and even their words, but they bear none of the responsibility for my remarks today. I should also point out that I will limit my remarks to the U.S. scene, although the European Community has conducted an impressive effort in parallel with the DOE project.

Before beginning, however, I need to say a word of background about the scope of the DOE project and about methodological framework it is using. The project itself aims to develop methods for estimating the full social costs of a new increment of electric generation capacity for several fuel cycles. "Full social cost" is the sum of the unpriced externalities associated with electricity production and the private costs -- fuel, labor, capital -- that are captured in market transactions. In principle, a fuel cycle with low private costs could have large externalities, and so a high social cost. Of course, the converse could also be true.

The object of the exercise is to select the fuel cycle that minimizes the total social cost. Developing a policy to do this requires three distinct steps, and it is crucial to any discussion of social costing to keep these distinctions clearly in mind.

1. The first step is to estimate the economic value of the physical impacts caused by building a new electric generating plant. Thus, the plant emits pollutants into the environment, these pollutants affect human health, and the health impacts have an economic cost. There can also be beneficial impacts, and their economic value can also be estimated. Although both benefits and costs may be involved, this first step is called damage estimation.

   Since I will not return to the question of estimating benefits again, let me note here that the analytic framework developed for social costing turns out to be extremely helpful in sorting out which of them are in fact real. I refer you to a paper by Douglas R. Bohi that will be presented later in this conference for a critical analysis of some of these benefits.

2. The second step is to determine how much of the damage is an externality. The externality is that part of the damage that is not already included in the market price, and it is common to find that some internalization has already taken place. For example, if environmental regulations require a power plant owner to install pollution control equipment, then the cost of doing so will show up in the price of electricity. In some cases, the damages may already be fully internalized, as is arguably the case with the tradeable emission allowance program for sulfur oxides in the U.S. Because damages and externalities are not equal, it is critically important to understand the difference between the two.

3. Finally, policy makers must choose the instrument that they prefer to use to internalize the remaining externality. Many factors other

than the economics of social costing will influence this choice. For example, emission fees and tradeable emission allowances are equally effective policy tools in economics theory. However, they are not equivalent politically, as anyone who has followed the energy tax debate in the U.S. can testify.

Most of the work on the DOE social costing project has thus far concentrated on the first of these steps, estimating the damage function. Moreover, the research has focused on the damages caused by a single new power plant. This approach helped make a very complex methodological problem more or less tractable, and so was a wise research strategy. However, this strategy necessarily leaves open a number of questions about the value of social costing as a basis for policy action.

It is my view as an interested observer that the research thus far completed has demonstrated that social costing can contribute usefully to policy. This conclusion is not as self-serving as it may sound, given that the research began with the expectation that it would be relevant to policy. The research has lived up to some of the original expectations, but has also suggested that it has much richer policy implications, as well.

The most direct contribution of research to policy is to help guide the state-level experiments in social costing that are already underway. A number of state regulatory commissions are using the concept to encourage electric utilities to make investment decisions in new generating capacity that more nearly reflect full social cost. Since large investments and long-term consumer welfare are at stake in these programs, it makes sense for economists to help ensure that the economic principles and methods of social costing are correctly applied.

There is, of course, considerable room for differences among state approaches, especially in the final step of designing policy instruments for implementing social costing methods. Nevertheless, it is possible to get the earlier steps in the process wrong. For example, damage functions are very hard to estimate, and so some have suggested using the cost of compliance with existing regulation as a proxy. Aside from heroically assuming that existing regulation is in some sense optimal, there are other serious problems with this alternative. Research can, and indeed has, thrown light on these problems and has helped regulators treat them with greater understanding.

Getting social costing right in regulatory proceedings is not simply a matter of using the correct methodology, however. There is also the potential for using social costing as a reason for promoting policy objectives that have larger implications. For example, because social costing is being applied only to electric utilities, the power to impose it is left to the states as a matter of economic regulation. This can result in state economic regulation being used to make environmental policy without participation by federal environmental policy makers -- or state ones, for that matter. Additionally, it is my observation that some advocates have learned that social costing provides a lever to get policies approved at state level that have not been enacted nationally. Imputing social costs to carbon dioxide control is one example. The conceptual framework developed for the DOE project is extremely helpful in spotting these excursions beyond the limits of social costing.

Helping to bring rigor and understanding to state-level experiments in social costing is an obvious and immediate application of research to a current policy issue. Indeed, doing so was an early motivation for conducting the DOE project. But the conceptual framework and applied methods that researchers have so far developed have also produced some less predictable insights into ongoing policy debates. For example, social costing research has illuminated some limitations of command-and-control regulation that may not have been widely recognized before now.

One contribution of social costing research to this regulatory debate lies in the location-specific nature of the damage estimate. The damages calculated for a new power plant at a particular site depend importantly on local conditions. In contrast, most current command-and-control regulation is nationally uniform, especially regulations that mandate best available technology. Locally specific damages estimates can be greater or smaller than the costs of implementing these national standards. Knowing this could help in the design of more cost-effective regulation.

In principle, of course, command-and-control regulation could require the owner of the power plant to spend exactly the right amount of money on pollution control equipment; that is, to spend exactly what would have been spent if the investment decision accounted for full social costs. But even for this unlikely outcome, command-and-control regulation would not have fully internalized the external costs. This is because there is always some damage left after pollution controls are put in place, and command-and-control regulation does not allow the residual damage cost to be reflected in market prices. If this cost were fully internalized, the price would be higher and the quantity of electricity purchased would be lower. This research result has important implications for the design of both environmental regulation and utility ratemaking.

I hope that I have now persuaded you that research into social costing has already produced results, some of them beyond original expectations, that are useful in current policy debates. For future research to proceed in an equally productive fashion, however, it seems to me important that the research agenda be guided by the needs of policy makers. It will be useful to forge a closer link between policy makers and researchers to ensure that the latter are working on the right questions. As important, policy makers need to understand enough about the research know what the right questions are. To illustrate this latter point, consider three methodological problems that future research could attack.

1. At the moment, social costing is being applied only to the electricity sector. Obviously, electricity generation is not the only source of some damages, notably pollutant emissions, and in some cases is not even the major source. Excluding other sources from the social costing analysis may therefore result in distortions. Nationally, there could be some efficiency losses if non-utility sources could reduce emissions more cheaply that utilities. Within the electricity sector itself, this piecemeal problem could disadvantage regulated utilities versus unregulated generators, and could lead to some large customers bypassing the grid entirely.

2. A second problem involves transboundary effects. These effects occur when an action taken in one regulatory jurisdiction creates damages in another; as, for example, when California buys power from a plant in Arizona. The issue is whether the California customer should pay for damages incurred in Arizona, and if so, how the residents of Arizona can be properly compensated.

3. Third is the issue of the scale of the policy being decided. Methods already developed can be very helpful in defining the social costs of building a single new power plant. However, a single plant has little bearing on national policy regarding, say, energy security. The methods for applying social costing to the analysis of national policy may be different from the methods have been developed for local analyses. In fact, the methods needed depend on the question asked. If the question is how the construction of a large number of new power plants of different social costs would affect national environmental goals, then the analyst needs to find a way to aggregate these costs. But if the question is how to implement a national policy on energy security by imposing a tax that reflects this social cost, then difficult issues of utility ratemaking present themselves.

Each of these topics opens up a challenging research agenda. However, whether the research proves useful for policy depends almost entirely on what question the policy maker wants to ask. Given the well-known tendency of researchers to do research, and of policy makers to ask questions of research that no one can answer, it seems to me essential that there be a close link between the two communities in defining the agenda for future social costing research. Happily, the social costing research already in hand has policy implications that are clear enough to permit a constructive dialog to take place.

A final but important word about asking the right question. It seems to me that the DOE project has demonstrated that "getting the prices right" is not a particularly useful goal for social costing. It would be better to aim to "get the policy right," whether the policy instrument involves price adjustments, tradeable allowances, or command-and-control regulation. The methods of social costing can help policy makers use any of these and other instruments more precisely and effectively. Thus, the benefits of this research can be richer and more immediate than the single-minded pursuit of the "right price".

Session 1 - Discussion

J. Gray (Chairman)

I would like to emphasize that when this symposium was first considered, those of us invited to help in its organisation concluded that it was not interesting to have a symposium that dealt singularly with the externalities associated with nuclear power. In effect, this was not to be a nuclear power meeting, but rather an electric power generation meeting, and that it would be much more interesting to see nuclear dealt with in context, along with all other fuel forms that might be available.

My question would be addressed to Yodasan: as you look ahead, say 20 years, what do you see the generation mix being in Japan, in terms of fuels, as compared to the generation mix at the present time?

S. Yoda

As I look at 20 years hence, I imagine that the share of the nuclear power in the Japanese energy supply picture is going to be as big as it is permissible, given the situation prevailing in that time frame. The reason is that the environmental cost of fossil fuel power generation will become very considerable, so that with further development of technology, the relative economic advantage of nuclear power would gain in importance in the time frame of twenty years to come. So, I would like to say again that the share of nuclear power would become bigger than what it is at this moment.

J. Gray

Let me ask if you would give us simply a guess as to the percentage level compared to today's percentage level?

S. Yoda

In terms of the percentage, while the share of nuclear power at this moment is close to 30 per cent of all the electricity generated in Japan, (it) is expected to increase to 40-50 per cent in 20 years; I think it should be infinitely closer to 50 per cent at that time.

S. Barabaschi

Yodasan, you mentioned the centralised power generation. Would you comment on the technologies that you foresee for this decentralised power generation in Japan?

S. Yoda

In Japan, distributed and decentralised co-generation will gain in importance. I believe co-generation can make a very important contribution to society because it can provide both heat and electric power to the community concerned. In this respect technological progress is expected to take place, in the area of co-generation as well as the development of a fuel cell, and solar cells.

M. Thibau

I would like to ask Mr. Yoda if Japan has unified the frequency or if it is still divided into 50-60 cycles?

S. Yoda

We still are divided into 50 hertz and 60 hertz in Japan at this moment. This is because of the tremendous cost involved in trying to unify this frequency into one.

K. Uetmatsu

I suppose the major interest on the part of many non-Japanese participants in this symposium, is, what would happen to the nuclear energy policy in Japan under the new Hosakawa administration? I would appreciate it very much if Mr. Yoda could share with us his personal view of this matter.

S. Yoda

I personally believe that the Hosakawa administration in Japan will have a positive effect upon the future of nuclear power generation policy in Japan. This is because those people who used to be in the opposition in the past, who were very critical of nuclear power development in Japan, are now in responsible positions in the government at this moment. I suppose they will take a more realistic view of the total Japanese energy picture. Therefore, they would speak more responsibly in the future vis-à-vis the nuclear power generation policy in this country. I suppose the existence of a new government in this country, in Japan, will have a very good effect upon nuclear policy.

J. Gray

I am reminded as we consider that kind of question, i.e. the impact of political change on nuclear power, that one of the early practitioners in the consulting world in the United States was Jack Hogerton, who was associated with a man named Sid Stoller who would be known to many of you. One of Hogerton's favourite questions, 25 years ago was "Can nuclear power survive democracy?" We would hope that the Japanese would help demonstrate that nuclear power can in fact survive more democracy, and we thank you for a great effort on that behalf.

J. Grawe

You (referring to S. Tierney) didn't elaborate very much on bidding procedures. If I understood you correctly, you said that in your opinion bidding procedures might be able to internalise externalities in the best way, or at least in a sufficient way. I was surprised to hear that remark. I am not sure that I understood you correctly because in my opinion every bidder will of course try to deliver the least cost option. He will, of course, regard the existing legislation and just do what the existing legislation would demand him to do, and probably not invest one dollar more on his own, because he has to fear that the competitors will offer cheaper solutions. So I did not understand why you are of the opinion that the bidding process, which may have a lot of benefits and advantages, would just solve this particular problem of internalising externalities.

S. Tierney

In my last job as a utility regulator, we struggled exactly with that question.

I did not mean to say that I believed bidding procedures fully internalised costs. I do think that many of the risks associated with changes of different parameters are internalised in the following case: in my State, we required, when bids were made, that there could not be contract re-openers. You had to bid at a contract price and there could not be contract re-openers in the changes in the environmental laws. We felt that the bidder backed up by the lender had to absorb some risk to accommodate changes in regulations that that project would be exposed to. Those who found out that there was a change in regulations, left the market place, and we are still faced with a lack of supply.

Then we decided that we would have to state what environmental equipment assumptions, what emissions, pollution emissions would have to be internalised to get a common bid. Other States still do include these adders on the residual emissions. You are right, it does not fully internalise it. I have spent many intriguing afternoons dealing with exactly the trade-offs that you are describing.

Unidentified

I think this would be addressed to Dr. Tierney. Can you identify for me some specific cases where a regulating agency or commission has actually mandated the selection of a higher real cost alternative that had lower social costs, thereby imposing on its ratepayers higher costs than the lowest real-cost alternative?

S. Tierney

There were numerous States that have proceeded with higher energy efficiency investments than they would have for the reasons we described, which is that if one were looking only at dollar costs and not internalising the externality value, you would find that X amount of energy efficiency investments were economic.

In many States that have decided either a 10 per cent premium for energy efficiency, there is a higher level of investment than would otherwise take place. There are many States, New York, Massachussetts, California and a long list that have found themselves in that situation.

Unidentified

I know that there are a number of States that have that. I guess what I am looking for is a specific case of where a new power generating plant, where the technology selected was the higher real-cost vis-à-vis the lower real-cost through the external environment.

S. Tierney

I believe that there are some, and I wish that I could give you a specific example. I am sure that there are some in Massachussetts for example.

D. Kopp

Both speakers (S. Tierney whose paper is unavailable and R. Fri) in their presentation used the term "residual pollution". I was rather interested in that term. One can accept the idea that the current level of regulated control is clearly not optimum, perhaps the entire debate we are having here is about that, but does this mean that in the United States, you are not accepting anything like threshold effects, that you are rejecting, for example, concepts of the outlying buffering of soils in certain areas regarding, say, acidification processes. You are automatically assuming a linear dose relationship in terms of human health impacts, and so you are not accepting that there is any threshold below which, in a sense, an emission is not pollutant, but simply an emission and therefore not, presumably, incurring any costs disadvantage. I am very interested in your apparently rejecting any threshold concepts in your attempt to evaluate externalities.

S. Tierney

There are wild debates going on among researchers and policy makers on the subjects you are raising in your question: whether there are thresholds, what is their character, whether and probably not, the effects are linear, associated with pollutants in dose response relationships.

I think that what you are hearing from me and Bob, is nomenclature that has developed in this country as we have tried to deal with the externality/internality issue. When people begin to use the terminology "externality", people who know about pollution control equipment say "well there are a lot of costs internalised". So I observe that we have developed language to deal with what we call the remaining amount of emissions. I think that I am probably sloppy in calling emissions the same as pollution and probably people get Ph.D.s in that debate, but I didn't. So my mind is probably sloppy, but we call residual to mean the remaining amount after the pollution control equipment that was put in place at the time that plant went into operation. You can have two plants that are both legal. One built 25 years ago, with yesterday's laws and one built with very strict standards, (but) which have different levels of residual pollution.

R. Fri

The term of art, I believe, although I didn't get my Ph.D. in it either, is residual damage. In some cases it has to do with threshold effects and that kind of thing, but let me give you an example that I have been given that helps me think about it. If there were absolutely strict liability imposed on transportation companies for oil spills, that is to say, if a damage occurred, the company involved (would be) responsible (and) would have to pay all the damages in full. The companies would act to prevent oil spills, double hull tankers and so forth. There would still, however, be the occasional oil-spill. As a matter of economics, the trade-off is simply that it gets too expensive to try to control it and you just pay off the few that happen. The oil spill that is left is the residual damage even though the costs are in that case fully internalised.

Now you can see why people get Ph.D.s in this.

Unidentified

I have one very specific question for Sue Tierney. In the Energy Act last year there was a provision for, I believe, $250,000 grants to individual States that wanted to hire people in the resource planning. But in its wisdom, as I understand, Congress did not allocate the money. I was wondering if there has been an update on that and whether the money has been appropriated?

S. Tierney

I don't think it has, but we have been trying to figure out how to work with States to deal with IRP (Integrated Resource Planning) even in the absence of appropriations. One of the specific things I did, with regard to that, was to do a customer survey with States with regard to their continuing interest in the kind of research that is in the EC/Oak Ridge DOE fuel cycle study. So we are continuing to do that kind of research, even though that is not part of the appropriations that you are talking about.

T. Suzuki

I have a question for Dr. Fri. When you try to calculate the cost of damage, it could be very difficult and extremely large. When you try to calculate the prevention cost, the prevention of such damage, it could be small. For instance, the cost of an airbag may be much smaller than paying hospital costs. Could you explain to us, could you give us ideas, which is better, to try to internalise external costs?

R. Fri

Well to take the airbag example, the cost of the accidents are in fact quite high. If the cost of adding airbags is low relative to that, then the benefit/cost ratio is very favourable. What happens is that in effect costs of the damage become fully internalised. The excess benefits are then what in economics are called an economic surplus, an extra benefit to the consumer that they don't have to pay for because it turns out to be cheaper to get the benefit then the value of the benefit, and that is a very good thing to have happen. You look for those, but at the margin, in theory at least, benefits equal costs, and that is where you stop.

S. Tierney

Can I add an example to show why I am not sure the question worked from my experience? In this country or any State with very high standards of pollution control on the margin, the marginal cost of control, the dollar spent to reduce the pounds of pollutants, is very, very high and therefore begins to exceed the damage function at some point.

R. Fri

There are two ways of estimating the costs: one is the damage function approach, the other is called an abatement option approach. At some point, they are the same. The trouble which Dr. Tierney just pointed out is that the sign of the first derivative is different at that point.

The abatement cost is going up because the marginal cost pollution control is going up, and marginal benefit is going down for each dollar spent on pollution. This is one of the problems, in my judgement, with using the abatement cost. In some cases, it may be a useful proxy because of the difficulty in calculation, but in cases, particularly where Dr. Tierney said the costs were high, what you really want to do is going in one direction, and what you are doing is going in another direction, and that can get you in very big trouble.

Unidentified

A question for Dr. Tierney. While various States are going ahead with either developing adders or some sort of mechanism to bring the externality into the bidding process (for example, Massachussetts, New York, now have some existing regulations), what is the Federal policy view on this issue and is it near-term or long-term in the sense of how soon the Federal Government would say "New York you can't do that alone, Pennsylvania, you should do that, Ohio, you should do that", because in boundary cases you would be shifting cost from one region to the other?

S. Tierney

I might give you two pieces of information in response to your question. One of them is our system of Federal/State joint regulation changes: the State regulates some things and the Federal Government regulates others. The Federal Government is not in a position to tell the States what to do because it is they who make the decisions in their jurisdiction and not the Federal Government. Because I come from a State, I am particularly wary of that approach. But the Federal Government regulates prices of wholesale transactions through the Federal Energy Regulatory Commission. To date the FERC has not moved to internalise externalities in the kind of way we have seen in States. There is one instance where the Federal Government just recently started on this, and that is on our authority to set up a plant efficiency standards. In the cost/benefit analysis that we will use to determine what level of standard to require, we have asked the public to guide and comment on how we should incorporate externalities that we know are not valued as zero. That is as far as we go at the moment.

REMARKS - GLOBAL ENERGY OUTLOOK AND EXTERNALITIES

J.E. Gray
Past Chairman, U.S. Energy Association

I would like to present a global energy outlook for the period 1990-2010, and views on the subject of externalities.

My global outlook is both current and dated. It is current because I've just updated it using the latest available OECD data. It's dated because I presented the same basic information before, at the WEC Conference in Madrid in 1992. In doing that, I there concluded, with the help of some of you here tonight, that there were no truly significant changes in the global outlook since the WEC Conference in Montreal in 1989, where I first summarised such an outlook.

So from a global perspective, the more things change, the more they are the same. This reflects both the stability and the ponderousness of the global energy supply and use systems, and their relative indifference to short term changes in national energy policies.

Let's look at the outlook for the period:

→ Resources are plentiful; distribution and price appear manageable.

→ Efficiency is a global imperative; to be driven principally by price and by regulation.

→ Oil demand is up by 39%, to 4200 mtoe, with attendant supply, price, transport and environmental issues.

→ Gas demand is up, by 70%, to 2800 mtoe, also with supply, price and transport problems.

→ Coal demand is up, by 44%, to 3300 mtoe, with attendant environmental and transport issues.

→ Nuclear power demand is up, by 24% to 690 mtie, with attendant financing and environmental issues.

→ Hydro demand is up 79% to 330 mtoe.

→ Alternatives, including geo-thermal, are up 320% to a level of 109 mtoe.

→ Electricity demand is up 71%.

Market pricing is seen as becoming pervasive and all energy related environmental and health and safety issues are seen as being manageable within the various national imperatives and trade-offs governing the balance between economic development and environmental priorities.

There will continue to be extensive analysis and consultation, as well as differences in opinion as to what's to be done on a global basis, regarding trans-national environmental and health and safety issues. Sorting out of issues, options and aggressive pronouncements will continue. Steps may be taken to further structure a global energy/environmental decision-making capability which is both serious and formidable. Such a structuring will go slowly in the absence of broader, more global perceptions of a "clear and present danger".

The situation may be typified by the old Chinese saying, "Big noise on stairs; nobody coming down".

Now a word on externalities -- related to energy supply and use. I first discovered and perhaps understood this class of issues in 1972 when I got involved with the Ford Foundation Energy Policy Project to do a study of decision making in the U.S. energy industry. Walter Mead of U.C. Santa Barbara was the staff economist. I adopted from Mead the understanding that the only thing to do with externalities was to internalize them, or stop talking about them as having a bearing on current economic driven decision making, re planning, investment, operations, or fueling. Above all, don't politicize what should be economic.

Subsequently, we have continued to learn that internalization carries with it costs which may or may not be acceptable to the suppliers and, most important, the users of energy. These are the people. We've also recognized that there are costs which are of a nature and magnitude that preclude considering internalization, such as the war in Iraq. There are other costs which gain national or regional acceptance as the result of the right mix of perception, science, politics, economics, and public acceptance, such as those associated with acid rain. These examples relate to defining desired results and the basis for paying and allocating the consequent costs. There is often an associated conviction that all other prospective costs of such actions are somehow considered and accounted for. Maybe. Sometimes they are ignored, sometimes unknown. This applies to both of my examples.

This has led to my own proscriptions regarding internalization of costs:

1) You can have any product, or service, with whatever characteristics you can afford. Short form is you can have any environment you can afford.

72

2)   Be careful to resist internalization unless the costs and the benefits are clearly understood, and are acceptable in the market.

3)   Be careful to insist on internalization where the costs and the benefits are clearly understood and are acceptable in the market.

4)   Be careful in internalizing costs which are not internalized by your national or corporate competition, unless the benefits enhance economic performance.

5)   Be careful in internalizing costs where you may, or may not, be able to capture the benefits, or the benefits may be illusory.

6)   You need not be careful in internalization if you know what you want, can afford it, can live with the consequences, and do no harm to others.

7)   Keep trying! Think big!

In closing, I note that a couple of major players in energy have moved towards managing the externalities, rather than internalizing them. One of these is the World Energy Council which appears to be thinking about changing from an industrialized, supply side orientation, towards one with a demand side, developing country, wealth transfer orientation. The other is Shell International whose strategic planners presented, in a briefing I recently attended, a serious vision of how to assure that the future world population, of many more billions, consists mostly of "middle class" citizens rather than having a small percentage of rich and a very large percentage of poor. One learns by questioning that in getting to that, yes, desirable world, the Shell planners presume that national governments should and must go away. They may or may not be correct in that view, but it certainly demonstrates thinking big in internalizing the externalities.

I shan't go further into the fabric of the meeting and will close by thanking you for your attention and again for choosing to be with us at this symposium.

Chairman/Président
J. Yasinsky, Westinghouse Electric Corp.

Power Generation Choices: A Canadian Perspective
E.A. Marriage, Ontario Hydro
L. Masson, Hydro-Quebec
G.E. Gunter, New Brunswick Power

How Generation Choices are influenced by Costs, Risks and Externalities: The Generation Planning Process in Ontario, Canada
E.A. Marriage, M.S. Rodgers, Ontario Hydro
(Additional paper, not presented orally)

Power Generation Choices: An International Perspective on Costs, Risks and Externalities
R. Carle, G. Moynet, Electricité de France

Generation Choices as Influenced by Costs, Risks and Externalities - Germany
L. Strauss, Bayernwerk Aktiengesellschaft

Generation Choices as Influenced by Costs, Risks and Externalities - Sweden
C.-E. Nyquist, Vattenfall

The Economics of Nuclear Power and Competing Technologies in the U.K.
B.L. Eyre, P.M.S. Jones, AEA Technology

Discussion

POWER GENERATION CHOICES: A CANADIAN PERSPECTIVE

E.A. Marriage, Ontario Hydro
L. Masson, Hydro-Quebec
G.E. Gunter, New Brunswick Power

Abstract

Canada has abundant reserves of all the major energy resources but regional variations in their distribution have shaped the development of the utilities serving these regions. Since the 1970's rising costs have reversed the earlier downward trend and the risks associated with very large projects have changed the approach to planning. Overall, however, environmental requirements have had the greatest impact on utility planning. The rising need to consider residual environmental impacts or externalities and the growing importance of demand side measures are becoming major influences.

LE POIDS DES COÛTS, DES RISQUES ET DES EXTERNALITÉS DANS LE CHOIX DES MODES DE PRODUCTION D'ÉLECTRICITÉ

POINT DE VUE DU CANADA

Résumé

Le Canada possède d'importantes réserves des principales ressources énergétiques, mais leur répartition géographique a joué sur le développement des compagnies d'électricité desservant les différentes régions. Depuis les années 70, l'augmentation des coûts a inversé la tendance autrefois à la baisse, et les risques associés aux très grands projets ont modifié les modes de planification. Toutefois, d'une façon générale, ce sont les impératifs d'environnement qui ont pesé le plus lourd dans la planification des compagnies. La nécessité de prendre en considération les impacts résiduels sur l'environnement ou les externalités environnementales, et l'importance croissante des mesures de gestion par la demande jouent aujourd'hui un rôle prépondérant.

INTRODUCTION

The generation and utilization of electrical energy has gone through some remarkable changes over the past century. Initially, the objective was to provide and distribute electricity to as many customers as possible at the lowest possible cost and until the late 1960's, the cost of the product steadily declined. Since that time, the trend has reversed. Cost increases have been due mainly to inflation and the need to reduce environmental impacts.

In the past, the economics of scale as well as favourable long-term cost projections led many utilities to commit to large generation projects. However, it has gradually become clear that schedule delays, cost overruns, regulatory changes and other factors can turn a large project into a financial burden. These risks have become major concerns in the planning process.

More recently, attention has moved to the area of residual environmental impacts, that is, the impacts that remain after the measures to comply with the regulations have been taken. These residual impacts have become known as externalities. The issue, of who pays the cost of these externalities is also becoming a concern.

Increasingly the selection of generation options is being based on rigorous consideration of not only traditional criteria such as cost and security but also financial and other risks and externalities.

THE CANADIAN UTILITY SCENE

This paper will address the topic of power generation choices from the perspective of three Canadian utilities: Ontario Hydro, Hydro Quebec and New Brunswick Power.

Table 1 compares sources of electricity generation by fuel type for the period from 1960 to 1992. Table 2 shows the distribution of electrical energy production by fuel type and by province for 1992.

DESCRIPTION OF CURRENT OPTIONS

Tables 1 and 2 show that while hydro dominates as an electricity source, fossil and nuclear are major contributors. Large fuel reserves, both fossil and nuclear, and major undeveloped hydro sites will ensure that these sources will continue to receive consideration to meet future generation requirements.

Table 1

Sources of Electricity Generation

Fuel Type	1960	1970	1980	1990	1992
Hydro	92.6	76.5	68.0	62.8	62.2
Thermal	7.4	23.0	22.3	22.4	22.6
Nuclear	----	0.5	9.7	14.8	15.2
Total	**100%**	**100%**	**100%**	**100%**	**100%**

Table 2

Electrical Energy Production by Fuel Type, 1992

	Coal	Oil	Natural Gas	Nuclear	Hydro	Total
Nfld	0	4.9	0	0	95.1	100%
P.E.I.	0	100.0	0	0	0	100%
N.S.	61.7	29.1	0	0	9.2	100%
N.B.	8.0	43.1	0	30.3	18.6	100%
Quebec	0	0.8	0	3.1	96.1	100%
Ontario	19.8	1.3	2.2	48.1	28.6	100%
Manitoba	1.0	0.1	0.2	0	98.7	100%
Sask	73.4	0.5	4.4	0	21.7	100%
Alberta	85.6	0	11.1	0	3.3	100%
B.C.	0	1.1	4.4	0	94.5	100%
Yukon & N.W.T.	0	30.1	0	0	69.9	100%
Canada	**17.2**	**3.1**	**2.3**	**15.2**	**62.2**	**100%**

Source: Energy, Mines and Resources Canada

The following are the principal options for new generation and some of the major issues associated with them.

Nuclear

Nuclear energy is a proven technology with high standards of safety and environmental protection and is used throughout the industrial world. In Canada, all operating nuclear plants use CANDU pressurized heavy water reactors fuelled with natural uranium. Ontario Hydro, which has over 90% of the nuclear capacity in Canada began its commercial nuclear program in the 1960's with its decision to build four 500 MW units at Pickering. Between 1971 and 1993, 5 large nuclear stations, each consisting of four units were placed in service. Hydro Quebec and New Brunswick Power each have one 640 MW CANDU-6 unit installed on their system.

The existing Candu 6 and 9 designs will continue to be available, probably in improved versions. They are joined by the 450 MW CANDU-3, a new design being offered by AECL. It offers modular design, passive safety features, and very short construction schedules.

The Candu 3 might be built for about the same unit cost as the 700 MW Candu 6 but it is a new design and while evolutionary in nature there are a number of departures from experience. The risk of regulatory action with the new design can affect cost and performance.

Nuclear plants are capital-intensive with relatively long project implementation periods, but low operating costs. Nuclear generation does not produce significant atmospheric emissions, however; the irradiated fuel must be stored for a long time. For the moment, in Canada, storage at nuclear plant sites either in pools or dry canisters is satisfactory, but work is underway to develop a permanent storage facility. Work is required to improve the public perception of nuclear power.

Hydroelectricity

Hydroelectricity is the least expensive future option in Quebec, as well as Manitoba and British Columbia. Nearly all of the economically viable hydro electric potential in the rest of Canada has already been developed.

In Quebec hydroelectric stations are divided into three categories: large, medium and small. In large river systems such as Grand Baleine and Nottaway-Broadback-Rupert, along with other smaller sites like Sainte-Marguerite, Haut Saint-Maurice and Ashuapmushuan, Quebec has an economically viable potential of over 14,000 MW. There is also the potential of developing Quebec's medium-sized and small rivers. As for small hydroelectric stations, preference is given to proposals from independent power producers for plants generating less than 25

MW and offering a purchase price comparable to the avoided cost of facilities in the Hydro Quebec system.

Ontario has about 12,000 MW of undeveloped hydro electric potential but most of it is unattainable because of economic and environmental considerations. About 400 MW could be developed in New Brunswick.

Hydroelectricity involves sizable capital outlays and long construction periods, but the operating costs are much lower than for any other means of generation. Hydroelectric developments have environmental and social effects and although plants are designed to maximize the net benefits, some effects such as the flooding of lands and the temporary increase in mercury in reservoirs, cannot be avoided.   On the other hand, hydroelectricity is a renewable resource and produces minimal atmospheric emissions.

Fossil Fuel

The main fossil fuel options are coal and natural gas. Oil will be used for combustion turbine peaking plants where natural gas is not available but new oil fired base load plants are unlikely. Other fossil fuels such as Orimulsion^{TM} will be attractive in certain cases. A 300 MW plant in New Brunswick is being converted to use this fuel.

Pulverized coal plants continue to be the base load technology of choice in a number of regions of Canada and a 165 MW fluidized bed plant is currently being commissioned in Nova Scotia. The newest pulverized coal plants employ flue gas desulphurization and low NOx burners. Future plants may require selective catalytic reduction or other high efficiency NOx control technology.

Advanced coal technologies, including various coal gasification processes and pressurized fluidized bed combustion offer superior environmental performance. In general, all emission streams, gaseous, liquid or solid will be of lower volume and more benign than for a conventional plant. The economics of the advanced technologies are, however, questionable at this time. Development will no doubt clarify and improve the picture but the conventional plants may retain their favourable economics. The advanced technologies will probably involve greater technical risk than conventional plants for some time. On the other hand, the environmental performance of conventional plants may expose them to the risk of stricter environmental requirements and result in higher costs.

The ability to control plant emissions and to deal with waste and residual impacts is arguably superior for the nuclear plant than for a coal plant. In the future this situation could become a risk for coal plants in that regulatory action may be taken to control carbon dioxide emissions and solid waste to the same degree as emissions from a nuclear plant are controlled.

Gas and oil fired combustion turbine units remain attractive for peaking capacity. Investment costs are low, implementation periods are short but operating costs are high. Environmental impact depends on the fuel selected but with their low capacity factors it is generally low.

Combined cycle plants have many attributes: low capital cost, short schedules, and good environmental performance. As long as gas prices remain low and the supply is assured, these plants will be an attractive option.

Fuel cells use hydrogen as fuel. It is produced from natural gas or syngas derived from other fossil fuels. They are at an early stage of development and will not be available as a central station option in the foreseeable future.

Alternative Energy Sources

Alternative energy sources, particularly wind and biomass are being considered by several Canadian utilities for demonstration or other small scale projects in specific locations.

Wind energy may offer some promise as a long-term option in Quebec and in western Canada. In Alberta work is proceeding on about 20 MW of wind energy development and Quebec is proceeding with feasibility studies and demonstration programs.

Biomass, mainly wood and wood waste are widely used in the pulp and paper industry for raising process steam usually with co-generation. There is also renewed interest in dedicated wood fired generating stations.

Interest in these options and others such as photovoltaics is strengthened by virtue of their being renewable and the fact that they offer supply security and local development advantage. Their potential is limited however and they are unlikely to make a significant contribution to Canadian electricity requirements in the near future.

Demand Side Management

Demand Side Management (DSM) initiatives to reduce consumption and better utilize existing low cost base load generating capacity and thereby reduce the need for additional capacity are being given high priority by most Canadian utilities.

DSM programs in Canada are directed towards the main customer classifications. In the residential sector, efficient lighting, heating, insulation and appliance programs are the most common and are typically encouraged through rebates or cost incentives. Commercial and industrial DSM programs are developed to provide incentives to cover the cost of converting to more energy-efficient equipment for lighting, space heating, ventilation, air-conditioning, water heating and mechanical drives.

UTILITY VIEWS ON PLANNING

Before dealing with some of the specific issues of planning, an overview of utility views is of interest. In general, the recent completion of new generation facilities coupled with low load growth means that few if any major capacity additions are required before the next decade. Looking to the future, new approaches to generation planning are being considered by all utilities.

Ontario Hydro

Ontario Hydro's electric power system has evolved from one which utilized only hydro-electric stations (at the end of the 1940s) to one which was dominated by coal-fuelled stations (by the mid-1970s). The shift in emphasis from hydro-electric stations to coal-fuelled stations resulted from the unavailability of further economically viable hydroelectric sites. By the early 1970's, Ontario Hydro again shifted its emphasis from coal-fuelled to uranium-fuelled stations because of the need to reduce dependence on imported coal, the economics which favoured nuclear stations, and the desire to utilize locally available nuclear technology.

In the past 10 to 15 years, much has changed in power system planning in Ontario and in the rest of the world. The changes can be traced to several major causes: a) the emergence of environmental issues as a primary consideration, b) increased awareness of the financial, regulatory and other risks associated with each generation option and, c) the uncertainty associated with load forecasts and the impact of DSM programs on electricity demand.

The largest impact on the planning process has undoubtedly resulted from the emergence of environmental considerations as a key decision factor. The planning process has become more open as the involvement of the public and of interest groups in the planning process has become the norm. DSM receives priority as an option to moderate future load growth. For new facilities Ontario Hydro plans on the basis that the most appropriate emission control technology (recognizing environmental impacts and costs) would be utilized to mitigate any harmful impacts of emissions to the environment from these units. Also, much effort is being expended on quantifying the environmental costs of the "residual emissions" and to "internalize" these costs in the cost of the generation option.

Environmental considerations, public preference and financial, regulatory and other risks favour alternative and renewable technologies such as solar energy

83

and wind turbines. Non-utility generation is also viewed as having lower environmental impacts due to renewable or natural gas fuel usage and in most cases, smaller plant size. Cogeneration is considered an attractive option due to the improved efficiency in fuel utilization.

Changes in the planning environment were reflected, in part, in the Demand/Supply Plan proposed by Ontario Hydro in 1989. In the plan, highest priority was given to the use of demand management to moderate future load growth. Next in priority was maximising the use of the existing generation through selected rehabilitation of existing units and retrofitting of environmental controls. The plan also included non utility generation, the orderly development of the remaining economically viable hydroelectric capacity in Ontario and the purchase of 1000 MW from Manitoba Hydro. The plan also recognized that there could be a role for alternative energy options in the future when these options become feasible. Only after all the economic options from the above sources had been fully considered did the plan require the building of new major fossil-fuelled and nuclear generating stations.

For the planning process primary criteria included customer satisfaction, reliability of supply, worker and public safety, environmental requirements and standards, cost, social acceptance, technical soundness and flexibility. Secondary criteria included resource preferences, resource smoothing, environmental characteristics (in addition to requirements and standards), economic impacts and other social considerations. The proposed plan represented a tradeoff among these factors. The most important criterion remained lowest long-term cost.

In January 1992, an update to the Demand/Supply Plan was issued. This update incorporated a revised load forecast and higher expectations for demand management and non utility generation, planning to the median instead of higher load forecast, a government moratorium on nuclear pre-engineering and an assumption that Ontario Hydro fossil stations would be life-extended. The requested approvals for hydro-electric facilities were reduced and the requested approvals for additional fossil-fuelled and nuclear facilities were withdrawn. In January 1993, Ontario Hydro withdrew from the hearing process altogether since changes in forecast conditions made it evident that the approvals for new generation options requested in the plan were no longer required in the next 3 to 4 years.

Ontario Hydro's planning process is adapting to the new role of providing electricity at lowest social cost and multiple criteria are now being considered when assessing the suitability of generation options. In addition to a long list of quantitative criteria such as feasibility, safety and environmental impact, Ontario Hydro is moving to additional measures to address the risks associated with generation options.

Hydro Quebec

In its 1993 Development Plan, Hydro Quebec has proposed a series of supply demand orientations it finds most beneficial for its customers and for society as a whole, in terms of both meeting future electricity demand and developing markets.

The impacts of resource and market options were analyzed according to a number of criteria: net economic benefits for Hydro Quebec and its customers, adhering to the principle of sustainable development, supplying services at the best possible cost, maintaining the utility's financial health, sustaining jobs in Quebec, and ensuring a margin of flexibility in planning.

For several years, Hydro Quebec has applied some of the principles of integrated resource planning. Such planning differs from more traditional planning methods in three ways: it includes energy efficiency in the range of options available to meet demand, it takes externalities into account and it incorporates public consultation.

In this context, to meet future electricity requirements, priority will be given to improving the existing system and to energy conservation in all cases where the cost of such measures, for equivalent service, is less than or equal to the cost of new generating facilities. These two solutions fully meet Hydro Quebec's criteria and have no significant environmental impacts. The addition of new generating facilities ranks only third as a priority.

Hydro Quebec has always assisted its customers in using electricity efficiently. In the past three years, however, efforts have been stepped up through such measures as promoting customer awareness, commercial and technical support, special rate formulas, financial assistance programs, research and development, and regulatory actions. The objective is to encourage all energy-saving measures that cost less than the marginal cost of supplying electricity. The energy-conservation target of 9.3 TWh, (or 5.4% of projected demand) by the year 2000 is comparable to those of the leading utilities in this field.

But despite energy-efficiency measures and improvements to the existing power system, new generation will be required to meet forecast demand between now and the year 2010.

Of all the options studied, hydroelectricity and certain gas-fired thermal generation (particularly cogeneration or combined-cycle plants) emerge as the most attractive. These options have very different features. Hydroelectric facilities require substantial capital and relatively long lead times of five to eight years, depending on the project. Gas-fired thermal stations have significant variable costs and short construction periods (about three years).

Hydro Quebec has given priority to hydroelectricity. This option, including small facilities operated by independent producers, remains the choice according to the latest development plan. This orientation in favour of hydroelectricity is based on the low real levelized costs and the most economic spinoffs in Quebec. Furthermore, hydroelectricity also offers the best guarantee of minimum long-term changes in costs of supply.

As with all generating facilities, and in fact as with all human activity, there are residual environmental impacts that cannot be completely eliminated. These include water, air and soil quality, perceptions and social changes, occupation and organization of the territory, and natural ecosystems. More specifically, large scale hydro projects involve the flooding of large areas of land with consequent loss of vegetation and wildlife habitats, the release of mercury compounds into the water and thus into the food chain, and the release of relatively low levels of methane. River flows are also modified, and in the case of the initial James Bay I project, this led to the relocation of a village inhabited by the Cree Indians.

Such impacts while undeniably present must however be put clearly in perspective. The hydro projects are located in a vast and virtually uninhabited area. The territory covered by the James Bay and Northern Quebec Agreement with the Native peoples has an area of over a million (I 088K) square kilometres i.e. twice the size of France or 20% greater than the area of the 13 founding states of the United States. A total of 43 000 persons live in this area, i.e. the population of a city suburb. These are 42% Native peoples, and almost all are located in coastal villages rather than the affected river basins. Because of the harsh climate and nature of the soil, there is no agriculture and vegetation is sparse.

The impact of the Hydro projects, grandiose as they are, is small relative to the size of the total area. The land flooded by all existing and envisaged projects amounts to 17 000 square kilometres, or 1.6% of the territory. In the case of the Great Whale project, the flooded area will be only 1700 square kilometres, or a mere 0.2% of the territory. Virtually nobody is living within the basin of the Great Whale river, and no settlements will be submerged by this project. As for the complete development, including the James Bay 1 and James Bay 2 projects, the only population displacement was the village of Chisasibi at the mouth of the La Grande River. Even here it can be argued that the environmental impact was positive, as the displacement involved a substantial improvement in housing conditions and social infrastructure.

It is fairly clear in fact that the issues surrounding these hydro developments are less environmental than social and political. The Native peoples here, as throughout North America, are challenging the legal base of land ownership and of the whole existing political structure of North America. At the same time, they are confronted with a major clash of traditional and modern cultures, in which hydro development is only one of several contributory features. Hydro projects may aggravate problems confronting Native populations who wish to retain traditional life styles while adopting at their own pace the fruits of new technology. In general however, these problems can be minimized by adopting a systematic approach of consultation with local peoples and a readiness to integrate the projects within a development plan for the region and thus insure that the native communities are partners in the economic development stimulated by hyrdo electric projects and that their aspirations are clearly taken into account.

To meet these concerns and so reconcile the 'white' interest for hydro development with the claims of Native peoples for material improvement, all development so far has been done under the terms of the James Bay and Northern Quebec Agreement signed in 1975 between the Canadian and Quebec governments, Hydro Quebec and representatives of the Cree and Inuit peoples. This agreement clearly recognizes and stipulates aboriginal rights over different classes of land in the territory, has resulted in massive financial contributions by Hydro Quebec to Native peoples, and has led to a tangible and startling improvement in living conditions. This is shown by the rise by approximately 25 years in life expectancy for the affected population. Furthermore the agreement has been a means of protecting and enhancing traditional ways of life and provides the Native peoples with the means to implant and manage their own institutions in the fields of education, health, social security, justice, economic development and income security.

Certain aspects of the treaty are contested by the Cree, but it remains the best ever signed with Native peoples in North America. The new disputes concern specific points, and Hydro Quebec is continuing negotiations with the Native peoples affected by projects. Overall, therefore, Hydro Quebec is convinced that hydro electric development is compatible with environmental protection and can be integrated within the principles of sustainable development.

Purchases of about 760 MW from independent power producers, mainly in the form of cogeneration, are also planned. Cogeneration and, to a lesser extent, combined-cycle generating stations, are the forms of fossil generation that cause the least damage to the environment. Direct economic spinoffs from cogeneration are fewer than from hydroelectricity, but may have a positive effect on the development of Quebec's main industrial sectors.

The nuclear option has not been retained. In addition to being much more costly and requiring longer construction lead times than the hydroelectric option, it is generally not well perceived by public opinion.

Wind energy continues to be very costly compared to hydroelectric generation. However, in-depth studies will be conducted to establish the economic value of a future wind-energy contribution to the main system. The situation is different in the isolated off-grid systems, since generating costs are much higher there. Demonstration programs will be launched in certain off-grid systems.

For a certain number of options, development of markets beyond the level of simply meeting requirements could be envisaged. These options involve relatively small amounts of energy over the next few years, but would require that commissioning of hydroelectricity projects be brought forward at the start of the next century.

This development comprises 3 elements: foster the use of electrotechnologies, promote limited and targeted establishment of electro-intensive industries, and favour export development. The latter is attractive because of its net economic benefits for Quebec and because export contracts provide revenues that help bring rate increases in line with inflation.

New Brunswick

In looking at the rather distant future requirements for new base load capacity, attention has been focused on coal-fired and nuclear options. Natural gas is currently not available in eastern Canada and combustion turbines for peaking service will use distillate oil.

NB Power's approach has been to choose low cost options while pursuing longer range goals of diversity and flexibility and strategies to maximize benefits from optimized use of interconnections with neighbouring utilities. In the past, electricity exports have made a substantial contribution to net income and have permitted the installation of large units.

Residual environmental and social costs have been internalized to some degree by doing better than the regulatory requirements and by spending for social/recreational infrastructure. The utility has also been able to implement programs to share economic benefits of projects with a larger group of local suppliers and manufacturers without in any way restricting participation by out-of-province or international companies.

NB Power has used an Integrated Resource Planning (IRP) process since 1990. As a result, some 110 MW of demand side management options equal to about 4% of the system peak have been targeted to be achieved by 1996-97.

IMPACT OF RISK FACTORS ON GENERATION CHOICES

While the fundamental objectives of electricity supply planning are tied to cost, reliability and the environment, risk is always a key issue. Every generation option and every planning decision contains elements of risk. Interest and inflation rates, capital cost over runs, fuel costs and availability, environmental and other regulatory requirements all present some degree of uncertainty.

Some risk situations are more intractable than others. The risks associated with high investment cost options can create situations which are very difficult to overcome. On the other hand, the major risks, for example, of a gas fired combined cycle plant are associated with the future fuel supply. A major price increase could possibly be dealt with by converting the plant to coal fuel using gasification technology. Alternatively the utility could change the duty of the plant from base load to intermediate by addition of new coal or nuclear capacity.

Ontario Hydro's Darlington station is an example of the risks of high investment cost options. Schedule delays, coupled with technical problems resulted in the station costing much more than originally forecast. Recognizing that while lowest long-term cost remains the primary criterion, other measures such as the proportion of fixed costs ratio and the crossover period, can be used to highlight risks associated with high capital cost and long lead time options.

Hydro Quebec's generation planning has viewed risks from both supply and demand sides. Supply side risks are incorporated into the planning process through probabilistic approaches or, where there is insufficient data on probabilities, through a scenario approach.

The variability of supply due to equipment failure and maintenance needs as well as hydro supply reliability, which depends especially on climatological factors, are handled by probabilistic methods. Probabilistic analysis is not adequate for risks such as possible delays or major modifications to projects and fuel price fluctuations.

The long lead times and capital intensity of certain types of electricity generation make the industry particularly vulnerable to unforeseen changes in demand. On one hand, a strong increase in demand, as was experienced towards the end of the 70's and again at the end of the 80's, creates a risk of power or energy shortages. On the other hand, a major shortfall of demand, provoked by economic difficulties or other changes in market conditions, can involve serious economic problems. These can be accentuated if, in response to its financial situation, the utility increases rates which may in turn lead to further erosion of markets. Once again these risks can be incorporated in planning through either a probabilistic or scenario analysis.

THE NEED FOR FLEXIBILITY

Faced with these risks on both the supply and demand side, electric utilities place great emphasis on attaining flexibility by incorporating the following measures:

a) Develop more flexible generation resources which can be installed with shorter lead times and lower capital costs even though the total cost may be higher.

b) Shorten lead times for all types of generating resources by undertaking the planning and environmental review phases in advance of final commitment.

c) Develop flexible markets including demand side management on a probabilistic basis and secondary markets which can be served in times of energy surplus. This is particularly appropriate for Hydro Quebec where hydrology factors are important. Markets for surplus power have been developed to serve New England and New York, and electric boilers have been provided to large industrial customers especially those in the pulp and paper industry.

EXTERNALITIES

The idea of placing a monetary value on residual environmental impact and other externalities was advocated more than half a century ago. Recently, the idea has resurfaced in Massachusetts in a Boston Edison submission to the Department of Public Utilities and in bills submitted to the U.S. House of Representatives by Rep. Pete Stark.

Many US utilities are now required to include estimates of environmental externality costs in their evaluation of generating options for new facilities. At present there is a wide range of numbers used, and there does not yet appear to be a consensus on externality values. Monetisation of externalities is limited, and does not apply for example to non-environmental externalities, nor to demand side management options. It is only applied in the planning of new options and externalities are not taken into account in the operation of existing plants nor in price signals given to customers.

As for Canada, in October, 1992, in a speech presented to the Independent Power Association of British Columbia and the Canadian Institute of Energy, the

Honourable Anne Edwards, B.C. Minister of Energy, Mines and Petroleum Resources said:

"...the province's objective is for domestic electricity requirement to be met at the lowest social cost, taking environmental and other impacts as well as financial cost into consideration... positive and negative features of the experiences in the United States will be carefully considered in developing social costing methodologies and processes for British Columbia..."

In Ontario Hydro's Demand/Supply Plan, environmental "control" costs were incorporated to the extent that the costs of the existing system included retrofitting existing stations with control equipment such as flue gas desulphurization (FGD) plants for SO_2 reduction, and the costs of new stations included the costs of control equipment such as FGDs and selective catalytic reduction (SCR) plants for NO_x reduction. However, costs of residual environmental effects were not quantified and, therefore, not explicitly included in the evaluation of plans and options.

Since 1989, Ontario Hydro has increased its efforts in attempting to estimate the environmental impacts and costs of the residual emissions from generation options. In 1991, a committee on Environmental Costs undertook to:

- Coordinate work within Ontario Hydro on environmental costs and benefits Identify the potential uses of environmental costs and determine the implications for Ontario Hydro
- Recommend an appropriate methodology for incorporating environmental impacts into resource planning, site selection, environmental dispatch, etc.
- Develop an Ontario database of damage estimates
- Communicate the corporate position on social and environmental externalities.

Recently attention has moved towards monetization of environmental externalities and incorporation into the planning process. Some of the topics studied by Ontario Hydro illustrate the complexity of the process:

- quantification of the dose-response relationships between human health and air pollutants
- development of dispersion models for pollutants
- estimating emissions or emission reductions related to demand management programs, non-utility generation and environmental dispatch
- quantifying the health impacts of PCBs, mercury, radioactivity and electrical and magnetic fields.

It should be observed here that the method involves making estimates of the direct damages rather than estimating the cost of control measures.

The current position on environmental costing is as follows:

- Ontario Hydro should, where possible, monetize the environmental impacts of its activities
- Ontario Hydro will develop a framework for monetizing and incorporating environmental externalities into the decision-making process
- Ontario Hydro will consult with the government to determine the best process for incorporating environmental costs.

At this time, Hydro-Quebec does not include externalities in its calculation of generation costs. It is noted, however, that renewable forms of energy, such as hydroelectricity, are generally favoured over fossil-fuel-based energy technologies when externalities are taken into account. Hydro Quebec intends to conduct studies on externalities and intensify research into multicriteria decision making aids; however, these studies will take several years to complete.

Environmental factors do however have a significant impact on internalised project costs, as they are taken into account in the initial choice, conception and design of all facilities. As a result, negative environmental impacts of large scale projects are greatly reduced and positive impacts realized.

At that same time, to find the best ways to mitigate the environmental and social impacts of hydroelectric projects, it is essential to work with the communities affected by such developments. Hydro-Québec is committed to ensuring that native communities are partners in the economic development stimulated by hydroelectric projects and that their aspirations are clearly taken into account.

NB Power has not incorporated externalities in the planning process. The fact, however, that many of its facilities incorporate environmental controls that perform substantially better than regulatory requirements means that some of the costs normally associated with externalities are already internalized.

There are significant difficulties in quantifying impacts of power plants. For fossil plants, there is the issue of CO_2 emissions. The links between power plant emissions and rising atmospheric CO_2 concentrations and the potential for global warming are controversial subjects. Neither the likelihood nor consequences of climate change are well known. Of course, it is possible to estimate the impacts and calculate the external cost but the range of plausible values that might be applied is very wide.

Nuclear plant emissions are arguably controlled to a higher degree than for other options. Nevertheless there are the issues of the accident risk and residual low level radioactive emissions. The consequences of a serious accident could be severe. The probability of occurrence is very low and the resulting value of the expected external cost is therefore also low. Low level emissions are continuous but are very small compared to natural background levels and therefore probably should have a low cost attached to them. On the other hand, public perception is that these are serious impacts and there can be pressure to assign high costs to them.

In view of the uncertainties associated with quantification, NB Power has chosen not to incorporate externalities in the planning process. Their impact is being examined however in sensitivity analyses.

CONCLUSION

Undoubtedly, the examination of multiple criteria when making choices between generation options is a practice that is here to stay. The process will likely be improved upon as greater efforts are made to quantify the value of characteristics such as short lead times and of residual environmental emissions over the next few years. On the environmental front, there will likely be a greater push towards considering the full life cycle impacts of each generation option and there will likely be considerable effort expended to attempt to quantify these impacts.

In the development of resource plans, there will likely be a greater push towards devising strategies which make plans flexible and able to respond to changing risks and opportunities. A step-wise approach combined with a feedback mechanism will likely become the norm. This approach consists of defining the problem (for example, a decision on whether or not to reinvest in an existing asset), developing scenarios, developing a flexible strategy to deal with the problem, then evaluating the problem and feeding the results back into the problem definition stage.

These developments will make the job of utility resource planning more challenging due to the range of uncertainties now included in the decision-making process. Utilities will continue to explore ways to improve its planning and decision-making process while balancing requirements to alleviate short-term financial/rate impacts, maintain good longer-term economics and to consider environmental and socio-economic concerns.

HOW GENERATION CHOICES ARE INFLUENCED BY COSTS, RISKS AND EXTERNALITIES: THE GENERATION PLANNING PROCESS IN ONTARIO, CANADA

E.A. Marriage
Director, Power System Planning, Ontario Hydro
and
M.S. Rogers
Planning Engineer, Power System Planning, Ontario Hydro

Abstract

Ontario Hydro is responsible for generating, supplying and delivering electricity throughout Ontario, Canada. Installed generation capacity of 32 GW consists of 20% hydro-electric (6.4 GW), 35% fossil (11.3 GW) and 40% nuclear (14.2 GW). Ontario Hydro's planning process has evolved significantly since its decision in the late 1970s to build the 4-unit 3500 MW Darlington Nuclear Station. The emergence of environmental issues as a primary consideration, increased awareness of financial and regulatory risks, and uncertainty about the load forecast and the impact of demand management programs on the load have all contributed to the changed planning process. This paper discusses Ontario Hydro's responses to these changes such as: increased public involvement in the decision-making process; the use of a broader range of options including demand management and non-utility generation; optimizing the use of the existing system; more complete risk analyses of generation options, and; recent attempts to incorporate externalities into the decision-making process.

COMMENT LE CHOIX DES MODES DE PRODUCTION EST-IL INFLUENCE PAR LES COUTS, RISQUES ET EXTERNALITES : LA PROCEDURE DE PLANIFICATION DE LA PRODUCTION EN ONTARIO, CANADA

Résumé

Au Canada, Ontario Hydro se charge de produire et de fournir de l'électricité dans tout l'Ontario. La puissance installée de 32 GW est composée à 20 pour cent d'hydroélectricité (6.4 GW), à 35 pour cent d'énergie d'origine fossile (11.3 GW) et à 40 pour cent d'énergie nucléaire (14.2 GW). La procédure de planification d'Ontario Hydro a considérablement évolué depuis la fin des années 70, où la compagnie a décidé de construire la centrale nucléaire de quatre tranches de 3500 MW de Darlington. L'apparition de problèmes d'environnement au premier rang des préoccupations, la sensibilisation aux risques financiers et réglementaires, les incertitudes concernant les prévisions de la charge et l'impact des programmes de maîtrise de la demande sur la charge ont contribué à modifier la procédure de planification. La présente communication expose les réactions d'Ontario Hydro face à cette évolution : participation accrue du public au processus de prise de décision, utilisation d'une plus large gamme d'options notamment pour la gestion de la demande et la production des autoproducteurs ; utilisation optimum des systèmes existants ; analyses plus complètes des risques liés aux divers modes de production et, plus récemment, tentative de prise en compte des externalités dans le processus de prise de décision.

How Generation Choices are Influenced by Costs, Risks and Externalities: The Generation Planning Process in Ontario, Canada

Introduction

Ontario Hydro, created in 1906 by special statute of the Province of Ontario, is responsible for generating, supplying and delivering electricity throughout Ontario. Ontario Hydro is a financially self-sustaining corporation without share capital. Bonds and notes issued to the public are guaranteed by the Province of Ontario.

Ontario Hydro sells electricity wholesale to 311 municipal utilities, which, in turn, retail electricity to approximately 2.8 million customers. Ontario Hydro also sells electricity directly to 108 large industrial customers and to approximately 940,500 small businesses and residential customers in rural and remote areas.

Ontario Hydro's existing generating system consists of 69 hydro-electric generating stations, 6 fossil-fuelled (5 coal- and 1 oil-fuelled) stations (four of which have four or eight units) and 5 four-unit nuclear stations. There are also two fossil-fuelled stations (coal and natural gas) which are mothballed. Power from these generating stations is transmitted to Ontario Hydro's customers over a transmission network of approximately 28,000 kilometres of high voltage transmission lines, a distribution system of approximately 107,000 kilometres, and many transformer and switching stations. As of August 1993, the dependable peak installed capacity of the Ontario Hydro system was approximately 31,900 megawatts, consisting of 6400 MW (20%) hydro-electric, 11300 MW (35%) fossil-fuelled and 14,200 MW (45%) nuclear-fuelled stations. In 1992, primary energy made available was 136 TWh. Of this total, 49% was supplied by nuclear stations, 27% by hydroelectric stations, 21% by fossil-fuelled stations and 3% by purchased power. Nuclear and some hydroelectric stations are used to meet base loads, while fossil-fuelled stations and the remaining hydroelectric stations are used to meet intermediate and peak loads.

Evolution of the Power System

Ontario Hydro's electric power system has evolved from one which utilized only hydro-electric stations (at the end of the 1940s) to one which was dominated by coal-fuelled stations (by the mid 1970s) to one in which the majority of the energy is now provided by nuclear power (1990s). The shift in emphasis from hydro-electric stations to coal-fuelled stations resulted from the unavailability of further economically viable hydroelectric sites. The shift in emphasis from coal-fuelled to uranium-fuelled stations resulted from a desire to reduce dependence on imported coal, economics which favoured nuclear stations, and the desire to utilize locally available nuclear technology and uranium. CANDU nuclear stations (pressurized heavy water reactors developed in Canada) are fuelled with natural uranium fuel, which is mined in Ontario and Saskatchewan and refined and fabricated in Ontario.

Ontario Hydro's major commitment to nuclear power was made in the 1960s with its decision to build its first commercial nuclear power station (four, 500 MW units) at Pickering. Between 1971 and 1993, Ontario Hydro placed 5 large nuclear stations in service, each consisting of four units, for a total of twenty units. The fifth and largest nuclear station, Darlington, was committed in the late 1970s and consists of four 900 MW units, the last of which was declared in service in June 1993. Ontario Hydro has not committed any major new generating stations since Darlington.

Changing Planning Environment

When the Darlington nuclear station was committed, the choice for Ontario Hydro was that of building either a large coal-fired generating station or a large nuclear station. The selection was made primarily on the basis of lowest long term costs, with the higher capital, but lower operating cost nuclear station being preferred.

In the intervening years, much has changed in the costs of options, the manner in which generation choices are made, and the manner in which planning of the power system is done in Ontario and in the rest of the world. The changes which have occurred over the past 10 to 15 years, in planning power systems, can be traced to several major causes, including: a) the emergence of environmental issues as a primary consideration in the evaluation of any proposed resource plan, b) increased awareness of the financial, regulatory and other risks associated with each generation option, and c) the uncertainty associated with load forecasts and the impacts of demand management programs on electricity demand.

Firstly, one major impact on the planning process has undoubtedly resulted from the emergence of environmental considerations as a key decision factor. The planning process has become more open as the involvement of the public and of interest groups in the planning process has become the norm. This has resulted in some significant changes in the approach to planning. For example, most utilities now give top priority to demand management as an option to moderate future load growth, thereby reducing the need for new generation and transmission facilities, and the environmental impacts resulting from these new facilities. The pre-eminence of environmental issues has also led to utilities maximising the use of the existing system before any new generation sources are considered. And when new generating stations are proposed, the available options must be carefully assessed to determine the environmental costs and benefits, both to the socio-economic environment and the natural environment, of the various options. For new facilities, Ontario Hydro plans on the basis that the most appropriate control technology, recognizing environmental impacts and costs, would be utilized to mitigate any harmful impacts of emissions to the natural environment from these units. Also, much effort is being expended on quantifying the environmental costs of the "residual emissions" and to "internalize" these costs in the cost of the generation option.

Environmental considerations can also be considered to be indirectly responsible for utilities considering a much wider range of generating options to meet future load growth. These options include not only the use of conventional generation technologies such as nuclear, fossil and hydro-electric stations, but also the use of alternative and renewable technologies such as fuel cells, solar energy and wind turbines. The driving force behind the public's expressed preference for these alternative technology options is the perception that these options have lower environmental impacts. Non-utility generation (NUGs) is also viewed as having lower environmental impacts due to renewable or natural gas fuel usage in most cases. Cogeneration is considered an attractive option due to the improved efficiency in fuel utilization which it offers.

The second significant impact resulted from a greater awareness of the risks associated with long-term planning decisions. In the past decade, utility after utility has been caught unprepared by changing load forecasts, a changing regulatory environment, cost increases and schedule delays on major generation projects and a changing financial environment. For many utilities, these uncertainties affected projects under construction, and often these were large nuclear projects.

In Ontario, changing load forecasts led, at first, to a deferral of the in-service date of some projects and then, later, to advancements of the in-service dates. The Darlington nuclear station was one project which was affected by these changes, leading to schedule delays. Regulatory risks can also be illustrated by Ontario Hydro's experience with the Darlington station. In the late 1980s, problems in proving the reliability of new shutdown system software to the regulators, led to licensing difficulties and further delays to the in-service date of this station. These delays, coupled with other technical problems led to the Darlington station costing much more than originally forecast. The experience with large mega-projects such as Darlington has, as much as any other factor, led Ontario Hydro to more explicitly consider the financial and regulatory risks associated with each generation option as it devises new resource plans. The long lead time associated with large projects and the resultant inflexibility in adjusting plans to fit changing conditions has brought home the disadvantages of large projects such as the Darlington station. This, in-turn, can be considered to be partly responsible for Ontario Hydro gaining a better appreciation for short lead time options, such as small hydroelectric projects, combined cycle stations and non-utility generation, as well as demand management programs.

Ontario Hydro's Demand/Supply Plan

Changes in the planning environment were reflected, in part, in the plan proposed by Ontario Hydro in 1989 for meeting the electricity needs of the province of Ontario until the year 2014. This plan, called the Demand/Supply Plan, was the culmination of a major planning effort and a long public consultation process started in 1984.

In the plan, highest priority was given to the use of demand management to moderate future load growth. Next in priority was maximising the use of the existing system; this included the rehabilitation of existing units and the retrofitting of environmental controls onto some units. The plan also included non-utility generation, the orderly development of the remaining economically viable hydroelectric capacity in Ontario (this included redevelopment of some existing sites) and the purchase of 1000 MW from Manitoba Hydro (to be supplied by a new hydro-electric station in Manitoba). The plan also recognized that there could be a role for alternative energy options in the future when these options become feasible. Only after all the economic options from the above sources had been fully considered did the plan require the building of new major fossil-fuelled and nuclear generating stations.

Before devising a preferred plan, many different plans were analyzed. Each plan was tested against a list of primary and secondary criteria. Primary criteria included customer satisfaction, reliability of supply, worker and public safety, environmental requirements and standards, cost, social acceptance, technical soundness and flexibility. Secondary criteria included resource preferences, resource smoothing, environmental characteristics (in addition to requirements and standards), economic impacts and other social considerations. The proposed plan represented a tradeoff among these factors. The most important criterion remained lowest long-term costs.

The plan was the subject of an environmental hearing for over two years. The sole purpose of the hearing was to demonstrate the need for new generation and related transmission projects. Many public interest groups participated in the process as well as representatives from various sectors of the electricity industry. In January 1992, an update to the Demand/Supply Plan was issued which recognized changing forecast conditions. These included a revised load forecast, higher expectations for demand management and non-utility generation, less conservative planning assumptions, a government moratorium on nuclear pre-engineering, and an assumption that Ontario Hydro's existing fossil stations would be life-extended. In this update, the requested approvals for hydro-electric facilities were reduced and the requested approvals for additional fossil-fuelled and nuclear facilities were withdrawn. In January 1993, Ontario Hydro withdrew from the process since changes in forecast conditions made it evident that the approvals for new generation options requested in the plan were no longer immediately required.

<u>Impact of Financial Risk Factors on Generation Choices</u>

Since the publication of the Demand/Supply Plan in 1989, changes have continued to occur in the planning environment. Although the plan considered many of the risk factors that exist in today's planning environment, the primary consideration still remained lowest long-term costs. Over the past few years, however, Ontario Hydro has had to request rate increases which were well above the inflation rate. These rate increases were required, in large part, to recover interest and depreciation expenses associated with bringing high capital cost assets, such as the Darlington nuclear station, into service. These new capacity charges were compounded by a lack of revenue growth due to a stagnant growth in electrical demand. Revenue requirements are driven by the need to recover expenditures on high capital cost, low operating cost assets such as nuclear stations, therefore, much of Ontario Hydro's costs, and hence its revenue requirements, are fixed. In business language, Ontario Hydro is a capital-intensive company.

This situation has led Ontario Hydro to more explicitly consider project characteristics such as: a) capital intensity; b) lead time/flexibility; and c) impact on rates and borrowing, recognizing that while lowest long-term costs remains the best measure, it could be supplemented by other measures, leading to more optimal decisions. Two of these measures, the proportion of fixed costs ratio and the crossover (or payback) period, are now being explicitly incorporated into Ontario Hydro's decision-making process.

Proportion of Fixed Costs Ratio

This measure provides an indication of the specific contribution of an individual alternative to Hydro's capital intensity. One method of calculating this is to take the ratio of the present worth of the lifecycle *capital* costs (including capital modifications, rehabilitations, etc) of a project to the present worth of *total* lifecycle costs. The application of this measure would require that a desired ratio be developed for the system. Capital work committed during a period would then be chosen so as to average down the system ratio over the period.

Crossover (or payback) Period

Ontario Hydro calculates the crossover period as the absolute number of years until the cumulative present value of the benefits exceeds the cumulative present value of the costs. Target crossover periods can be recommended for new projects so as to average down the "system crossover". The crossover period also recognizes the advantages provided by options with short lead times. The faster a project can be placed in-service, the faster it starts recouping revenue, and thus the shorter the time that the initial capital expenditure is at risk.

Economic evaluations of projects, instead of focusing exclusively on lowest net present value of costs, would use judgement to trade off the economic assessment factors including the lowest net present value of costs, the proportion of fixed costs ratio, the crossover period and the impacts on rates and borrowing.

Other financial measures can also be used, such as project specific discount rates (assigning higher discount rates to perceived higher risk projects) and negative preference adders (adding a premium to the present value of costs of higher risk projects). These measures all provide additional information and would allow factors other than lowest long term costs to be taken into account in the decision-making process, but are not employed by Ontario Hydro at this time.

Impacts of Environmental Considerations on Generation Choices

The Demand/Supply Plan published in 1989 included an Environmental Analysis which summarized the advantages and disadvantages of alternative generation options and demand/supply plans. The analysis was done using a consistent set of natural and social environment criteria as follows:

Natural Environment Criteria	*Social Environment Criteria*
Resource Use	Socio-economic effects
Non-renewable resources	Regional employment
Land use	Regional economic development
Water use	Local community impacts
Emissions/Effluents/Waste	Societal Considerations
Atmospheric emissions	Social Acceptance
Aquatic effluents	Special/Sensitive Groups
Solid Waste production	Lifestyle Impacts
	Distribution of risks and benefits

Figure 1 indicates how the Environmental Analysis process fits into the overall planning process.

The Environmental Analysis of the 1989 Demand/Supply Plan included an analysis of each alternative plan against these criteria. Both the environmental benefits and the environmental costs of each alternative plan were evaluated. It was noted that achieving acceptable environmental effects for each alternative plan would require careful siting, design, construction and mitigation measures for the various plan components. In evaluating alternative demand/supply plans, environmental "control" costs were incorporated to the extent that the costs of the existing system included retrofitting existing stations with control equipment such as flue gas

102

Figure 1: How Environmental Analysis Fits into the OverallDemand/Supply Planning Process at Ontario Hydro

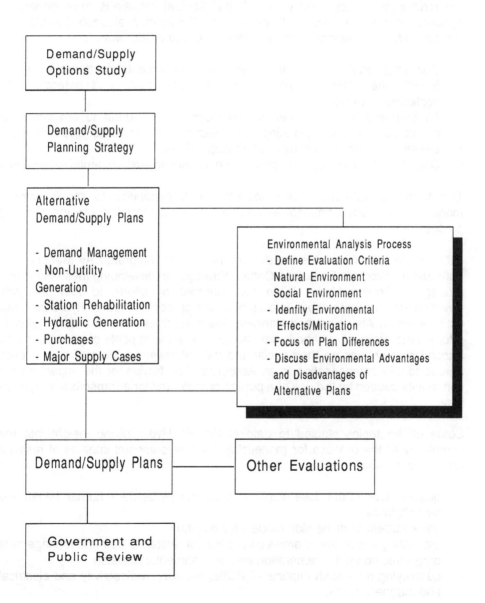

desulphurization (FGD) plants for SO_2 reduction, and the costs of new stations included the costs of control equipment such as FGDs and selective catalytic reduction (SCR) plants for NOx reduction. However, "damage" costs of residual environmental effects were not quantified and, therefore, not explicitly included in the evaluation of plans and options.

Since 1989, Ontario Hydro has increased its efforts in attempting to estimate the environmental impacts and costs of the residual emissions from generation options. In 1991, a Steering Committee on Environmental Costs (SCEC) was established. The original terms of reference of the SCEC were to:

- Coordinate work within Ontario Hydro on environmental costs and benefits
- Identify the potential uses of environmental costs and determine the implications for Hydro
- Recommend an appropriate methodology for incorporating environmental impacts into resource planning, site selection, environmental dispatch, etc.
- Develop an Ontario database of damage estimates
- Communicate the corporate position on social and environmental externalities

The terms of reference have recently been expanded to move towards monetization of environmental externalities and incorporation into the planning process.

Other related work includes the development of an Integrated Air Management Plan and a Corporate Emissions Control Strategy. In developing these plans and strategies, Ontario Hydro is in the forefront of efforts to fully integrate environmental considerations into its planning process. Ontario Hydro has also established an Aboriginal and Northern Affairs Business Unit in order to directly address the concerns of Aboriginal peoples about the impacts of generation and transmission options on their social and natural environment, and has recently agreed to pay compensation to an Aboriginal First Nation for the impact on the community caused by the use of a portion of their land for a transmission right-of-way.

Some of the topics studied to date in Ontario Hydro giving insight into the complexity of the process for monetizing the environmental impacts of residual emissions include:

- quantification of the dose-response relationships between human health and air pollutants
- development of dispersion models for pollutants
- estimating emissions or emission reductions related to demand management programs, non-utility generation and environmental dispatch
- quantifying the health impacts of PCBs, mercury, radioactivity and electrical and magnetic fields.

The debate continues over the appropriate manner in which to "internalize" these externalities. Suggested methods for incorporating externalities in the planning process include recommendations for specific adders for each generation technology based on the cost of environmental controls and specific adders based on estimates of "damage" costs. Ontario Hydro prefers, and is pursuing the "damage" cost approach. Some utilities use externalities costs in making resource acquisition decisions and one method of doing this is to apply percentage adders to the purchase price for preferred options (renewable and high efficiency options). Several intervenor groups have advocated the use of specific adders to the cost of each technology in order to "level the playing field" between conventional and alternative technologies. These groups argue that, although there are uncertainties about the exact value of specific adders, this should not preclude their use in the planning process. To bolster their argument they point out that uncertainty in the load forecast does not preclude the use of load forecasts. In line with a strategy established in 1989, and without doing a detailed analysis, Ontario Hydro has been applying up to a 10% "preference premium" to the purchase price of renewable and high efficiency options in order to recognize the public's and Ontario Hydro's preference for these generation options.

Ontario Hydro's current position on environmental costing is as follows:

- Ontario Hydro should, where possible, monetize the environmental impacts (both costs and benefits) of its activities
- Ontario Hydro will develop a framework for monetizing and incorporating environmental externalities into the decision-making process to the extent possible.
- Ontario Hydro will consult with the government to determine the best process for incorporating environmental externalities

Integrating Costs, Risks and Externalities into the Planning Process

Ontario Hydro's planning process is adapting to the new role of providing electricity at lowest social cost, as against lowest cost. Instead of depending on lowest long term costs as the sole measure, multiple criteria are now being considered when assessing the suitability of generation options. Figure 2 is an example of an analysis sheet for one of the nuclear options considered in the 1989 Demand/Supply Plan for future system expansion, which qualitatively evaluates the characteristics of this generation option against a long list of criteria. In the future all generation options will be subject to at least this level of analysis before being considered for incorporation into resource plans. In addition to the qualitative analysis shown in Figure 2, as previously mentioned, Ontario Hydro is moving towards the adoption of additional measures such as the proportion of fixed costs ratio, the crossover period and the short-term impacts on rates and borrowing when assessing generation options.

In a related manner, overall resource plans are also being analyzed from a variety of points-of-view to gain an understanding of the manner in which uncertainties in forecast conditions affect the results of planning studies. "Point" forecasts are being replaced by "bandwidth" forecasts for almost all of the important inputs to the planning process, such as: load growth; fuel prices; capital and operating costs of generation options; allowable emission levels; the degree of success of demand management; and the level of non-utility generation which will be available. This approach results in a large number of scenarios being analyzed, but it provides information which allows decision makers to devise plans which can be easily adjusted to changing forecast conditions. Figure 3 is an example of one process under consideration for managing uncertainty in Ontario Hydro.

What does the future hold?

Undoubtedly, the examination of multiple criteria when making choices between generation options is a practice that is here to stay. The process will likely be improved upon as greater efforts are made to quantify the value of characteristics such as short lead times and of residual environmental emissions over the next few years. On the environmental front, there will likely be a greater push towards considering the full life cycle impacts of each generation option and there will likely be considerable effort expended to attempt to quantify these impacts.

In the development of resource plans, there will likely be a greater push towards devising strategies which make plans flexible and able to respond to changing risks and opportunities. A step-wise approach combined with a feedback mechanism will likely become the norm. This approach consists of defining the problem (for example, a decision on whether or not to reinvest in an existing asset), developing scenarios, developing a flexible strategy to deal with the problem, then evaluating the problem and feeding the results back into the problem definition stage.

These developments will make the job of utility resource planning more challenging due to the range of uncertainties now included in the decision-making process. Ontario Hydro will continue to explore ways to improve its planning and decision-making process while balancing requirements to alleviate short-term financial/rate impacts, maintain good longer-term economics and to consider environmental and socio-economic concerns.

FIGURE 2: Example of an analysis sheet for a Generation Option

Description: 4 x 881 MW CANDU

Feasibility:
- Well established and proven technology.
- Able to be licensed in Ontario; may require layout changes for later need dates.
- Suitable sites exist in Ontario. Selection/approval of new sites extends schedule.
- Availability of design, construction and key operating staff and of suppliers of major equipment requires special attention due to discontinuity in nuclear program.

Cost:
- High total capital investment cost but low specific power cost.
- Low operating and fuelling costs. Lowest LUEC of the options studied.
- Rate impact is high initially but low in long run.

Planning Flexibility:
- Limited flexibility because of long lead time, large size and integrated nature of the station.
- Large capital commitment required. Initial capital commitment could be lowered by committing in pairs, but there is a risk of higher lifetime costs if the second pair is not built.

Health and Safety:
- Public safety concerns have been extensively reviewed, most recently by Dr. F.K. Hare in the Ontario Nuclear Safety Review. It was found that Ontario Hydro reactors are being operated safely and at high standards of technical performance.
- Excellent safety record .

Natural Environment Considerations:
- Meets current environmental regulations and Hydro standards.
- The CANDU system is a closed system for which historically radioactive releases have been less than 1% of the Derived Emission Limits of the AECB.
- Relatively low volume but long-lived high and low-level radioactive waste produced.
- A method for the long-term disposal of used fuel is undergoing public review by the federal Environmental Assessment Review Panel.

Socio-Economic Considerations:
- Significant direct and indirect Ontario GDP and employment benefits.
- At 80% ACF, over 85% of life cycle cost is spent in Ontario.
- Community impacts moderate for existing sites, potentially higher for new sites.
- Utilize and possibly expand the existing infrastructure of Ontario nuclear industry.
- Public support for nuclear is split 50/50 - main concern is emissions and safety.

Performance:
- Integrated CANDU stations have demonstrated reliable performance in Ontario. Bruce B station has achieved high lifetime capability factor to date. Early setbacks at Darlington are temporary; potential root causes identified and solutions are progressing. Forecast 80% lifetime capability factor.

Figure 3: Model for Managing Uncertainty

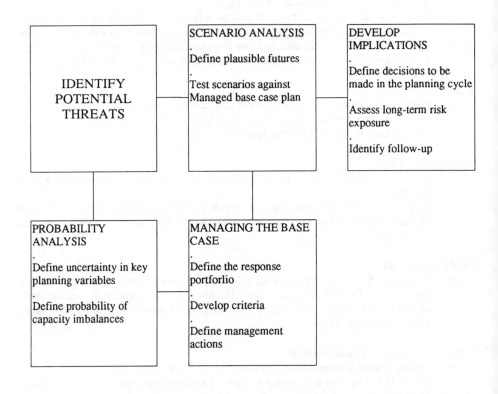

POWER GENERATION CHOICES:
AN INTERNATIONAL PERSPECTIVE ON COSTS, RISKS AND EXTERNALITIES

Rémy CARLE
Deputy General Manager
and
Georges MOYNET
Electricité de France

Abstract

In France, as in other countries, it is no longer possible to discuss issues relating to the power or electricity generating sector without taking account of a wide range of sociological, political and environmental factors. Given the uncertainty regarding either the short- or long-term impact of such factors, analysts can no longer rely on forecasts and need to adopt an approach based on the use of extremely wide-ranging scenarios. Many researchers are currently examining ways in which these factors might be incorporated into economic calculations in the form of externalities. Such externalities cannot be properly taken into account, however, until we have reliable methods and databases with which to quantify them. At present they are taken into account implicitly. Thus the extensive nuclear power programme launched by the French Government in the 1970s was aimed at securing advantages in both economic and environmental terms. These advantages may be seen today in the structure of France's energy supply.

LE CHOIX DES MODES DE PRODUCTION: LE POIDS DES COÛTS, DES RISQUES ET DES EXTERNALITÉS - ÉTUDE INTERNATIONALE

Résumé

En France comme dans les autres pays, il n'est plus question de parler d'énergie ou d'électricité sans tenir compte des multiples facteurs sociologiques, politiques, environnementaux qui interviennent dans ces questions. Tous ces facteurs introduisent des incertitudes soit de court terme soit de long terme. Il convient donc de ne plus raisonner en prévisions mais bien à l'aide de scénarios très largement ouverts. L'introduction de ces facteurs dans les calculs économiques en termes d'externalités fait l'objet de nombreuses réflexions. Ils ne pourront être sérieusement pris en compte que lorsque nous disposerons de méthodes et de bases de données solides pour les chiffrer. Aujourd'hui il en est tenu compte implicitement. C'est ainsi que la France a lancé dans les années 70 un vaste programme nucléaire sur la base d'avantages aussi bien économiques qu'environnementaux. Ces avantages peuvent être appréciés aujourd'hui dans la réalité de l'approvisionnement énergétique français.

1. Since the 60s, the human race has realised that the world in which it lives has its limits. Man's impact on nature and the environment, formerly thought to be marginal, suddenly appeared fraught with danger. Twenty years have passed since the alarm was sounded by the Meadows report published by the Club of Rome on the future of our planet. The pessimism of its forecasts was much criticised. It was due to the use of insufficiently soundly based development models. The announced shortages and catastrophic situations have not come about, but this is partly thanks to the very fact that the report was published. The authorities and, more generally, all those taking part in economic decisions reacted to it. They realised that no growth could continue at a high rate for very long, that our world was not boundless and that we were all interdependent. The notion of exponential growth gave way to that of sustainable growth.

With the slowing of growth, and more or less correlatively (it is not the aim of the present paper to discuss this), came the calling into question of existing technologies through environmental considerations, the emergence of new technologies and products, a certain relocation of production to the developing countries (notably in South East Asia), political and then economic instability in many under-developed countries, the collapse of the Soviet empire, etc.

The age of continuous economic growth has thus given way to a period of increasing uncertainty in a world which is changing, technically of course, but even more socially and politically, blurring the reference framework for development and even making the very nature of the decisions to be taken unclear.

This is particularly the case in the field of energy, where the response times to the challenges which have arisen are long. Medium to long-term energy supply remains, despite the structural changes which are affecting growth, one of the important factors determining this growth, and will continue to be.

The earlier criteria for electricity generation choices now appear simple as compared with today. It was essentially a matter of determining the most economical mix for generation technologies and adjusting the amount of annual investment to the expected rate of growth. This does not mean that it was not necessary to develop sophisticated econometric methods, but they were applied to a future that was virtually certain or in which the uncertainties could be clearly delimited.

That age is now past. Simultaneously to the economic criteria there are now others, both qualitative and quantitative, through which we seek to capture the now uncertain long-term, medium-term and sometimes even short-term context. There is no question here of embarking upon a complete analysis of all the factors which may influence electricity generation choices, but we must mention the most significant:

- Medium to long-term strategic options;
- Short-term context: political, structural, financial, etc.;
- Comparison of the relative costs and competitiveness of the different generation technologies;
- Evaluation of the specific risks: technical, financial, regulatory, strategic (dependence on primary energy resources), etc.;
- Externalities: local or global parameters in the environmental field (air and water pollution, global warming) or the political field (acceptability, national policies, international agreements), etc.;

2. In this context of uncertainty, it now appears essential to explore not an increasingly unknown future but a number of possible futures and to examine their implications for the construction and operating conditions for the generation units envisaged.

This leads to the establishment of a fairly limited number of scenarios but sufficiently differentiated to give a reasonably broad representation of the situations with which we may be confronted.

The development of a scenario whose prime quality is to be coherent within itself, requires taking into account the interplay of a certain number of influencing variables to be selected from:

- *macroeconomic variables* (world economic growth prospects, growth/employment correlation, action by the political authorities, economic revival, évolutions in the energy and electricity-intensity of growth, energy and electricity demand, dependence on primary energy resources);
- *structural variables* (evolution in electricity generation/transport/distribution systems, regulation/deregulation aspects, breakdown of public/private capital, development of new structures: independent producers, cogeneration);
- *technical and environmental variables* (technical and technological evolution of generation equipment, evolution in plant safety, nuclear or other, evolution in regulations governing air pollution, water pollution and land use, greenhouse effect and its impact on climates, possible international measures);
- It is also necessary to take account of *actors' strategies*, since they do not remain neutral but industrial companies, electricity companies, research and development organisations,

All these considerations lead to a necessarily limited number of medium to long-term options which we shall call strategic options, which form the framework for the action to be taken to make the best present and future choices.

3. The definition of medium to long-term options must absolutely not lead to failure to examine the short-term situation: it is good to know where we are going, but we must not stumble on the first step.

The short term is full of political and structural constraints, but it should be noted that the financial aspects are those most often mentioned:

- financial balance, profits and losses;
- taxation on profits and capital;
- return on capital, shareholders' dividends;
- investment pay-back period;
- interest rates on short and medium-term loans;
- debt/cash flow ratio;
- debt structure;
- regulatory constraints, in particular concerning tariffs.

The political and structural constraints appear through changes in the regulations, political incentives or pressures relating to the structure of the electricity system or the capital structure, the more or less marked maintenance of the public service nature of the industry, etc.

Depending on present and anticipated financial performance, capital market investment yield requirements, pay-back periods aimed at, requirements of banking and financial circles for granting more or less favourable credit conditions, and finally taking into account the structural factors mentioned above, the decision-makers' choice may tend towards solutions with low or less capital intensity, or which offer prospect of a more rapid return, more attractive for investors, notwithstanding the definition of significantly different longer-term options.

Lastly, the existence of oppositions by pressure groups and their echo in political circles or purely local opposition (nimby - not in my back yard - effect) may lead to the postponement or even cancellation of certain projects despite their economic interest and the fact that they fit into the strategic lines pursued.

A trade-off, not always quantifiable, between the short and long terms may be necessary and can be clarified by iterative analyses concerning generation cost comparison, risk analysis or the taking into account of externalities.

It is clear that a strong lamp is necessary to light up the immediate area, but it is also well-known that full headlights are no use for driving in fog. If the prospective exploration described in the first section has not been carried out or has not been adequate then driving will be dangerous, because a brick wall or a sharp bend may lie ahead of the curtain of fog and surprise even the best of drivers.

4. Although the expected "technical" lives of generation plants are somewhat variable, from 20 to 40 years depending on the type of equipment (from combined cycle to nuclear), they are in any event long-term investments.

Calculating generation costs including capital, operating and fuel costs thus requires a short-term/long-term intertemporal evaluation of the different components of total cost.

This evaluation is generally made by using a discount rate in order to calculate an average long-term discounted cost or levelised cost. The method used and standardised by the UNIPEDE for the European Communities in the late 70s has since been adopted by the OECD (NEA and IEA) and most of the bodies interested in electricity generation cost comparisons, thus making this method virtually universal.

An important factor in calculating average discounted costs is examining the price trends which can affect the different cost components throughout the lifetime of the plant, i.e. operating costs (in particular maintenance) and fuel costs. This is especially important for natural gas fired combined cycle plants for which the fuel component can amount to 60-80% of the total cost. This cost thus becomes highly sensitive to the hypotheses used for future gas price variations.

Two recent studies have updated the generation cost calculations for installations to be commissioned in 2000 and slightly beyond:

- UNIPEDE Copenhagen 1991 Electricity Generating Cost Report of the 007 Cost Working Group;
- OECD (NEA-IEA) Projected costs of generating electricity from power plants for commissioning around the turn of the century (to be published in 1993).

5. In France, cost comparison calculations have been made regularly for some decades now. They were formerly the prerogative of the PEON Commission, but they are now updated by the Directorate for the Gas, Electricity and Coal Industries of the Ministry for Industry. A new edition was published in 1993.

Three basic generation technologies are considered: nuclear, coal, and natural gas in combined cycle plants.

The comparison between the three sources has to be made on a full cost basis, i.e. integrating investment expenditure, operating and maintenance costs, and fuel costs.

Two scenarios have been used for each fuel: high and low scenarios reflecting two very different market situations, in order to cover the field of different possible trends during the period considered.

The cost of the nuclear kWh also takes into account of the complete dismantling of installations (Level 3), waste disposal and research and development. It is the only generating technology for which dismantling costs are taken into account.

In France, thanks to the standardisation of equipment and the experience acquired, nuclear power remains the most competitive mode for base and intermediate load capacity, as shown by the following table.

Table 1

Base load electricity generating costs in France
(for an utilisation of 6 000 hours per/year)
(1993 french francs)

COST (centimes/kWh)	NUCLEAR	COAL (fluidised bed)	NATURAL GAS COMBINED CYCLE
Investment	13.6	11.3	5.9
Operating	5.6	5.8	2.4
Fuel	4.5-6.2	11.7-17.7	21.1-27.4
R&D	0.4		
TOTAL	24.1-25.8	28.8-34.8	29.4-35.7

We would stress:

- first, the soundness of these results, which have been maintained for many years despite the fluctuations seen on the raw materials markets, and this in all the scenarios considered;
- second, that the nuclear costs include a certain number of "externalities" not taken into account for the other energy sources.

These results explain and justify France's constant commitment in favour of nuclear power since 1974, this constancy having permitted in return the development of long-term programmes, based on standardised models, substantial industrial investments and a policy of adapting Electricité de France to its new generation tools which were necessary to the technical and economic success of the enterprise.

We all know that the figures in the above table make it possible to determine the operating conditions under which there is equivalence between high investment cost modes, here nuclear, and those with high operating costs, here coal and natural gas. This equivalence, in fact, lies in the 4 500 to 5 000 hours range.

Nuclear on the one hand and coal or gas on the other are thus in fact complementary and there has never been any question in France, or anywhere else, of an "all nuclear" solution. Only the existence of coal-fired power stations inherited from the past explains the lack of investment in this field in recent years. When consumption growth requires it, it is clear that new investments will include both nuclear power stations and coal- or natural gas-fired plants.

In addition, France has a substantial hydroelectric plant stock, generating 60-70 billion kWh a year, or some 15% of total production today. The characteristics of these installations (in fact different according to whether they are running water or reservoir plants) require particular methods of economic calculation, though they depend on the same general philosophy as that outlined above. They also have advantages (notably rapid start-up) which also need to be monetised. On the other hand, their disadvantages in terms of environmental impact and water resource management, so important today in our countries, also need to be taken into account. The scarcity of sites and local opposition mean that the development of hydropower will be problematical and certainly limited, at least in Europe.

Be this as it may, Electricité de France can but be pleased at having three modes of generation: nuclear, conventional thermal and hydro. Quite apart from the economic calculations, one of the virtues of electricity is that it can use different energy sources and hence if necessary compensate for any unfavourable evolutions in one of them. Was this flexibility of electricity generation not largely used in the energy crises of 1974-79? It must be preserved whenever natural conditions permit it.

6. Quite apart from the global risks mentioned above, a certain number of risks directly associated with the building or operation of a new plant may modify the base conditions or values of certain factors entering into choices:

- Technical risks

Often evoked in connection with nuclear plants, they can affect new or not yet fully tried and tested fossil fuel technologies such as fluidised bed combustion (FBC) in coal boilers, or combined cycles associated with coal gasification (IGCC) and to a lesser extent natural gas (CCGT), notably as regards the real lifetime and maintenance costs.

- Financial risks

These are all the events which threaten to compromise the financing of a given project or modify its profitability, notably vis-a-vis investors. The financial criteria were reviewed in section 2 and will not be returned to here.

On this subject it seems necessary to mention the doubts and reservations expressed in certain circles regarding the significance of economic comparisons of generation costs based on the average discounted long-term cost method (section 3). According to these criticisms, the practice of discounting over the lifetime of the installation would be insatisfactory for taking into account the reality of the financial constraints and associated risks and would leads to dubious or inadequate comparisons.

The recent methodological work reassessed by UNIPEDE, to be made public at the Birmingham Congress in 1994, shows that this methodology remains perfectly suited to taking account of different and contrasting economic and financial contexts, provided that the discount rate is made a parameter.

Financial simulations permit us to say that a rate of 5-10% a year in constant money is largely sufficient to cover all the economic and financial contexts reasonably applicable to investments with a lifetime of over 20 years and for a product which, whatever the context considered, public or private, regulated or competing, remains one of the important factors for progress and economic growth.

- Regulatory risks

A project is drawn up within an existing regulatory framework and as a function of officially announced new measures. This framework is constantly moving however, with repercussions on structures, tariffs or technologies.

- Strategic risks, notably vis-a-vis resources

This is a greater or lesser risk according to the degree of dependence of the country concerned.

Generation costs are often calculated on the basis of the fuel prices known at the project study stage, which is likely to lead to a completely erroneous evaluation, notably in installations with a low capital intensity where the fuel component represents over half of the total cost. In fact, and according to the dependence of the country concerned vis-a-vis primary energy resources, such a solution which now looks attractive could lead to a deficit or extra economic cost in the short or medium term.

While uranium supplies, in view of present nuclear programmes, and coal supplies, thanks to fairly well-distributed world coal reserves, look as if they should not cause any problems during the next few decades, there could be problems with natural gas supplies in the case of large-scale investment in combined cycle plants using this fuel.

Keeping to the European OECD countries, for which self-sufficiency in natural gas (all uses) is at present in the order of 75%, it is easy to show that a policy of massive investment in natural gas fired combined cycle plants (for example 50-100 GWe installed by 2005-10) would lead to radical changes in the pattern of supply for this same OECD area.

It can be said that any new investment in combined cycle after 2000 would mean that all the gas consumed would have to be imported from Russia, from Siberia or other countries of the former Soviet Union, the Maghreb (a necessarily limited share), the Middle East or Central Africa, all areas where there are problems in terms of political or economic stability.

This would also require substantial investment: thousands of kilometres of gas pipelines, liquefaction plants, LNG transport facilities, storage capacities in user countries.

All this gives reason to study and take into account several gas price trend scenarios, scenarios which can very significantly affect the generation cost comparisons and in certain cases reverse the choices which would be dictated by short-term considerations only.

It is tempting to say that the natural gas combined cycle option appears all the more attractive so long as it does not develop.

7.   The problem of taking the **externalities** into account in the choice of mode of electricity generation brings us back to what we said in the introduction about the limits of the world in which we live and the global conditions for sustainable growth.

In 1991 world (marketed) energy consumption was about 8 billion tonnes oil equivalent. The average individual consumption of 1.6 tonnes oil equivalent conceals flagrant inequalities which are going to become increasingly unsustainable: the average North American consumes over 7 toe a year, the European a little over 4 toe, ten times more than the African. The developed

118

countries are slowing the rate of increase of their consumption and trying to reduce it. The priority for all the poor countries is first to survive, then to industrialise and raise the standard of living of their populations. Third World energy consumption is exploding; assuming only that average individual consumption stabilises at 1 toe a year, world energy demand in 2020 will be 13-15 billion toe, almost twice as much as today.

Given these needs, we are faced with two dangers: one is being unable to meet them, the other is damaging our environment so severely that the survival of part of the human race would be affected. In the medium term, on the scale of one or two generations, there is no danger of a shortage of fossil fuels. World oil and gas reserves can meet demand, but with two ineluctable consequences: a considerable price increase as soon as the most accessible deposits are exhausted and increased dependency on a small number of producer countries with all the risks which this implies. Coal resources are better distributed and their abundance means that they will not be exhausted for a very long time to come, but their use brings us face to face with the problem of air pollution.

With rising energy consumption, pollution levels have considerably increased: dust emissions, oil spills, land flooded by hydraulic schemes, acidification of air and water by sulphur and nitrogen oxides, forest die-back, impairment of the protective ozone layer and lastly, the most recent fear, the greenhouse effect due to carbon dioxide and methane whose concentration in the atmosphere is also growing.

We discuss below the two issues which are most often cited in connection with the choice of power station technologies: atmospheric pollution due to the acid emissions in the flue gases and the contribution to the greenhouse effect.

Atmospheric pollution and acid rains

The question of the damage caused by acid rains has been agitating public opinion for about fifteen years. The original problem was the forest die-back seen in Western Europe and the extreme situations in the Czech Republic and East Germany; there is also the acidification of lake waters in Sweden and Canada. The cause is the emission of gases containing sulphur and nitrogen oxides in the combustion of coal and hydrocarbons. Certain coals and lignites can have sulphur contents as high as 7 or 8%, while crude oils can contain up to 2%. The nitrogen oxides are due to reaction of the nitrogen in the air with the high temperatures of fuels of all sorts.

Bringing these polluting emissions down to acceptable levels is a necessity accepted by all countries; for this it is necessary to limit the use of fuels rich in sulphur and fit thermal plants with fume scrubbers which fix the sulphur oxides.

It is also necessary to develop less polluting car engines. All this is good, but it must be remembered that no depollution system is 100% effective.

Nuclear power generation can make a vital contribution to reducing atmospheric pollution, as shown by the example of France. Thanks to the construction of nuclear power stations, sulphur oxide emissions from thermal plants was divided by about 5 between 1980 and 1990 (see Appendix). Overall SOx emissions in France are now at less than 30% of their 1980 level. Particulates due to fly ash from thermal power stations are now insignificant. At national level, though NOx emissions from power stations are now much lower than they were, the total has changed little because of the dominant role played by motor vehicle engines.

The greenhouse effect

The greenhouse effect is above all a natural phenomenon which guarantees the earth a temperate climate. Without this effect the temperature would be 20-30°C lower than it is today. However, we are now seeing an unprecedented increase in the concentration of "greenhouse gases" in the atmosphere, due to human activity. Thus the concentration of CO_2 alone has increased by over 25% since the beginning of the century, with a strong acceleration as from the 60s. The energy sector itself is responsible for 80% of CO_2 emissions, almost 70% of the greenhouse effect and will be responsible for over half its growth in the future.

The emission of CO_2 into the atmosphere due to the combustion of any product containing carbon is in fact inevitable. It is no longer a matter of eliminating toxic products connected with a few per cent of impurities present in the fuel. For every tonne of carbon burned (whether it be coal or the carbonated fraction of natural gas or any other heating fuel or motor fuel), a little over three tonnes of carbon dioxide are produced. Direct capture cannot be envisaged because it would require equivalent masses of reactants whose manufacture would necessitate energy consumption of the same order of magnitude as that produced in the combustion and which would have to be produced in addition, which would lead to a vicious circle. Therefore only indirect methods of capture can be envisaged and reafforestation is one.

The risk run with the greenhouse effect is global warming, which could be in the order of +1.5 to +4.5°C by horizon 2030-50. There is great uncertainty about the amplitude of this warming and the scientific world is divided on the point. Certain essential factors, such as the complete carbon cycle (role of the oceans in CO_2 absorption) or the existence of possible stabilising mechanisms, still remain very little known.

Combatting the greenhouse effect requires concertation on the world level, since no country can combat it effectively in isolation. The present state of the debate has shown above all the difficulty of introducing any such regulation in the face of national interests. Thus the North-South confrontation has found a new field: on the one hand the population explosion and on the other the wastage on the part of the rich countries.

The problem of combatting the greenhouse effect is a typical example of the collective decision-making in the face of uncertainty. The probability of the risk is unknown, but its impact is potentially considerable. Information will go on improving, but the phenomenon is irreversible and certain commentators speak of the "gamble of climatic risk".

In such a context, associating uncertainty with irreversibility, decision theory teaches us that we should give preference to actions which keep open or increase the possible future choices.

Nuclear power can play a significant role in this problem of the greenhouse effect, provided that the other external effects associated with its development are treated seriously and effectively.

We once again cite the example of France, where CO_2 emissions have diminished by one-third since 1980. Two-thirds of this reduction were obtained thanks to nuclear energy (see Appendix), the other third resulting from the energy conservation policy.

Comparison with other countries is instructive with regard to this policy. Expressed in tonnes of carbon per person per year, **CO_2 emissions in the main industrialised countries** are as follows:

France	1.9
United Kingdom	2.8
Germany	3.2
Eastern Europe	almost 4
United States	over 5

It would obviously be interesting to be able to calculate, in economic terms, the value to be placed on a given energy chain because it makes it possible to reduce the emission of a given type of pollutant or a given greenhouse gas. Beyond this, we could imagine quantifying the risks associated with the different possible choices. The problem is obviously very delicate because it would be a matter of quantifying entities of different natures, and certain subjective elements may come into play.

Many research bodies and universities are working on this problem. UNIPEDE recently set up a working party on the broader subject of the economic evaluation of environmental factors. This would enable us to introduce a certain rationality into controversial areas; we must therefore hope that this work will bear fruit. In the other hand it would be unwise to anticipate the results.

******

Economic calculation has been contributing for many years to making our energy choices and strategies more efficient. But it is clear that it needs to become even more sophisticated to be able to take account of all the factors to which our society is quite justifiably attaching increasing importance. This will make it possible, we hope at least, to consider these factors on a more rational basis and in a less subjective fashion than at present.

However, the actual quantification of the externalities and the weight to give them still remain and in many cases will remain difficult problems. In the meantime, it seems important to keep all the energy chains we have available - and even to develop and improve them. Let us neglect none of them. This is the only way to ensure that our children will have the energy supplies they need.

APPENDIX

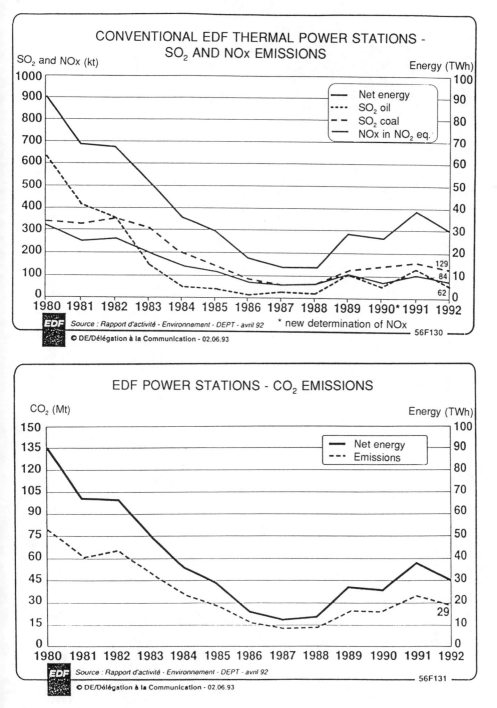

CONVENTIONAL EDF THERMAL POWER STATIONS - SO_2 AND NOx EMISSIONS

SO_2 and NOx (kt)

Energy (TWh)

Legend:
- Net energy
- SO_2 oil
- SO_2 coal
- NOx in NO_2 eq.

129
84
62

Source : Rapport d'activité - Environnement - DEPT - avril 92   * new determination of NOx

© DE/Délégation à la Communication - 02.06.93

56F130

EDF POWER STATIONS - CO_2 EMISSIONS

CO_2 (Mt)

Energy (TWh)

Legend:
- Net energy
- Emissions

29

Source : Rapport d'activité - Environnement - DEPT - avril 92

© DE/Délégation à la Communication - 02.06.93

56F131

GENERATION CHOICES AS INFLUENCED BY COSTS, RISKS AND EXTERNALITIES - GERMANY

Ludwig Strauss
Member of the Board of
Bayernwerk Aktiengesellschaft

Abstract

The pattern of power generation in Germany is the result of cost-effectiveness considerations, governmental decisions on energy policy, environmental protection legislation and questions of social acceptability. The relative importance of the various factors has changed over the years. While in the past, the economics of cost accounting provided the decisive criteria for the types of power plant chosen, nowadays and most probably in the future as well, political factors and public acceptance are becoming more and more important. The increasingly frequent discussion of the concept of inclusion of external costs and the present plans to alter the fundamental framework of the power generation industry will have lasting effects on the make-up of the power generation mix. There is discussion, for example, of the possible consequences of introducing a CO_2/energy tax, of amending the German Atomic Energy Act and of introducing new elements of competition into the electricity generation industry. Effectively the most significant factor in the decision as to the future pattern of power generation in Germany, however, is the social acceptability of the energy sources concerned. Thus the shape of the future development of power generation in Germany to a significant degree on the outcome of the attempts at present in progress to find a consensus on energy - in particular with regard to the continued use of nuclear energy.

LE CHOIX DES MODES DE PRODUCTION : LE POIDS DES COÛTS, RISQUES ET EXTERNALITÉS - ALLEMAGNE

Résumé

En Allemagne, les modes de production d'énergie sont choisis en fonction de considérations de rentabilité, des décisions gouvernementales en matière de politique énergétique, de la législation sur la protection de l'environnement et de critères d'acceptation par le public. L'importance relative de ces facteurs a évolué au fil des ans. Si les critères économiques jouaient autrefois un rôle décisif dans le choix du type de centrale, aujourd'hui, les facteurs politiques et l'adhésion du public prennent une place de plus en plus importante, qu'ils conserveront très certainement à l'avenir. L'idée de plus en plus fréquemment évoquée de tenir compte des coûts externes, et les projets actuels visant à modifier la structure fondamentale du secteur de la production d'énergie auront des effets durables sur la répartition des divers modes de production d'énergie. On s'interroge par exemple sur les conséquences possibles de l'introduction d'une taxe sur l'énergie/CO_2, de l'amendement de la loi allemande sur l'énergie atomique et de l'introduction de nouveaux éléments concurrentiels dans le secteur de la production d'électricité. En Allemagne, le facteur pesant le plus lourd dans le choix des futurs modes de production d'énergie est en réalité l'acceptation, par la communauté, des sources d'énergie envisagées. Ainsi, le développement futur de la production d'énergie dépendra dans une large mesure des résultats des actions menées actuellement pour parvenir à un consensus sur l'énergie, en particulier sur le maintien de la production d'énergie nucléaire.

One of the most decisive prerequisites for positive economic development is the worldwide provision of energy by reliable, cost-effective and environmentally safe means. Due to its wide variety of possible areas of use, electricity as a consumer energy has a leading role to play in this process.

It is of particular interest against this background to take a closer look at the factors affecting electricity generation options and the ways in which they can be influenced, taking the Federal Republic of Germany as our example. In order to be able to do justice to the dynamics of this subject, I will divide my observations into two sections. First of all I will attempt to illuminate the factors which have determined the make-up of the power generation mix up to now. In the second part I will look at aspects which can be expected to be decisive factors in determining the pattern of electricity generation in Germany in the future. I will attempt to show that decisions regarding the power generation mix in Germany are increasingly being made not so much on the basis of cost-effectiveness as under the influence of political demands, environmental aspects and social acceptability. I should say before going any further that overall, the factors which determine the decisions reached by energy suppliers with regard to the types of power station operated will remain basically the same in the future - the relative importance attached to each one will, however, alter decisively.

Let me first of all briefly outline the present pattern of power generation in Germany. Generation of electricity is currently characterized by a relatively well balanced overall mix of the different types of primary energy. The largest share of power supplied is provided by nuclear power stations which account for a good 30%. Since re-unification of the two Germanies, the second position is now occupied by the lignite-burning power stations which have a share of just under 29%. Lignite was far and away the most important power generation fuel in the former German Democratic Republic, being used for 90% of the electricity produced. The over-emphasis on this one source of energy in former East Germany was a consequence of the desire to avoid importing energy which would have to be paid for in convertible currency (DM, dollars).

The third most important source of primary energy for electricity generation in Germany is hard coal, which has a share of around 27%. Natural gas and oil follow a long way behind with shares of 6% and 2% respectively of the total electricity generated. The regenerative energy sources account for 6 ½%, two thirds of which is accounted for by hydro-electric power generation.

These figures for the country as a whole of course hide the fact that the make-up of the groups of power stations operated by the individual regional utility companies differs considerably from one region to another due to differences in geopolitical factors. Thus in the Rhineland and Central German coal-mining regions, hard coal and lignite are the most significant energy sources, in Bavaria, where water abounds but raw materials are scarce, most power comes from hydro-electric plants and the quasi-indigenous nuclear power stations.

Let us now examine the individual factors which have affected the structure of power generation in Germany.

The primary energy structure I have just outlined is in the first instance, and always has been, the result of the **economics of cost optimization** according to the following scheme: the power utilities calculate the proportion of base-load, medium-load and peak-load power stations required to cover the level of demand on the basis of the consumer structures of their supply regions. The availability of the various sources of primary energy, or as the case may be, the cost of bringing them to the required locations, i.e. **geopolitical factors**, then determine by way of cost comparison the types of power station to be used for basic, medium and peak demand electricity. Thus, in the past, these economic considerations were the most important factors in deciding which sources of primary energy to use for the generation of electricity.

Nevertheless, we must not overlook the fact that there have always been **political requirements**, especially in the areas of **structural and environmental policy** which have exercised an influence on the power generation mix. This becomes clear when we look more closely at the power generation structure in Germany.

Thus although the present collection of power stations with its well-balanced mix and minimal dependence on oil imports quite adequately satisfies the requirements of security of supply, from an economic and ecological viewpoint, the proportion of medium-load power stations fuelled by - primarily domestic - hard coal is too high, while the proportion of nuclear power stations in the base-load bracket is lower than would be required for an optimum mix. The fact that this situation exists must be seen as a consequence of the influence of politics in the energy industry.

The high proportion of medium output plants is a result of the obligation to use domestic hard coal as a fuel. Since 1965, a whole series of laws and agreements has subsidized the use of hard coal in power stations and required the utility companies to purchase fixed quotas of German hard coal for fuelling electricity generation. Behind this policy of "priority for hard coal" lies the desire to safeguard local jobs, to provide a solution to structural problems and to contribute to the security of the supply of electricity. However, the implementation of such structural policy measures by way of energy policy dictates is bought at a high price. Since the price of domestic hard coal is at present a good $125 per tonne above the price of imported hard coal, the subsidy requirement runs to around $6.25 billion a year. These funds are obtained essentially by means of a special levy - the so called "Kohlepfennig" - added at a specific rate to the price of electricity. There is, however, a remaining shortfall which has to be made up by the power utilities themselves. The protected status of German hard coal is further enhanced by the requirement to obtain approval for oil or gas-fuelled power stations with a maximum capacity of over 10 MW.

This degree of political influence on the power generation mix and the higher electricity prices associated with it are today becoming more and more difficult to justify from the point of view of national **energy policy**. In particular, the argument that the use of German hard coal safeguards the supply of electricity has lost a lot of weight since re-unification and the availability of lignite reserves in Eastern Germany.

The fact that the contribution of **nuclear power stations** to basic output is **too low** is a consequence of difficulties in gaining **public acceptance** for this type of fuel in Germany. Since the mid-seventies, the power utilities in Germany have been faced with a level of increasing opposition experienced in very few other countries. The sometimes violent demonstrations at building sites for nuclear power plants contributed to the termination of construction of the proposed reprocessing plant in Germany, and the massive obstacles and sometimes extreme upgrading requirements demanded by approval procedures have in effect put an end to the building of new nuclear power stations. This crisis of public acceptability today exerts a greater influence on the planning of power stations than the economics of fuel costs or environmental considerations.

Since the end of last year, some movement has been brought back into the gridlocked debate on nuclear power in Germany. In the spring of this year, a high-powered "negotiating committee" made up of politicians of all persuasions and representatives of industry, trade unions and ecological pressure groups was constituted. It was clear from the start that we could not expect any substantial results before the beginning of 1994, the year that promises to be one continuous election campaign. The lead-up to the election does not, in my opinion, create the necessary conditions for this to happen. But I will return to this point later.

For many years now, electricity generation considerations have been influenced not only by cost accounting factors and political intervention but increasingly by **environmental aspects**. Environmental acceptability is today the prime objective of the supply of energy alongside the security of the supply. In order to achieve this goal, the environmental protection authorities have, in the past, made virtually exclusive use of statutory regulations. The economic basis for government restrictions is the theory of negative external effects or the inclusion of external costs.

The costs seen as external are those which are imposed on society without being accounted for within the individual cost accounting systems. The most significant of these are the costs for off-setting or eliminating environmental damage (for example air, water and soil pollution, and health hazards). In order to quantify precisely which types of effect on the environment should be restricted or compensated for, the external costs would have to be precisely ascertained. There are, however, considerable problems of evaluation associated with such an undertaking, which as yet have not been properly solved. As a result, what happens at present is that environmental protection demands are put forward and politically sanctioned without any precise knowledge of how such demands would stand up to a cost-benefit analysis from the point of view of the national economy.

In practice in Germany, numerous and far-reaching air pollution restrictions were introduced, for the electricity generating industry in particular, primarily under the terms of the "Bundesimmissionsschutzgesetz" [Federal Immission Control Act] and the associated statutory ordinances (e.g. the "Großfeuerungsanlagen-Verordnung" [Ordinance on Large-Scale Furnaces]) for new plants but also for existing plants. These government-imposed limits did not directly alter the relative numbers of the various types of power station, but did create considerable additional burdens for the utility companies. The German electricity generating industry invested approximately $8.9 billion in the reduction of sulphur dioxide emissions and at least $4.2 billion in limiting nitrogen oxide pollution. To that we can add annual operating costs of at least $3.1 billion. The resulting additional electricity generating costs, which can be looked upon as internalized external costs, amount on average to at least 2 cents per kWh. The measures introduced reduced sulphur dioxide emissions by 87% compared with 1982. Average reduction of nitric oxide emissions nationwide was approximately 72%, again compared with 1982.

The "internalization" of the external cost of air pollution by means of the imposition of new conditions for existing plants thus created a considerable additional burden for the power generating industry, without initially changing the pattern of electricity generation in West Germany.

The lasting effect of these types of regulations can be seen in the case of existing plants for which upgrading to the new standards is not worthwhile and in the decisions regarding the building of new plants in the future. Thus in the former GDR, where power stations are now subject to West German environmental protection laws, lignite-fuelled power stations with a total installation capacity of around 8000 MW will be taken out of service in the next few years.

The introduction of the **"Stromeinspeisungsgesetz"** [the law governing the supply of power to the national grid] in 1991 brought with it additional financial burdens for German electricity companies for reasons of environmental protection and energy policy. The new law was justified on the grounds of the need to safeguard resources and limit effects on global climate. The proportion of electricity produced using regenerative energy sources was to be increased by offering a price for this type of electricity supplied to the national grid which is well above the acquisition costs of the alternatives. A statutory minimum price was decided upon for electricity generated at plants using solar, wind or water power, or methane or sewage gas. The legally guaranteed prices are not related to the cost of the energy being replaced. Instead, they are based on the average earnings from the supply of electricity to all electricity consumers nationwide. It is my opinion that this price fixing policy on the part of the legislature represents a type of government intervention which goes against both the principle of free market prices subject to controls on monopolies and the principle of remuneration for electricity supplied to the national grid on the basis of the costs avoided. The remuneration for electricity supplied to the national grid produced by regenerative energy sources is not based on commercial value in accordance with the level of demand in each case. In addition, levels of remuneration which are based on sales revenue also include the electricity company's proportional costs for availability and distribution. In order to be economically justifiable, any system of payment would have to be based on the costs actually avoided by the electricity supplier. Distribution and availability costs should not be included in this calculation.

The departure from the principle of prices based on costs represents an attempt to make regenerative energy sources profitable to the sole detriment of conventional energy sources in the electricity generation industry and on the pretext of the avoidance of external costs.

I would like to mention two further factors which have affected the structure of power generation in the past and which were relevant not only to Germany, but also had repercussions in the rest of Europe and the world. I am referring firstly to the oil crises of 1973 and 1978/9 and secondly to the limitation of the use of natural gas in power stations by an EC Directive of 1975. The oil crises, especially the first one, caught the world economy unprepared. The sudden jump in the price of oil could only be gradually off-set in the power generating industry by changes in the make-up of the power station mix. My company in particular, Bayernwerk, was hit especially hard by the sudden hike in prices, having invested heavily in oil-fired power stations in the mid-sixties only shortly before the first oil crisis. As a consequence of the oil crises, the pattern of power generation across Europe was restructured. The proportion of electricity produced using oil dropped sharply in Germany and - as I previously mentioned - is now down to around 2%.

The second influencing factor was a Directive of the Council of the European Communities which made approval necessary for the building of new natural gas-fired power stations and the conclusion of new agreements or the renewal of expiring agreements to purchase natural gas for power stations. The reason for the Directive was the assumption, in the aftermath of the first oil crisis, that reserves of natural gas were insufficient with regard to safeguarding the supply of electricity. It wasn't until 1991 that this Directive of 1975 was rescinded. The reasons given by the Council of the European Communities included the fact that, due to technological progress, the use of natural gas in electricity generation was now to be viewed more favorably from an environmental, technical and commercial point of view.

My exposition up to now has attempted to show that in the past, the power generation mix in Germany was determined by **commercial cost considerations** and **political restrictions**, while **environmental protection measures**, although representing an additional financial burden on the power generation industry, did not have any lasting effect on the pattern of power generation.

I would now, in the second part of my paper, like to examine the current developments which might influence the pattern of power generation in the future. Let me start with a particularly ambitious environmental protection proposal which is being strongly promoted in the European Community in particular: the introduction of a CO_2 tax or a combined CO_2 and energy tax. The intention of such a measure is to attempt to internalize the external costs of the use of energy in the broadest sense - that is the costs consequential on CO_2 emissions - at source, in other words to transform them into business or internal costs and use the effect on prices to initiate changes in patterns of behavior which will reduce environmental effects.

With regard to the CO_2/energy tax which is the subject of much discussion particularly in Germany, I will not pretend that I personally do not take a more positive view of voluntary commitments on the part of industry to reduce CO_2 emissions than the tax option. Nevertheless, should the tax option prove ultimately to be politically unavoidable, then based on the intention of the proposal, i.e. the reduction of global climatic changes, a pure CO_2 tax would naturally be preferable in principle to a combined CO_2 and energy tax. If the aim of the tax is to reduce CO_2 emissions, then that tax should be levied at the point of emission and should not lead to across-the-board taxing of any type of energy use and the questionable effectiveness that might result from it. In my opinion, it might well be possible in the long term to bring about a shift in power generation toward low-CO_2 or CO_2-free fuels. All other things being equal, this would mean a shift of the power generation mix toward low-emission or zero-emission fuels, for example toward nuclear power, gas and greater use of regenerative energy systems.

However, a strictly "cause-and-effect" approach such as this for reducing CO_2 emissions and the simultaneous internalization of external costs would not appear to have much prospect of success at present. Both on a national level in Germany and internationally within the EC, the discussions at best involve a combined CO_2 and energy tax and in some cases only an energy tax. The reasons given quote competitive disadvantages which would result from a pure CO_2 tax. But in actual fact, this argument is counter-productive. Because if you want to reduce CO_2 emission and therefore you raise a tax on it, it is only natural and entirely intentional that the various competing operations will be affected to differing degrees according to the level of CO_2 emissions they currently produce. It is especially those countries which make almost exclusive use of fossil fuels that are vehemently opposed to the "fundamental theory" tax option.

At present, therefore, the discussion is tending toward a combination tax with a CO_2 and an energy component. Such a compromise would - for all its shortcomings - at least be preferable to the pure energy tax also under discussion, not to mention the fact that there are widely diverging views on the necessity of such a tax at all.

The absolute precondition of any tax option must be that it is introduced internationally. This is due on the one hand to the simple fact that CO_2 is not a local or national problem, but a global one. Any measures introduced by individual countries in isolation would from the outset not have any hope of achieving any level of success. And furthermore, unilateral moves to introduce a CO_2/energy tax at national level, in Germany for example, would significantly impair the country's industrial competitiveness. For this reason, the EC has stated that parallel moves are necessary in the USA and Japan as well.

At national level in Germany, the proposed amendment of the German Atomic Energy Act could be a major factor affecting future decisions on investment in power stations.

As indicated earlier, in consideration of CO_2 emissions, greater emphasis ought to be placed on nuclear energy since it is a zero-emission fuel. The extent to which this will be possible in Germany in future will be partly dependent on the nature of the intended changes to the Atomic Energy Act. The primary purpose of this law up to now has been to promote the peaceful use of nuclear power, to provide protection against its hazards and to redress the damage caused by nuclear energy. The amendments to the law planned by the Federal Government, and especially the initiatives put forward by the opposition unfortunately represent moves which at the very least will not make continued operation of existing plants and most certainly the expansion of nuclear power any easier.

Let me illustrate this by itemizing a few points: one of the aspects to be revised, for example, is the conditions for approval of "damage prevention measures". Opponents of the use of nuclear energy will doubtless take the opportunity offered by the amendment to once again test the limits of their rights of action. This would make the requirements for new plants unquantifiable. The legal security of operators and investors would be severely impaired.

Further sources of uncertainty for existing nuclear power stations in Germany could also result from the planned proposals to make it easier to impose supplementary conditions at a later date and the total abolition of compensation for such new conditions. Since there is no consensus in Germany at present on the future use of nuclear energy, the proposed amendments provide certain regional authorities led by local governments keen to opt out with the ammunition to put a stop to the use of nuclear power altogether. The inability to predict the extent of future supplementary regulations represents a considerable risk in the face of the high cost of investment in the nuclear energy industry.

In view of the discussions taking place at present at the highest political levels in Germany on the subject of a consensus on energy and in which the question of the future use of nuclear power plays a central role, the planned amendment of the Atomic Energy Act is sending out the wrong signals. The threat to the future of existing plants and the uncertainty overshadowing any plans for new plants not only could but most certainly would have a significant effect on plans for the shape of power generation in the future. The Federal Government's draft proposals are more likely therefore to be detrimental to the use of nuclear energy in future.

To a certain extent, the proposed amendment to the Atomic Energy Act can also be placed under the heading of internalizing the external costs of electricity generation - not least in view of the increase in the required insurance cover for possible damages claims also under discussion.

I am not saying anything new when I tell you that there are enormous uncertainties with regard to assessment and allocation in this regard and that the doors are thrown wide open to political whim. Nevertheless, in this connection I would like to refer briefly to a study on this subject performed by the Swiss company Prognos AG. I do not intend to go into the considerable problems of assessing external costs particularly with regard to the uncertainties in the assessment of altered fatality risks, instead I would just like to draw attention to one aspect which seems to me to be particularly important in this context. The study I referred to carried out by Prognos AG, which was entitled "Identification and Internalization of the External Costs of Power Generation", calculated the external costs for nuclear, conventional and regenerative energy systems. The external costs of a core meltdown are set at $6,600 billion. Based on the likelihood of a core meltdown occurring, the annual damage costs work out at $4 billion. If these costs were to be internalized, the price of a kWh would increase by approximately 2.7 cents.

What appears to me to be significant is that damage of the level of the estimated $6,600 billion can in practice not be insured against. If for this reason the whole future of nuclear energy were brought into question, then this would also have to apply to numerous other large-scale technical installations and plants for which not all possible eventualities can be insured against or for which the insured liabilities are exceeded.

There is another point too: doing without nuclear energy would mean replacing it with fossil fuels. In that case, however, the external costs of CO_2 emissions would have to be taken into account. The consequences that this might have for the price of electricity is made clear by another investigation within the Prognos study which looked at the estimation of damage caused by CO_2/CH_4 accumulation. Various different authors came up with calculations based on certain assumptions and conditions of between $0.003 and $16.25 per kWh for the specific damage cost of fossil fuels. An average level calculated using risk analysis methods gives an increase of $1.5 per kWh.

In spite of all the uncertainties, this comparison of figures illustrates what internalization of external costs would mean if the principle is applied to all types of fuel. An overall view of costs of this type would not automatically stamp nuclear energy as an uneconomical system, and would at least give it new impetus in comparison with the fossil fuels. This is especially remarkable because of the fact that in Germany the claim is often made that if complete internalization of external costs were put into practice, nuclear energy would no longer be able to be used for electricity generation.

It must still be said, however, that a comparison of the possible external costs of nuclear power and fossil fuels of the type I have just mentioned can hardly be taken as an accurate guide due to the immense problems associated with estimation but does, nevertheless - if the internalization of external costs is put to one side - allow a judgment to be made about the different fuels from the point of view of environmental policy.

In addition to the influence of **environmental protection initiatives** including schemes for the inclusion of the **external costs** of the various types of fuel, I would like to mention two other developments which in my opinion will have a decisive effect on the future pattern of power generation in Germany. I am referring firstly to another item to be included under the heading of **"international political restrictions"**, namely the EC initiative aimed at introducing more direct competition into the European electricity generating industry. Secondly, there is the question I touched on earlier of the national discussion regarding the **social acceptability** of the individual power generation options and in particular the future use of nuclear energy in Germany.

Let us look first of all at the possible consequences of more competition in the European electricity market. The EC Commission has put forward a draft directive which would fundamentally interfere with the established structures of electricity supply. Among other things, it envisages the abolition of exclusive rights in the areas of generation, transmission and distribution as well as free access to the grid (so-called third party access or TPA) initially for all electricity producers and large-scale industrial consumers over 100 GWh per annum as well as for regional and local suppliers above certain threshold levels. Whether or not these proposals will be enacted and how they could be put into practice is still to a great extent unclear. Since the spring of this year, the European Parliament has come up with more models, which I can not go into any further at this stage, however.

But it is possible to make some statements about the possible consequences of implementation of the EC Commission's proposals even at this early stage. We can gain some insights from the electricity supply system in the UK which was reorganized a few years ago. This system has many similarities with the EC proposals. The initial reaction to the political initiatives in the UK led to the planning/construction of - to the best of my knowledge - 16,000 MW of new capacity to be fuelled almost exclusively by gas. The preference for gas-fuelled power stations is understandable from the point of view of minimizing amortization periods for investment in power stations in a pure seller's market. I will not offer any opinion on this aspect - especially not with regard to future developments in the UK model. But it is at least fairly obvious that investment-intensive, but in the long term economical basic output nuclear or coal-fired power stations have little chance in a system which looks no further ahead than 15 years at a time. It is my conviction that the resulting pattern of power generation will thus move away from the possible optimum with regard to the long-term quality of the supply.

In addition to the legal framework, the conclusions reached on the future of the supply of energy in the discussions taking place at present between politicians and the relevant industrial associations and social pressure groups are of crucial importance for business decisions on the structure of the future power generation mix in Germany. The question at the center of the negotiations set in motion in the spring of '93 is whether lasting consensus among the broader public on the future use of nuclear power can be found. The present lack of agreement in the public debate on the acceptability of the use of nuclear power has led to the previously mentioned de facto moratorium on the building of new nuclear power stations in Germany. In one case, legal technicalities are still preventing a reactor completed 6 years ago from going into service. The last nuclear power station to be taken into service in Germany started operation in 1988. In the absence of a clear political decision in the near future in favor of the continued and increasing use of nuclear power in Germany, its share of the electricity generated will steadily shrink, since a number of companies have stated that without a broad political consensus they are neither able nor prepared to risk investment in the building of new plants. In that case, the uncertainty of the political situation would at the very least result in a significant reduction of the number of nuclear power stations in Germany when the reactors operating at present reach the end of their service lives. Environmental considerations as well as those of business economics, which would tend to support the continued and even increased use of nuclear energy, would then have been pushed into the background by the political situation.

A Franco-German co-operative venture is currently attempting to prepare the ground for a re-assessment of the position of nuclear power in Germany by the development of an advanced type of reactor. The project involves working together with France on the development of an advanced European pressurized water reactor (EPR). At the center of this project is an advanced safety concept. The aim, among other things, is as far as possible to limit the effects even of theoretically conceivable core meltdowns to the immediate reactor site. The EPR is intended to meet approval criteria in both France and Germany.

It would appear unlikely that we can expect any conclusions in the near future from the consensus negotiations on the future of nuclear energy in Germany. I would like to emphasize at this point, however, that opting out of nuclear technology in Germany is not a step that could be taken with the agreement of the electricity generating industry. If the politicians decide nevertheless to bring an end to the use of nuclear power in Germany, they must also be prepared to take full responsibility for their actions.

Allow me to briefly summarize the points I have made.

The present pattern of power generation in Germany is the result of numerous factors. The **cost-effectiveness** of the individual types of power station plays a decisive role. But **political restrictions, environmental protection considerations** and the **question of social acceptability** are exercising an ever greater influence on the make-up of the power generation mix of the power utilities in the Federal Republic of Germany. The intensity of effect varies widely. Although the environmental protection legislation introduced up to now has increased the financial burden, it has not had a lasting effect in the sense of restructuring the pattern of the types of power station in use.

The obligation to use expensive domestic hard coal for electricity generation has, however, not only pushed up the price of electricity, it has also resulted in the use of hard coal for generating a large proportion of the electricity produced. For many years now, nuclear energy has faced the greatest problems. Not only do the strictest safety regulations in the world create an enormous financial burden, but the present social acceptability debate makes the building of new nuclear power stations extremely difficult.

The **future pattern of power generation** in Germany therefore depends to a significant degree on the decisions made by the politicians. Stricter environmental protection regulations, and in particular further inclusion of the external costs of electricity generation - for example by means of a CO_2 tax - can be expected to exercise an increasing influence on the decisions made by the electricity suppliers regarding the type of power stations they operate. The same can be said of the socio-political restrictions, such as the present stalemate in the discussion about the future use of nuclear energy and the building of new nuclear power stations.

The way in which the pattern of power generation in German will develop against this background remains to be seen. It is my view that in the medium to long term the need to drastically reduce CO_2 emissions will have a significant effect on the types of power station in operation. From this aspect, nuclear power ought to be of central and growing importance. As I have shown, greater consideration of the external costs of all energy systems would tend to strengthen the competitive position of nuclear power in comparison with fossil fuels. In the long term, we can also expect the relative importance of fossil fuels in electricity generation to diminish, if for no other reason than the finite nature of the reserves, at least in the developed nations of the world. In its place we should see regenerative forms of energy supplementing nuclear power. The effects of the levelling off of the growth of demand for electricity which is already foreseeable will also makes itself felt, however.

But there is a long way to go before we reach that point. As far as the situation in Germany is concerned, the most pressing matter in the immediate future is the achievement of public acceptance of the continued use of nuclear power.

GENERATION CHOICES AS INFLUENCED BY COSTS, RISKS AND EXTERNALITIES - SWEDEN

Carl-Erik Nyquist
President and Chief Executive Officer,
Vattenfall

Abstract

Sweden's electricity supply is a low-cost and efficient power generation based on hydro power and nuclear power - 50 percent each. Good performance and an ambitious waste handling system is crucial for the public acceptance of nuclear power in the future. To achieve this power companies which own nuclear power must have all the expertise which is necessary to operate and maintain ageing nuclear plants. Cross border trading of electricity - especially the cooperation betweeen the Nordic hydro power system and north-European thermal power system opens new commercial opportunities.

LE CHOIX DES MODES DE PRODUCTION : LE POIDS DES COÛTS, DES RISQUES ET DES EXTERNALITÉS - SUÈDE

Résumé

La production d'électricité suédoise est efficiente et peu coûteuse et se divise à part égale entre hydroélectricité et énergie nucléaire. Pour accepter l'énergie nucléaire le public de demain devra être assuré de ses performances et de l'existence de programmes ambitieux de traitement de déchets. Dans cette perspective, les compagnies propriétaires des centrales nucléaires doivent être en mesure d'assurer pleinement l'exploitation et la maintenance des vieilles installations. Les échanges transfrontières d'électricité, notamment dans le cadre la coopération entre le réseau nordique de production d'hydroélectricité et celui de production d'énergie thermique d'Europe du nord, offrent de nouvelles perspectives commerciales.

Do you know of a country where nuclear power has a very high availability, where there is virtually a complete program for handling nuclear waste, yet where the decision has been taken to phase out nuclear power by the year 2010? The country in question is Sweden, and that is where I come from.

My company, the Vattenfall Group, accounts for 50 percent of Sweden's entire electricity generation and is a wholly owned limited company by the Swedish Government. We are also Sweden's largest distributor of electricity.

I will comment on two areas where externalities are at reality in Sweden i.e.

- the nuclear issue
- the deregulation of the electricity market.

My opinion is that many externalities could be turned into possibilities for the power industry if the questions are treated in a correct manner and with sensitivity. After all electricity has been the winner on the energy scene since electricity was introduced on the market.

Electricity supply in Sweden (slide 1)

Sweden is a substantial consumer of electricity – utilizing approximately 15,000 kWh per capita. This derives from the fact that we enjoy low-cost electricity – with 50 percent generated by hydro power and 50 percent by nuclear power. Hydro and nuclear power both generate extremely favorable operating returns and are, in addition, environmentally friendly with no emissions at all.

Following the incident at Three Mile Island, Swedish public opinion turned against nuclear power and a general referendum was implemented in 1980. Based on the result of the referendum, Parliament decided that the nuclear power plants, then under construction, should be <u>continued</u> and <u>completed</u>, but that nuclear power should be phased out by 2010 at the latest. Following the event at Chernobyl, which had a serious effect on parts of Sweden, Parliament decided to initiate the shut-down of two reactors as early as 1995/96. This decision was changed in 1991 by a political agreement among three political parties. The development of alternative environmentally-friendly electricity generation and the country's economic situation are <u>now</u> the two factors most crucial to decide when the phase-out of nuclear power can be commenced. However, the parliamentary decision that nuclear power will be phased out by the year 2010 remains unchanged.

The alternatives in the area of renewable environmentally friendly electricity generation are limited. Political decisions are hindering the development of additional hydro power facilities while other renewable energy resources, such as biofuels and wind-power, will provide only minor contributions to the electricity supply. Accordingly, nuclear power must be replaced mainly by fossil-based fuels.

I will start to talk about nuclear power as this is the most crucial issue in Sweden.

Nuclear power (slide 2)

In Sweden, we have a total of 12 nuclear power reactors at four sites. Combined, these units generate up to 74 TWh. That is 15 TWh or 20 percent more than projected when the blocks were first placed in operation.

Vattenfall operates seven blocks, which currently generate 48 TWh – at Forsmark and Ringhals. Our goal is to operate these plants with high availability and good economy up to the year 2010. If required, we can increase capacity within the blocks by an additional 10 TWh. Part of this increase can be achieved at a relatively modest cost, compared with new electricity generation. For the time being there is no demand for new capacity and the year 2010 sets limits for what can be done commercially.

(slide 3)

Availability at Sweden's nuclear power plants is very high, seen in an international context. An availability level of 90 percent was recorded in 1991 in the nine Boiling Water Reactors, with 86 percent achieved in the three Pressure Water Reactors. These figures compare favourably with the world average of 70-75 percent. Availability declined temporarily on the BWR side in 1992, as a result of the five oldest BWR reactors having to be shutdown during the final four months of the year to rectify problems in the emergency core cooling system. Following this year's audit, this weakness have been corrected and I am hoping that we can then reattain and even exceed 1991's peak figures.

Operating economy is highly satisfactory. Variable costs are well under 1,0-1,2 US cent per kWh, of which 0,25 cent is allocated to nuclear waste-handling operations and for the dismantlement of the facilities. In Sweden, the owner of the nuclear power plant is responsible for all waste-handling and in the future for the dismantlement of any nuclear power plant that has been permanently shut down.

During recent years, we have noted cracks and materials faults in the PWR:s. The increasing number of faults is in part due to not having made an optimal selection of design and materials when the blocks were constructed, and in part due to our now having new, more advanced measurement equipment to analyze material problems more efficiently. Naturally, we also have problems that are directly related to aging phenomena.

We have experienced that access to skilled engineering resources is vital in the handling of these design-phenomena and problems with the materials. During recent years, we have successfully developed techniques for countering these faults.

Nuclear power represents a great deal of capital and is associated with many sensitive issues in terms of public opinion. It cannot be underlined too often. The importance of being able to operate nuclear power facilities safely, cost effectively and in a way that creates public confidence.

In this area, nuclear waste handling is a very important issue. As I mentioned during my opening remarks, Sweden has virtually a complete program for taking care of nuclear waste. Our program is illustrated on this slide.

(slide 4)

Sweden's nuclear power companies have jointly formed a company SKB (Swedish Nuclear Fuel and Waste Management AB) to handle all nuclear waste operations in the country.

The low- and medium- level radioactive waste is stored in a special final storage facility adjacent to the Forsmark nuclear power plant. We estimate that in its current and expanded form, this facility will be capable of storing all such waste, including material that will emerge of the future dismantling of nuclear units.

The spent nuclear fuel is initially stored for a period of one year at the nuclear power plants. After that, it is transported to the SKB facility at the Oskarshamn nuclear power plant – known as the CLAB (Intermediate Storage Facility). It is a wet storage. Here the spent fuel remains until it becomes due for final storage, which is expected to occur in about 30 years. In brief, the final storage solution that we are currently working with is based on the spent fuel being securely sealed in steel and copper capsules, which are then insulated some 500 meters down in the Swedish bedrock. In that respect the power industry has a lead compared to other industry sectors in funding cost for the total process.

During the 1990s, we are making studies to establish the most suitable locations for final deep storage. We plan to make detailed investigations on 2 sites and in about 15-20 years time, we are planning to implement a demonstration facility for deposition in a first step. After 10-15 years of experiences from this unit, the nest step in the final storage facility will be built to contain all of Sweden's spent nuclear fuel plus other long lived waste. It is estimated that deposits will be made in the final storage facility during the period 2020-2050 that means 30-60 years from now.

The nuclear waste is transported in a specially built maritime vessel, since the power stations and storage units are so far located on Sweden's coasts.

This system has been accepted by the Government authorities, who have also stipulated that the waste material shall not be converted or reprocessed. The costs of all this are being financed through a fee on current nuclear electricity generation. For each kWh generated in a nuclear power plant, the power companies pay 2,5 mills/kWh, or about 15 percent of the total generation costs, into a special fund.

Just now we are in a process to gain acceptance among some municipalities in Sweden to be the host of the final storage location.

Here, we can expect to meet the attitude, "not in my backyard".

(slide 5)

This picture shows how people have responded to the question "If it is assessed that the best place for storing highly radioactive nuclear waste is your municipality, would you accept or reject that such materials are stored in your municipality?"

The proportion of people prepared to accept this proposal has exceeded 50 percent since 1989. The number of "don't knows" has increased and is currently 12 percent.

(slide 6)

Public opinion relating to nuclear power on the whole is highly volatile. Resistance increased during the 1970s, particularly after the incident at Three Mile Island. The Chernobyl accident in 1986 had a profound effect on Sweden with radioactive rain, but the negative effect on public opinion was short-lived.

The shutdown of five blocks in the fall of 1992, did not have any long term effects on opinion. During the past year, public opinion has been relatively stable, with some 50-60 percent believing that nuclear power should be used beyond the year 2010, if there are good technological and economical reasons for doing so and I would also say environmental reason.

The Nordic electricity system

Up until now, I have focused exclusively on the Swedish electricity generating system and in particular on nuclear power because nuclear is so important for Sweden's electrical future.

For many years, Sweden's power companies have been cooperative with each other. A key part of this is the production optimization. In practical terms this means that power is generated whenever possible in those plants offering the lowest varable costs and that the customers as well as power companies share the profit generated by this strategy. I estimate the value to more than 100 MUS$ annualy.

(slide 7)

We use the same power exchange concept on a Nordic basis, in the NORDEL cooperation where the individual production structure of the various countries is utilized. Norway, for example, has only hydro power, Sweden has hydro and nuclear power, Denmark has only fossil-fuel power and Finland a mixed system with roughly similar proportions of hydro, nuclear and fossil-based power. We thus have excellent opportunities to combine hydro power and thermal power in these countries.

This means also that hydro and nuclear power usually eliminate fossil-based power, resulting in improved environmental performance overall in the Nordic region.

New rules applying to the electricity market, such as deregulation, increased competition and internationalization of the electricity supply, are factors that make it uncertain whether the production – optimization measures I have described will be able to continue – or I would say They will not be! Instead, they may be replaced by more bilateral agreements between power companies.

Here ambitions to create a more competitive market will reduce other economical advantages. The impact of the economy of the power companies is difficult to estimate.

Future electricity supply (slide 8)

When I think about Sweden's future electricity supply, I think beyond the national electricity systems. A number of factors indicate that a much larger geographic area will be considered electrically. Sweden will be more interconnected electrically to neighbouring countries than today.

(slide 9)

The following trends are of vital importance to Sweden's electricity supply:

* stagnating electricity demand in Europe
* increased competition as a result of deregulation
* increased internationalization
* rehabilitation requirements in Eastern Europe
* Linking of power grids between Western and Eastern Europe

Electricity consumption is currently stagnating in Sweden and in many other countries in Northern Europe. Even though we can count on an increase of the electricity share of the total energy use, I believe that in the longer term in the industrialized world we shall see stagnation continuing as electricity consumption becomes more efficient. Thus the need for new capacity will stabilize or even decrease in future.

Increased competition is another trend in the electricity markets of Europe. Great Britain initiated the process by breaking up its monopoly. Our neighbour, Norway, has followed, and Finland is planning to introduce a new system, with Third Party Access, next year. In Sweden, the results of two recent government studies have resulted in proposals designed to increase competition in the Swedish electricity market. One of the proposals is based on opening the electricity network so that consumers will have the opportunity to choose their own suppliers. The other study cover the establishment of a public power pool. The study proposals are expected to result in a decision by the Swedish Parliament to open the entire electricity network. I think Januari 1, 1995 if the politicians will reach consensus.

The Swedish power industry is positive to the concept of increased competition and opening the electricity network but is anxious that the present, highly efficient system not be replaced by anything less efficient.

Looking at the rest of Europe

Within the EC Commission, there are also aspirations toward open markets. The EC has introduced rules in the area of common-transit and price transparency. However, since a number of countries remain doubtful about taking additional steps towards Third Party Access, it is unclear whether and how the movement towards more open electricity markets will continue. My opinion is that we will have to consider an open electricity market in the future.

The electricity system in Eastern Europe is in need of a major overhaul and modernization. A general review of the plans for Poland, Germany, the Baltic States, the Czech and Slovak Republics, Hungary and Russia reveal that substantial additional capacity (of approximately 25,000 MW) will be required up to the end of the present century. If financing problems can be resolved and institutional frameworks created, major business opportunities will exist for, primarily, from the western hemisphere power companies.

We envisage the increased internationalization of the power industry, with increased cross-border trading in electricity and a greater number of alliances and cooperation agreements between power companies in different countries.

One factor that will contribute to increased trading activities is the linking of the electricity networks of Western and Eastern Europe. The estimate is that the network in Eastern Germany will be linked with that of Western Germany in 1994, while Poland, the Czech and Slovak Republics and Hungary will be connected with the Western European system in 1997.

Conclusions

Among the most important conclusions that can be drawn regarding future electricity supply is the fact that the elimination of the monopoly will lead to increased uncertainty in terms of how the electricity will be sold – and at what price.

This will in turn result in greater caution being observed with regard to plant investments. There will be less heavy capital investment in new power plants.

Instead, the focus will be on measures designed to retain and increase production capacity in existing power plants.

Should new power facilities be built, interest will mainly be directed on plants requiring only modest capital investment, and where the licensing process can also be implemented fairly quickly that is shorter lead times. Moreover, low environmental impact will be required.

For us in Sweden, this means the following objectives:

> (slide 11)

> * for hydro power: renewal in combination with environmental measures
> * for nuclear power: efforts to keep maximum efficiency, a sustained operating performance and increased public acceptance in nuclear power as an energy source
> * for other thermal power operations: increased flexibility thus enabling different fuels to be used, and reduced environmental impact.

In terms of new electricity generation, first of all we intend to use very opportunity to attain combined electricity and heat generation where mostly biomass or natural gas will be used.

Towards that background, I do not believe that we shall see any new nuclear power capacity in Sweden during the foreseeable future even if I myself is very much in favour of nuclear power.

Bearing in mind the slow increase in demand for electricity, it is doubtful that new nuclear power plants will be considered necessary in many industrialized countries.

With little demand for new nuclear power in the future, a central issue will eventually be: how can we maintain sufficient competence within the industries supplying nuclear power technology and equipment to the power companies? We must have access to such competence inorder to handle renewal and maintenance operations. At the same time, the power companies must ensure that we have the necessary qualified technological expertise to staff the facilities we operate.

And, if our own base is too small, we should seek cooperation with other power companies. As an example, just such in-depth cooperation is currently being established between the owners of ABB's BWR:s. It can also be noted that on the PWR:s side, the Westinghouse User Group is already in existence and creates a good base for exchanged of experience.

For the fully developed nuclear power sector, the future will be determined by our own skills in the handling of this technology – technically, administratively and in terms of public acceptance.

Public acceptance will be the major externality for power generation in the future.

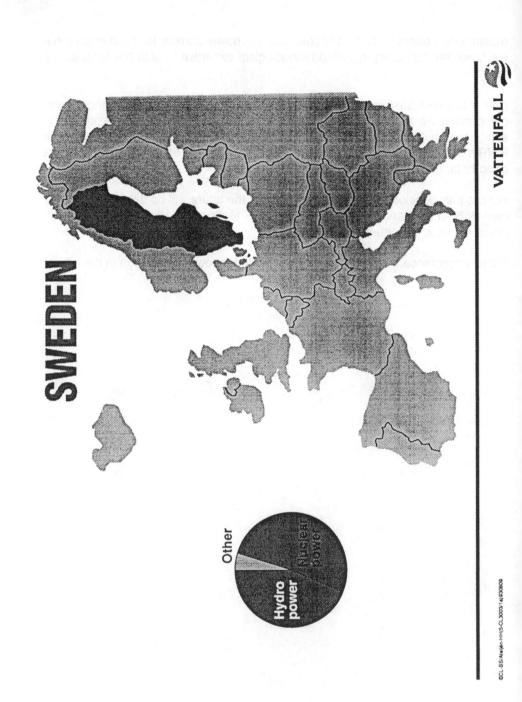

Slide 1

NUCLEAR POWER IN SWEDEN

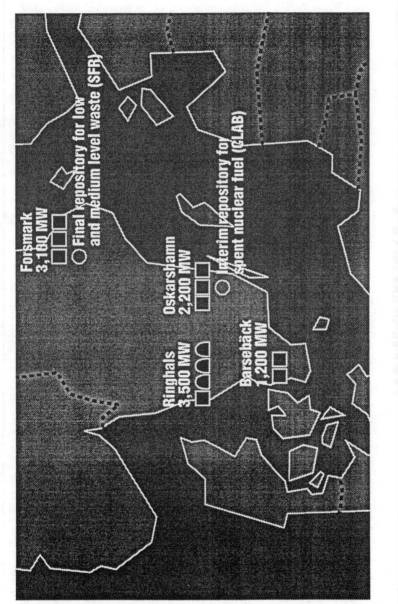

AVAILABILITY

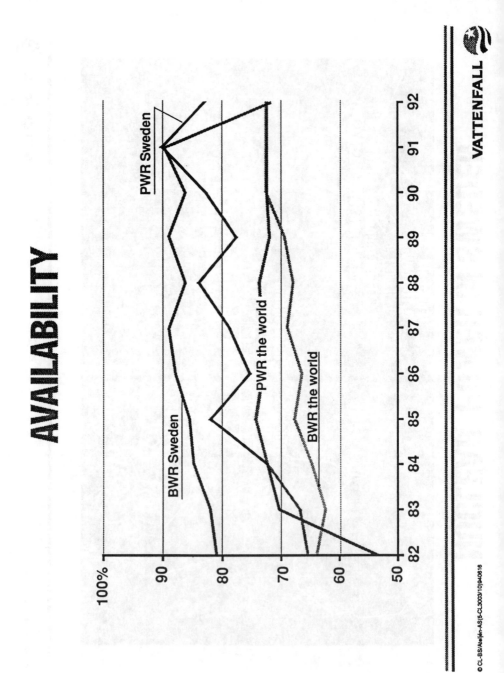

VATTENFALL

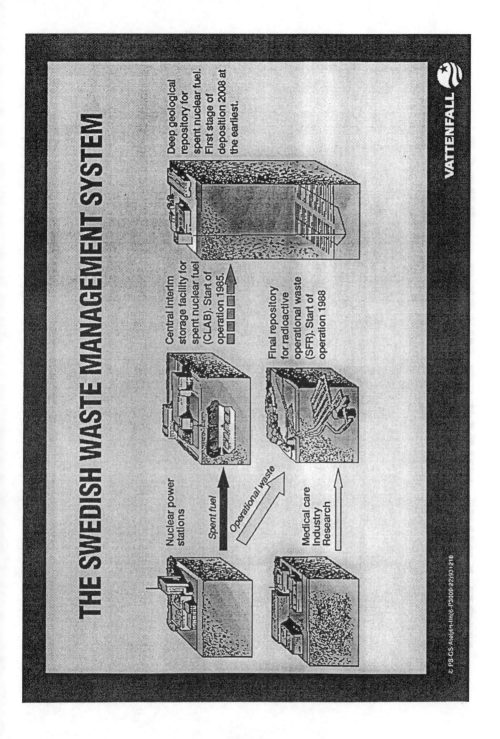

Slide 4

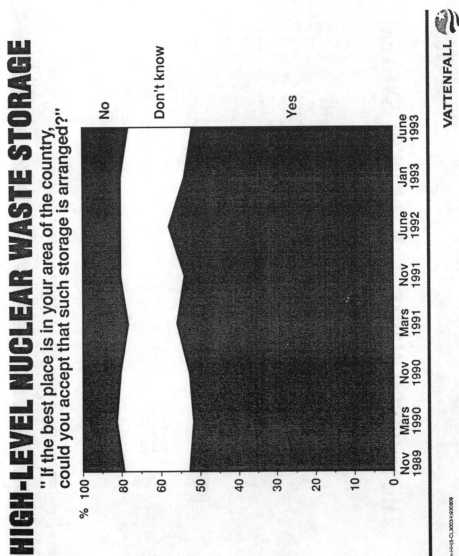

Slide 5

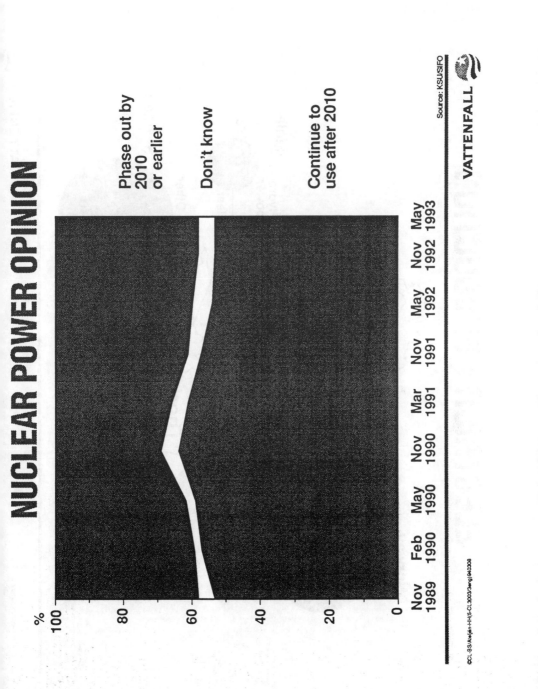

Slide 6

ELECTRICITY PRODUCTION

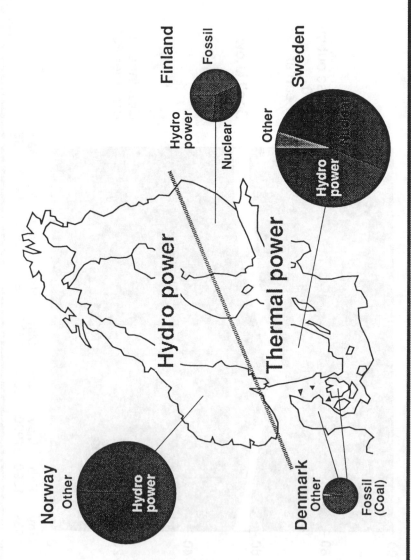

FUTURE ELECTRICITY SUPPLY

TRENDS

- Stagnating electricity demand

- Increased competition by deregulation

- Increased internationalization

- Rehabilitation requirements in Eastern Europe

- Linking of power grids between Western and Eastern Europe

VATTENFALL

Slide 8

POWER EXCHANGE WITH NEIGHBOURING COUNTRIES

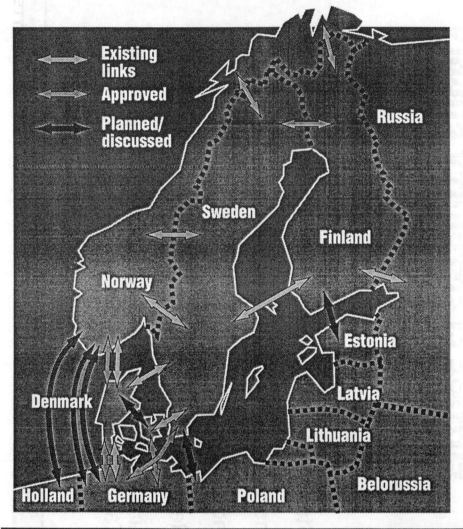

Slide 9

FUTURE ELECTRICITY PRODUCTION

CONCLUSIONS

- Greater uncertainty of electricity markets and prices

- Less heavy capital investment in new power plants

- Greater interest in retaining and increasing production capacity in existing power plants

- New power plants with modest capital investment, shorter lead times and low environmental impact

VATTENFALL

© CL-BS/Ateljén-HH/(5-CL3003/7)930812

Slide 10

CONCLUSIONS FOR SWEDEN

OBJECTIVES

- Hydro power: renewal in combination with environmental measures

- Nuclear power: maximum efficiency, sustained operating performance and increased public acceptance

- Other thermal power: increased flexibility enabling the use of different fuels and reduced environmental impact

VATTENFALL

© CL-BS/Ateljén-HH(5-CL3003/8)930812

Slide 11

THE ECONOMICS OF NUCLEAR POWER AND COMPETING TECHNOLOGIES IN THE UK

BL Eyre
Chief Executive, AEA Technology, United Kingdom

PMS Jones
Consultant to AEA Technology, United Kingdom

Abstract

The paper discusses the development of civil nuclear policy in the UK with reference to the factors that have influenced the past programme. The privatisation of the UK's electricity supply industry has radically altered the way in which decisions on new capacity are made, resulting in a dash for gas and problems for both coal-fired and nuclear generators. Analyses undertaken for the 1993 Government Review show new nuclear plants to be economically competitive, but market choices based on financial criteria may not reflect this. Wider factors will be considered in the review but whether and how the Government will allow for them in its market based philosophy remains unclear.

ASPECTS ÉCONOMIQUES DE L'ÉNERGIE NUCLÉAIRE ET DES TECHNOLOGIES CONCURRENTES AU ROYAUME-UNI

Résumé

La présente communication analyse l'évolution de la politique civile nucléaire au Royaume-Uni, à la lumière des facteurs qui ont influés sur le programme passé. La privatisation du secteur de la production d'électricité au Royaume-Uni a radicalement modifié les processus de prise de décision concernant les nouvelles capacités et s'est traduite par une ruée vers le gaz et par des problèmes pour les générateurs nucléaires et à charbon. Les analyses entreprises dans le cadre de l'étude gouvernementale (Governement Review) de 1993 montrent que les nouvelles installations nucléaires sont concurrentielles d'un point de vue économique, mais que les choix du marché fondés sur des critères financiers ne reflètent pas cette réalité. Bien d'autres facteurs seront pris en compte dans l'étude, mais il reste à savoir si les pouvoirs publics les intégreront dans leur philosophie de marché, et comment.

INTRODUCTION

The United Kingdom's civil nuclear programme was launched in the 1950s when there were concerns over the long term security of supply of fossil fuels and an expectation that nuclear costs would in the future be lower than those of competing technologies. Initial development was based firmly on indigenous capabilities and a desire to avoid the costs and potential imports that would have arisen if light water cooled reactors requiring enriched fuels or reactors using heavy water moderator had been adopted. The choice of the graphite moderated gas -cooled reactor systems led to the technically simple and very reliable Magnox plants, the first of which, Calder Hall, is still operating without problems after 36 years.

The advanced gas-cooled reactor (AGR) was a natural progression to higher power density, more compact plants with higher operating temperatures and thermal efficiencies. This, combined with on-load refuelling, was expected to make them competitive with the light water cooled reactors being developed in the United States.

Security of supply, balance of payments and the development of an exportable technological capability figured highly in the policy decisions of successive UK Governments[1]. These were implemented through the investment programmes of the publicly owned Central Electricity Generating Board, the South of Scotland Electricity Board, the Atomic Energy Authority and, from its formation in 1973, of British Nuclear Fuels plc.

A combination of factors contributed to confound the nuclear industry's and Government's expectations. The premature establishment of a multiplicity of design consortia forfeited the potential benefits of replication. Labour problems and design changes during construction led to delays and cost escalation for many of the plants. Technical concerns delayed the introduction of on-load refuelling of AGRs. These problem, together with the worldwide decline in fossil fuel prices in the mid-1980s, have resulted in nuclear generation from existing plants looking unattractive in full resource cost terms in the United Kingdom. The selected technology has also failed to win a niche in export markets against the widely deployed light water-cooled reactors.

The switch to the PWR was intended to revitalise the UK's nuclear programme and bring it into the mainstream of world development. Sizewell B was seen as a forerunner of a series of replicate plants with, at the time of its planning inquiry in the early 1980s, an expectation that it would provide power far more cheaply than new coal or oil fired plants. Gas was expensive and its use for baseload power generation proscribed.

Following the Sizewell Inquiry, the situation has changed dramatically in ways that had not been predicted. The plans to follow Sizewell B with a series of PWRs have been put on hold pending the outcome of a Government led review of nuclear power originally scheduled for 1994. Although the terms of reference for the review have not been announced it is clear that a key aspect will be the ability of nuclear power to compete economically with other technologies.

This paper presents the results from a UK nuclear industry Task Force, set up as one of seven by the Nuclear Utilities Chairmen's Group. Although the work of the Task Force is now drawing to a conclusion, it will not be completed until the terms of reference for the review have been announced, probably this autumn. Thus this text must be seen as a progress report which is subject to further refinement as our work proceeds, rather than as a definitive statement.

BACKGROUND TO THE UK NUCLEAR REVIEW

The UK Government embarked on its plans for privatising the electricity supply industry after approval had been given for the Sizewell B PWR and during the planning Inquiries being conducted for its proposed successor at Hinkley Point.

Between the Inquiries fossil fuel prices and expectations concerning their future movement had declined dramatically. The re-evaluation of the back-end spent fuel management costs and decommissioning costs for the existing and ageing gas-cooled reactors, necessary to establish the initial capital structure of the intended privatised generation companies, had caused alarm in financial circles and led Government to withdraw nuclear power from the privatisation package.

Estimates of the future price of PWR electricity from Hinkley C (over 6p/kWh) provided by the prospective private sector owner were based on high profit levels and short pay-back times. They were far higher than the generation costs derived in earlier public sector analyses for Hinkley C (under 3p/kWh) and higher than those for competing coal technology. This led to the withdrawal of Sizewell B from the privatisation plans and, when the Hinkley C plans were approved in principle, its construction and further nuclear development in the UK were made subject to the Government led review, then scheduled for 1994.

The privatisation proceeded with the creation in England and Wales of two privatised fossil-fuelled generation companies, 12 Regional Electricity Companies (RECs) with distribution franchises, and a separate National Grid company owned by the RECs. Two vertically integrated non-nuclear generator/distributor companies were formed in Scotland. The nuclear capacity was retained in that country in Scottish Nuclear and in England and Wales in Nuclear Electric, both of which remain in the public sector.

Despite existing overcapacity in the UK, the termination of the ban on gas fuel for generation, the availability of new combined cycle technology and the ambitions of the RECs and independent power producers to enter the generation market, have led to the so-called dash for gas. There were two consequences; the prospective margin of total capacity over peak winter demand in the late 1990s rose to nearly 60%[2] unless existing operating plants were closed prematurely, and the projected requirements for coal, particularly British deep mined coal, plummeted.

Existing nuclear capacity and Sizewell B (under construction) were protected under the Government's transitional arrangements, as are contributions from renewable sources, which obliged the RECs to take a specified amount of non-fossil fuelled electricity. The non-fossil fuel obligation provides assured electricity markets to 1998. Beyond that time the nuclear generators will have to be fully commercially competitive and any new plant will have to be financially viable.

An immediate threat to British Coal arose with the renegotiation of supply contracts as the initial transitional coal contracts neared their end earlier this year. The privatised generators preferred gas or lower cost coal imports and were reluctant to sign coal contracts while the RECs were unwilling to enter into long term contracts to purchase electricity. They are unwilling because their captive markets are threatened as the size of their franchise is progressively reduced to encourage competition in distribution.

British Coal's response was to plan extensive and irreversible mine closures. This sparked a political furore and led to an extensive Government review of coal markets, the published findings of which set down the framework of current Government energy policy[3].

This is based on a belief that markets offer the most efficient and effective means of meeting energy needs and that they will assure security and diversity of supply. Key elements of the policy are summarised in Table I. Those relating to the wider non-financial aspects and externalities are highlighted.

A number of ministerial statements have stressed the importance Government attaches to the need for nuclear power to be economically competitive before new programmes can be endorsed. There has been no clear recent indication of how this should be interpreted although the then Secretary of State for Energy, John Wakeham, when announcing the intent to hold the review[4], stated that nuclear would need to be economic taken over its life as a whole and taking broader factors like its contribution to security of supply and environment into account.

Table I - Key Policy Elements

- To encourage competition and choice

- To establish a legal and regulatory framework to enable markets to work well

- To ensure ... consumers pay full costs of the energy resources they consume

- To ensure the discipline of capital markets is applied to state owned industries by privatising them where possible

- To monitor and improve performance of remaining state owned industries while minimising distortion

- To have full regard to impact of energy sector on environment including measures to meet Government's international commitments

- To safeguard health and safety

- To promote wider share ownership

- To promote energy efficiency

NUCLEAR OPTIONS

The United Kingdom has eight operating Magnox stations with a combined capacity of 3200 MWe and seven operating Advanced Gas Cooled Reactor stations with a combined capacity of 7500 MWe. The first UK PWR, the 1188 MWe Sizewell B is nearing completion within the agreed budget and ahead of schedule. It is expected to begin producing electricity in 1994.

Since the two new public sector nuclear generation companies were vested in 1990 they have improved their performance very considerably through both substantially increased output (Figure 1) and reduced operating costs, so that the accounting costs of nuclear electricity generation have declined dramatically. Nuclear Electric's return on capital has risen from 4.5% to 10.1% per annum in three years and over the same period (1990-1993) output has risen 29% and productivity per employee by 55%[5]. Scottish Nuclear have a similar record of improvement[6] with net output increasing by 17% from 1991 to 1993 and unit operating costs decreasing by 7% in 1992-93.

Special arrangements had to be established when the companies were formed to fund the long term liabilities for spent fuel management and for reactor decommissioning which had been inherited from their predecessors. The previous provisions for these had been invested in both nuclear and coal-fired plants or absorbed by the Treasury. These special arrangements are scheduled to phase out by 1988 and both nuclear generators are expecting to operate as viable, profitable and competitive companies in the UK electricity supply market without subsidy or special assistance within a few years[5][6].

Existing gas-cooled nuclear plants produce some of the cheapest electricity for the UK grid. Only hydropower is cheaper and its contribution is limited and virtually fully exploited. The objective of the generators is therefore to keep the plants in operation for as long as possible, consistent with maintaining safety standards. The AGRs output has improved markedly in recent years so that nuclear's market share in England and Wales has increased from 16.5% in 1990 to 21.6% in 1993. With the commissioning of Sizewell B nuclear's contribution to UK electricity supplies will reach 25% but beyond the late 1990s/early 21st century it will decline as the older Magnox and later AGR plants are retired, unless new capacity is constructed (Figure 2). Nuclear electricity's market share in Scotland has risen to 47% (1992-93).

The Nuclear Review is important from the industry's point of view since it is likely to be influential in determining the timing of the new capacity additions and the scale of the nuclear contribution to UK supplies in the first decade of the next century.

The options for new plants for commissioning in this period are limited. Nothing other than a PWR could now be considered and there are obvious advantages in repeating the Sizewell design. Sizewell B has carried all the launching first-of-a-kind costs (FOAK) of design and tooling and UK contractors are now experienced in the techniques and materials used in its construction.

A repeat station located at the Sizewell site would offer the fastest restart to the UK's nuclear programme, and, if it were built as twin units, substantial savings could be expected compared with the costs of Sizewell B. Not only would FOAK costs be avoided but there would be gains from shared services and replication. Additional gains would be achieved by enhanced output and improved thermal efficiency, which lead to a 1288 MWe output, up by 100 MWe on Sizewell B.

The UK safety case for the design has already been established and this gives it an edge over potential international competitors. Nuclear Electric has been developing proposals for a Sizewell C twin plant and will be hoping that the Nuclear Review will provide the basis for it to proceed to a formal planning application and ultimate construction of the station, with a view to commissioning it early next century.

Other options open to the UK are international designs such as those listed in Table II and, in the longer term, AP600 or other small reactors or the European NPI design.

Table II - Possible PWR Options

Repeat Sizewell B
Extended core Sizewell
Framatome N4
Siemens Konvoi
ABB System 80+
MHI Standard 4-Loop design
MHI APWR
NPI — EPR
Westinghouse — AP600

At this time neither the AP600 nor the European NPI design is sufficiently advanced to justify serious consideration for deployment in the UK. Each of the remaining designs (Table II) is a possibility, although all would need to be subjected to a lengthy and detailed safety review and possibly tailored to match UK requirements. This would introduce delays and increase their basic costs but, subject to their eventual availability and the existence of lead plants in their countries of origin, they could be adopted in the UK for commissioning from about 2005 on.

Our assessments have therefore been focused on a detailed review of Sizewell C costs and on contractors quotations against an outline specification for the overseas designs. The latter quotations have been analysed in detail and appropriate additions made to cover owner's and site costs, with allowances for design modification where these were considered to be necessary.

For reasons of commercial confidentiality it is not possible to give details of the cost breakdowns in this paper but the most recent estimates for the imported designs range from £1160-£1420/KWe net compared with £1350/KWe for Sizewell C, all on the basis of a twin unit plant. There is inevitably greater uncertainty at this stage over the imported designs since their ultimate costs would be dependent on the requirements of the UK's independent Nuclear Installations Inspectorate.

The remainder of this paper will focus on the economics of Sizewell C since the results for the other options will not diverge greatly from it.

COMPETING NON-NUCLEAR OPTIONS

For our review we have examined the full spectrum of competing options including fossil-fuelled plants and renewable sources. Although some of the latter are potentially competitive, the contribution they can make at low costs is resource limited so that the main competition to nuclear for large scale base load power generation comes from gas and coal.

Gas-fired combined cycle generation with a thermal efficiency of around 50% was able to undercut all competing options for new construction at the gas prices prevailing in the UK in 1991, viz 15p/therm or £1.4/GJ. Not only were the capital costs of gas-fired plants significantly lower than those of coal and nuclear plants but their construction lead times were short (2 years) and the fuel price was well below the price of UK deep mined coal (£1.9/GJ). Higher gas prices now prevail and coal prices have been reduced but gas remains a prime competitor for new investment for the coming decade at least.

New pulverised fuel combustion coal-burning plants would have to be fitted with flue gas desulphurisation and fluidised bed combustion or integral coal gasification combined cycle plants (IGCC) with higher thermal efficiencies now look more attractive. Circulating fluidised bed plants could be available for commissioning, around 2000 but pressurised fluidised bed plants would be somewhat later. IGCC plants are still at an early stage of development, whilst the British Coal developed Topping Cycle, which could be even cheaper, has not yet been demonstrated and its costs are necessarily speculative.

For this paper the generation costs for coal and gas-fired plants are based on published investment costs[7][8] and an assessment of the likely range of fossil fuel prices. The latter are assumed to be at projected international prices delivered to the UK since this is the target set for British Coal by Government[3] and the UK is expected to be linked to the European Gas Grid by the year 2000[3].

The cost of imported fuels to UK generators is sensitive to the exchange rate for sterling. Forecasting exchange rates in the decades after 2000 is even more difficult than forecasting fuel prices. During 1992 alone the US$/£ sterling rate fell from nearly 2 to below 1.4. We have adopted a wide range to reflect the true uncertainties that face generators relying on imported fossil fuels.

ECONOMIC COMPARISONS

The basic methodology we have used to derive costs is that of lifetime levelised cost[9] which gives the constant money price that would need to be charged for each unit of electricity sent out by a plant, to exactly match its investment and operating costs over its working life, plus the costs of waste management and decommissioning. It also provides a return on capital equal to the discount rate used in the calculations. Based on analysis of UK investment markets we conclude that a rate of return of 8% per annum post tax in constant money terms is appropriate for a private sector generator in the UK's competitive environment.

The above method is the same as that adopted by NEA and IEA for their analyses[10]. We differ in two respects however. We provide for long term liabilities for spent fuel management, decommissioning and waste disposal through a notional provisions fund earning 2% per annum pre tax in real terms. We also include pro-rata allowances for central overheads rather than using incremental costs. Both practices slightly increase our costs relative to those that would be derived on the standard NEA/IEA methodology at an 8% discount rate.

For existing plants (and Sizewell B) we derive the economic resource cost of their future operation, ie their avoidable cost. This is the levelised cost difference over their planned lifetime between their continued operation and their immediate closure. Sunk investment costs and, for Magnox and AGR, costs of dealing with existing spent fuel, wastes and decommissioning are unavoidable and have to be met whether the plants operate or not.

The avoidable costs of generation in Magnox and AGR plants and Sizewell B are given in Table III together with the operating costs of existing coal and gas combined cycle plants.

Table III - Avoidable Generation Costs April 1993 money values	
Plant	Average avoidable costs at 8% RoR p/kWh
MAGNOX AGR Sizewell B PWR	1.2 1.3 1.3
	Operating costs at 8% RoR p/kWh*
Coal-Fired CCGT	2.2 - 2.6 1.8 - 2.1

*   The ranges are based on sterling prices of fossil fuels: coal £1.2 to £2.0/GJ and gas £1.5 to £1.9/GJ at the plant.

On this basis it is clearly economic to continue running the existing nuclear plants to the fullest extent possible for as long as possible, and to complete and operate Sizewell B.

The situation for new plants is less clear cut. Table IV sets out the costs for electricity from a new twin reactor PWR station, Sizewell C, at the 8% rate of return. Plant life is taken to be 40 years and the levelised load factor 85%. The latter has been derived on the basis of detailed analysis of the plant's technical features and the actual achievements of similar well managed modern plants. It is not inconsistent with the expectations in other countries[11].

Costs for coal-fired and gas-fired plants are taken from published papers[7][8] but they have been converted to our conditions and fuel price ranges. They are included in Table IV (coal 40 year life, 85% load factor; gas 25 year life, 90% load factor).

Table IV - Generation Costs For New Plants 8% per annum real return, 1993 money values p/kWh				
Plant	Investment	O&M	Fuel	Total
PWR	1.8	0.7	0.45	2.8-3.0
Coal	1.3	0.7	1.0-2.0[a]	3.0-4.0
Gas	0.6	0.45	1.7-2.5[b]	2.8-3.5

a £1 to £2/GJ at the power station
b £2 to £3/GJ at the burner tip

The range quoted for PWR reflects the residual uncertainty over constituent costs at the time of preparing this text. The fuel costs are insensitive to most of the underlying assumptions, including the choice of spent fuel management strategy in UK conditions. They are based on 49GWd/tonne uranium oxide fuel.

The range quoted for fossil-fired electricity reflects uncertainties in fuel prices to UK generators. The published investment cost data used for coal-fired plants[7] relate to 200 MWe plants. Some scale benefits may be achievable at larger sizes although the practicability of this has not been demonstrated for circulating bed or the longer term IGCC plants. The investment costs for gas cover larger plants and the range derives from uncertainties about fuel prices to UK generators with, at the bottom end, prices that might prevail in a UK market isolated from the European Grid, and at the top end international agency views on likely border prices in Europe after 2000[10].

SENSITIVITIES

The basic cost estimates presented in Table IV represent best central judgements whilst reflecting the very real uncertainty that surrounds future fossil fuel prices in the UK post-2000. This is a consequence of the existence of a large number of different ways in which fuel markets could develop and the ways in which exchange rates could change. Both are sensitive to national and international political and economic developments.

Nuclear costs are particularly sensitive to the interacting factors that determine the unitised capital component of generating cost (Figure 3). These include, in addition to the investment cost itself, the load factor the plant will achieve and the rate of return on capital.

They are however relatively insensitive to fuel prices (Figure 4) and, contrary to popular perception, to the back-end costs of the fuel cycle and decommissioning (Figure 5), which contribute less than 4% and well under 1% to the overall generation costs, respectively.

Fossil-fuelled plants have smaller specific capital costs and are less sensitive to this parameter and, in consequence, to load factor. This is shown for the three technologies in Figure 6, which brings out the crucial importance to nuclear plants of their filling a base load role - one they should logically enjoy as a consequence of their low overall operating costs.

The greater sensitivity of fossil-fired generating costs to fuel prices was indicated in Figure 4. This is extremely important in view of the major uncertainties that surround this parameter in the post-2000 period.

The levelised price of fossil fuels will effectively determine the rate of return investors can earn on nuclear investment in a competitive market (Figure 7).

Overall fossil-fuelled generation costs are most sensitive to the factors outside the control of the electricity industry (fuel prices) whilst nuclear generation costs are sensitive to factors that can be established (in principle) before investment commences and are within the industry's control.

UNCERTAINTIES AND RISKS

The decisions faced by Government in the Nuclear Review, and by potential sources of finance for future investment in new generating capacity in the UK, are not easy ones. The three main options have quite similar economic costs and the option favoured on financial grounds will be quite sensitive to changes in world fuel prices, exchange rates and required rates of return on capital, even if these are within the variations experienced over recent years.

Several characteristics of private sector investment choices could act to nuclear power's disadvantage in such a situation. A preference for quick returns leads to the adoption of high screening rates of return and short amortisation periods. Risk aversion leads to the addition of extra contingency factors to investment costs and the downrating of expected performance, as well as a preference for low front end costs and short construction lead times. All of these factors would add disproportionately to nuclear costs and could make it difficult to raise private finance.

Contrary to popular perceptions the technical and financial risks of investment in modern PWRs are well known and potentially controllable. Fuel costs are only 15% of overall generation costs and are hardly affected by changes in the economic parameters or spent fuel management choices. Back-end costs contribute only 3% of the total so that residual uncertainties in waste repository costs, etc, have little financial significance. Uranium contributes 5% to costs so that changes in market conditions do not have a major impact on nuclear electricity costs. PWR decommissioning costs, unlike those of Magnox reactors, add only one half of one per cent to overall costs on our conservative provisioning assumptions. Thus they present no real financial risk. Investment costs are probably better defined in the UK for the PWR than for most other energy technologies, although investors will still fear setbacks during construction, problems with sales of electricity and the effects of retrospective regulatory changes. A major counter to these difficulties is the fact that the generation costs from nuclear plant will be extremely stable.

With regard to fossil-fired plants they face uncertainties over investment costs and performance which are probably greater than those for nuclear pro rata, although smaller in absolute terms. The major financial risks attach to fuel prices and environmental concerns, both of which can be subject to political influences outside the UK's control. There is concern that fossil-fuelled plants, once built, will have generation costs that are volatile and subject to escalation. As the UK market is structured however, investors' concerns may be reduced by an expectation that such risks can be passed through to consumers by price increases.

There is therefore an imbalance in the way the risks and uncertainties are perceived which currently acts to the detriment of nuclear power. This represents a major challenge to the nuclear industry which has to get its message across and to find means of providing reassurance to potential investors.

WIDER ISSUES

As this conference is well aware, the direct financial costs to plant operators and consumers do not reflect the full impacts of the choice of generation technology on society at large. The biggest externalities appear to be environmental and the analyses conducted for the UK Nuclear Review have left us in no doubt that these costs are far larger for fossil-fuelled power generation than for PWRs. Our principal comparisons are between new plants for future construction.

For nuclear plants built and operated to the UK's strict regulatory requirements the external costs associated with routine operation and the hypothetical incidence of low probability high consequence accidents are minute compared with the direct generation cost, as shown in Table V. The acid emissions from even the most modern coal-fired plants and the effects of greenhouse gas emissions from coal and gas-fired plants, on the other hand, are significant even when no allowance is made for the more catastrophic consequences that could arise from global warming.

Table V - External Costs Of Electricity Generation* p/kWh			
Technology	Normal operation	Severe accidents	Global warming
Coal (new)	0.6-1.3	0.01	?
Gas	0.3-0.5	0.003	?
PWR	~0.01	0.001-0.1	Negligible

* These costs are based on the <u>total</u> fuel cycle and the Sizewell risk of major releases of under 1 in 10^6 per annum.

The sums shown in Table V for normal operation are comparable to the levels of carbon/energy tax (or pure carbon tax) that would be imposed if the \$10/barrel of oil equivalent discussed by the EEC were adopted (1p/kWh for coal and 0.6p/kWh for gas).

Fuel diversity and its contribution to security of electricity supplies has been given considerable weight in past Government decisions. It is however hard to attach a meaningful economic value to a benefit that is largely dependent on the speculative scenarios that have to be proposed as justification for a diverse fuel base, especially because the probabilities of their occurrence are no more than guesses. Reasonable estimates could be provided only when the risks are recognisably high, and by then it is too late to change the technological base of a national power network. The long planning and construction lead times for nuclear and coal-fired generating plants are paralleled by the times taken to develop new gas or coal-fields.

The one thing that is clear is that the enhanced short term security afforded by fuel stockpiling can be gained more cheaply using nuclear fuels than either coal or gas.

171

A one year fuel stockpile costs 0.03p/kWh for nuclear fuel, three times this for coal and between 3.5 and 6 times more for gas, even when the special storage requirements for the latter are excluded.

Balance of payments is another area where, for fuel importing countries, nuclear power offers savings. To the extent that uranium can substitute for imported coal or gas (or release these for export) it offers balance of payments gains of some 1p/kWh and 1.5p/kWh or more respectively.

Based on price elasticities for imports and exports it has been argued that the net cost to the economy will be 166% of its nominal cost[12]. The effect will be revealed through a decrease in sterling's exchange rate or the need for higher interest rates, both of which impact on the real wealth of the population at large just as much as environmental externalities.

We have considered other forms of externality associated with the macroeconomic effects of energy choices but, like the OECD[13], we have concluded that these do not represent significant external benefits attributable to nuclear power deployment in the UK at present.

CHALLENGES FACING NUCLEAR

The principal issue facing the nuclear industry in the UK at present is the uncertainty concerning the future structure of the electricity market and its implications for financing new investment. The private sector in the UK is notoriously risk averse and committed to short-term returns. Government is faced with a large public sector deficit and committed to reducing the public sector borrowing requirement.

The nuclear generators are publicly owned but are competing with the private sector generators. They are vulnerable to Government intervention in pursuit of specific policy objectives in a way that the private generators are not, and they are constrained in the extent to which they can compete in non-franchise markets. Unless they can invest in new capacity in the relatively near future, their overall share of UK electricity supply will begin to decline as the older stations come to end of life and this will have an adverse effect on their overall productivity. In the near term there is a window of opportunity for embarking on construction of a licensable plant which offers sizeable savings. If this opportunity is not seized future costs are likely to be higher and indigenous skills will be lost.

The outcome of the Nuclear Review is therefore critical to the future development of nuclear power in the UK. Its decisions will have far reaching and prolonged effects.

HOW WILL IT TURN OUT?

We cannot know how Government will respond to the case the nuclear industry will put forward. On the positive side it has already accepted the logic of the economic arguments in favour of continued operation of existing plants and the completion and operation of Sizewell B, despite strong anti-nuclear and pro-coal pressures to the contrary.

We believe that PWRs commissioned in the early years of the next century will be economically competitive with coal and gas-fired plants, but we recognise that they cannot offer a guaranteed sizeable financial gain to potential investors whose perceptions of the market risks may lead them to favour less capital intensive options.

PWRs do however have sizeable external benefits in terms of avoided environmental damage and, in the longer term, for the UK's balance of trade. They also provide stably priced electricity that is not subject to the vagaries of international economic and market developments. Without a significant nuclear contribution the UK Government's environmental commitments will be hard to maintain beyond the early years of the next century.

The failure of existing electricity markets in the UK to reflect these nuclear benefits is a matter the Government will need to reflect upon, alongside concerns about the anti-competitive features of the existing mechanisms set in place at privatisation to encourage price competition in generation.

There are market based methods by which the disadvantages faced by nuclear power could be redressed. Internalising external costs or benefits is one. Moving competition from spot electricity markets to new capacity markets with subsequent assured sales contracts is another. The provision of cost-free guarantees that remove the divergence between perceived and real financial risks of nuclear investment could also be attractive.

The UK Government has acknowledged recognises the wider benefits that nuclear power confers and this would suggest that it should be receptive to the nuclear case for new capacity, given the close comparability of the direct financial costs of the principal generating technologies. It will not be long before we see whether this is the case.

ACKNOWLEDGEMENTS

The authors wish to thank colleagues on the Plant Economics Task Force for the inputs and comments that underpin this paper.

REFERENCES

1. Jones PMS, Nuclear Power: Policy and Prospects, Wileys, Chichester, 1987.

2. National Grid Company, Third Seven Year Statement, March 1992, National Grid Company, London.

3. Department of Trade and Industry, Prospects for Coal, HMSO, London, 1993.

4. Wakeham J, Ministerial Statement, Department of Energy, London, September 1990.

5. Nuclear Electric, Annual Report and Accounts, Nuclear Electric plc, Barnwood, 1993.

6. Scottish Nuclear, Annual Report and Accounts, Scottish Nuclear Ltd, East Kilbride, 1993.

7. Dawes SG, Cross PJI, Michener AJ, Topper JM, Advanced Coal Burning Systems for Power Generation, paper to IEA/OECD conference on Coal, the Environment and Development, Sydney, Australia, November 1991.

8. McCloskey Coal Information, The Gas and Coal Debate, Coal UK, October 1992.

9. Nuclear Energy Agency, Projected Costs of Generating Electricity, OECD, Paris, 1983.

10. Nuclear Energy Agency/International Energy Agency, Projected Costs of Generating Electricity, OECD, Paris, 1989 and 1993.

11. Glorian D, Nuclear Power Station Performance Worldwide, Situation and Prospects, UNIPEDE, Stockholm Conference, June 1992.

12. Glyn, Memorandum to Trade and Industry Select Committee Review of British Energy Policy and the Market for Coal, HMSO, London, 1993.

13. Nuclear Energy Agency, Board Economic Impacts of Nuclear Power, OECD, Paris, 1993.

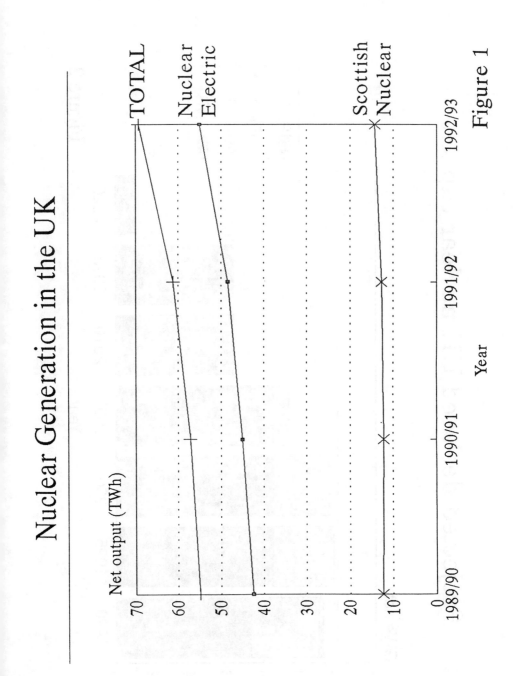

Nuclear Generation in the UK

Figure 1

Projected declared UK nuclear capacity

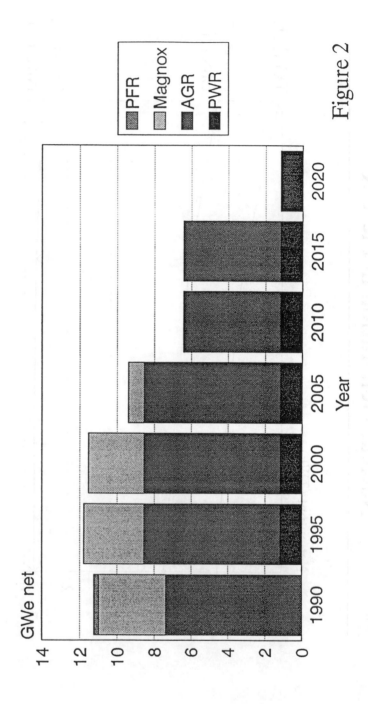

Figure 2

Sizewell C Twin Station Cost Sensitivity

p/kWh

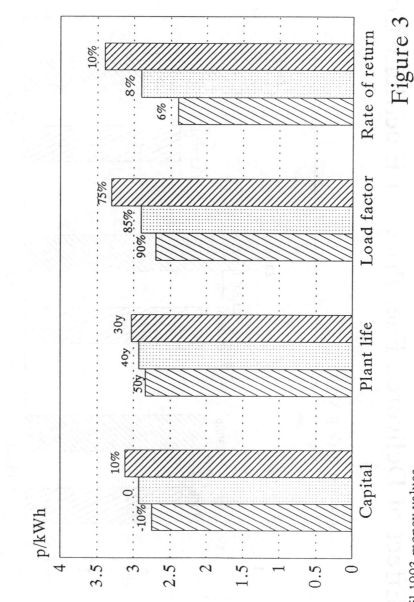

April 1993 money values

Figure 3

Effect of Delivered Fuel Price on Electricity Cost

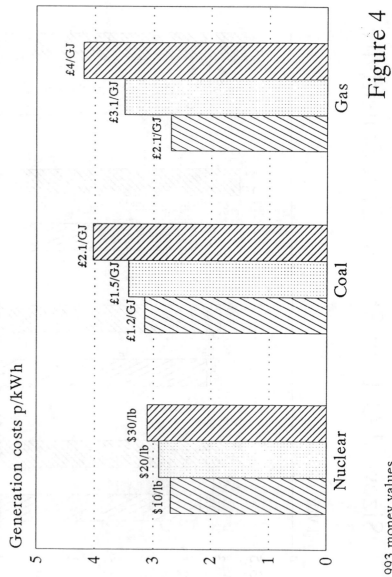

Figure 4

Sensitivity of PWR Generation Costs to Backend Costs

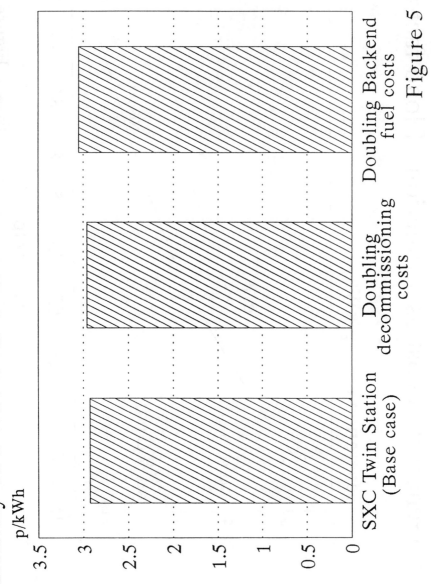

Figure 5

Generation cost as a function of load factor

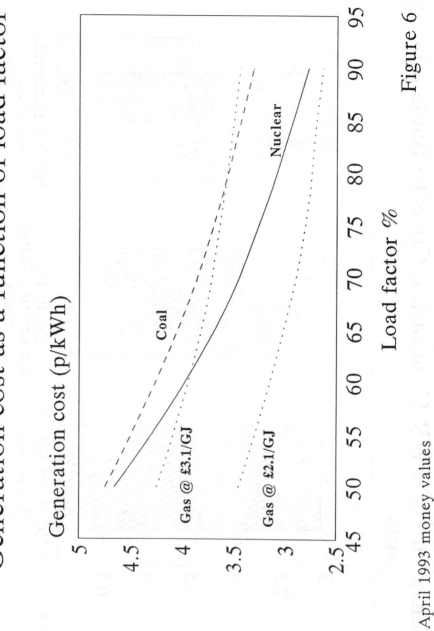

Figure 6

April 1993 money values

Relative Generation Costs
as fuel price and rate of return vary

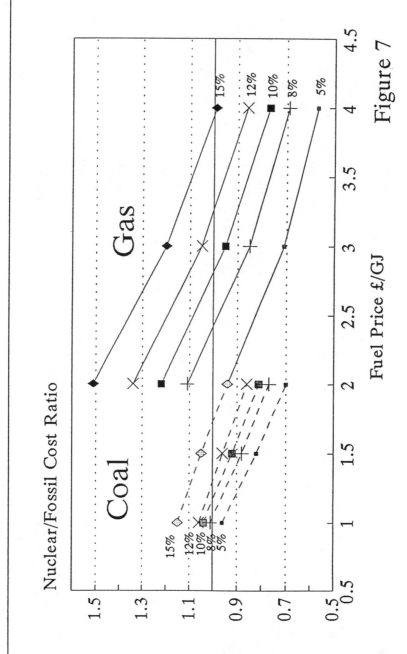

Figure 7

Session 2 - Discussion

J. Foster

I would like to ask Brian Eyre a question. Why is the Government doing this review or planning to do this review? What does the Government have at stake?

B. Eyre

A review was announced in 1989 when nuclear generation was withdrawn from the privatisation of electricity supply. I think the situation has changed quite a lot even since then. Now it is our intention to have the review and as you know we had a lot of problems about a year ago over the closure of coal mines. The Government then had the coal review and, at that time, they stated their intention to bring the nuclear review forward to 1993. We have been trying to understand what they want as the terms of reference for the review and have found it impossible to get an answer so far. It is obvious that they are having some difficulty in defining what the terms of reference should be.

There are two particular issues that I think we will address. One is whether a case can be established for further nuclear stations to be ordered and the background to that. That would take into account the present capacity in the UK, the position that is resulting from the dash for gas, and so on. And of course the position concerning coal fired stations, a number of which are closing down. So there will be a general review of whether the UK's electricity supply is going to be met properly and the role that nuclear will play in that. I think inevitably whether it will come in directly or not, but the review will focus on the question of whether and under what conditions nuclear can be moved to the private sector. I suspect that that will be the more important underlying agenda item.

J. Yasinsky (Chairman)

I have a question that any of our European participants could answer. To what degree do any of you anticipate a closer degree of co-ordination on your power generation decision-making activities in the future, assuming that the process of continual closeness among the EC nations continues? Do you expect to see more in the way of close co-operation in generation planning, especially as you look at externalities and especially those externalities that involve trans-boundary effects?

R. Carle

We certainly expect some changes for the creation of a big market for electricity and energy in general. We shall have more exchanges and certainly the goal for the utilities I think is to optimise the general prediction system in Europe which is not optimised to date. For example, we see today some plants operated at capacity where some other cheaper plants are stopped. We have to improve every use of source of electricity, hydro power, coal and nuclear and all these could be certainly much better optimised.

The question is to know whether this will be optimised by more co-ordination or by more competition? Probably by both and certainly the utilities want to continue to co-operate. They did already in the past, and probably in a more homogeneous market they could do more. If the system is not optimised to date, it is (because) of political constraints in this or that country. If these constraints can be removed and some more homogeneity can be created, certainly the utilities are ready to co-ordinate their efforts. We have already decided, more or less in advance of the European Commission, to exchange information about our new investments, to meet together and say "well, do we need more investment or is it really necessary? Is there any capacity existing in another country?" and to try not to make useless investments. We shall also, I think, see more competition because of the introduction of new actors on the European scene, whether these are called IPPs or not, autoproducers and so on. There will certainly be more companies involved in the business of producing electricity. We must accept that with some satisfaction because it will certainly create better conditions for the clients, for the customers.

J. Yasinsky

Do any of your European colleagues disagree?

C. Nyquist

I agree completely with Mr. Carle. I think looking at, for instance, the Nordic power systems which has been optimised now for at least thirty years. Now the Nordic hydro power will go down further south into the North European system. That means also a new type of optimisation where we utilise hydro power more efficiently. I think the Swiss have done this for several years already. This interconnection will be good for everybody.

B. Eyre

I should just like to comment. I agree with Remi (Carle). I think that there is no turning our backs on the market approach to electricity generation and actually opening up the markets Europe-wide will assist that process and introduce more competition. One of the dilemmas that we have had in the UK is that because of the way nuclear was withdrawn, at the end we just had two major private sector generators. There's been a lot of pressure to break that open even further, to get a greater degree of competition in. We have seen in our other energy industry, British Gas. There is now a move to break up British Gas to introduce more competition; one way of getting the competitive pressures to increase is actually to open up the market to European utilities and so the two would come together.

J. Gray

What would it take in changes in the U.S. system for French investments to be interested in new nuclear power in the United States?

R. Carle

I think nuclear investment is possible only if the rules of the game are very clear and are industrial. They may be clearer now but they are not industrial at the moment. Nuclear commitment is a long-term commitment, it is a commitment for forty years or more, and has to be based on a solid foundation. Which plant can be really realised for construction and for operation in this country? On which assessment? That has to be very clear if anybody wants to invest in these countries. Certainly that would be a very important point.

I am not sure that introducing new investors would change the problem, and certainly for the moment EDF is ready to join any other company to develop new ideas, to maybe build some first of a kind systems, but that has to be considered very carefully.

J. Yasinksy

I think we have a window of time over the next five years or so to complete what we have already started and that is the elimination of the various barriers that existed to bringing back the next generation of nuclear power. Clearly the action last year to codify a single step licensing process was a very, very important step. I believe that another remaining barrier we have is fair decisions on a workable approach for eventual high-level waste spent fuel disposal. I think that has to be created, but a credible plan and a consensus on how that plan is to be implemented has to be put in place. I believe that the new NRC's process on standardization and certification of a standardized designs is a strong and excellent move. Also, it may not surprise you that in the past we never started to build a nuclear plant where the engineering was all completed. It will be different as we plan it this time. We are now into a programme to complete what we call first-of-a-kind-engineering. Other than site-specific issues, all design activities will be complete and detailed design drawings will be available on every component plant before there is initial sale. The new designs that have been developed here in the United States have been to a large degree dictated by the US utilities who believe in nuclear power, telling us what they wanted and that involved a smaller and a simpler plant. That was reflected by the sixteen utilities who committed their funds to partially support the completion of this first-of-a-kind engineering for a design such as the Westinghouse AP600.

So some of the barriers are down. I would say the key barrier remaining right now is to reach a consensus on a workable solution for high-level wastes and spent fuel. I am optimistic that in the period beginning about 1996, we will see a new order in the United States. I do not think we will see an order that reflects the previous way in which a single utility would contract for a nuclear plant. In other words, going out with an NSSS supplier, with an AE, with a constructor. I think that it is highly likely that you will see some more innovative approaches to utilities themselves combining perhaps non-regulated subsidiaries into a new company which could involve suppliers and other investors. There are utilities here in the country today who were talking about these kinds of approaches. So is it a slam-dunk? No. Does the possibility exist that we will see a new generation of nuclear power in the United States? Yes, but there are still barriers that have to be removed.

Unidentified

It does seem to me as an independent observer of nuclear power that in terms of its conventional operation, in comparison to alternative sources of electricity generation, in terms of environmental impact, it is an open and shut case. The figures that have presented this afternoon have done nothing to change my opinion about that. So it all comes down to questions about assumptions you make about non-routine operations, accidents. For that reason I was somewhat interested to see the figures that appear, the first one Dr. Eyre's paper and then Herr Strauss' paper, where, unless I have missed a trick, Herr Strauss talks about the Prognos study and talks about core meltdown, and produces a figure of a damage equivalent of 2.0 cents per kilowatts/hour to such a situation. Dr. Eyre, referring to the nuclear chairman's study, has a figure where the upper bound is 0.1 of a pence, i.e. 0.16 cents/kilowatt hour.

I may have missed a trick here, but I could I ask the two authors to consider why there is an order of magnitude difference in their estimates of the per kilowat/hour cost of what in one place is called a core meltdown, and in the other a severe accident?

L. Strauss

Because I only reported on the study made by the Prognos Institute. I only converted the figures into a figure for a specific cost. I cannot imagine what other reasons for such a great difference. I also mention that perhaps assessments made by the two studies are very different and that is the reason I mentioned in my paper. At the moment it is too difficult to take measures from various studies about the external costs of nuclear.

J. Yasinsky

Excuse me, I think Mr. Strauss' colleague would like to supplement that response.

Unidentified

Mr. Chairman, I could give some more details on the Prognos figure. First of all there are no Prognos figures. I must stick to that. Prognos had the task to examine the possibilities, the consequences, the issues connected with the whole subject of externalities. They were not entitled, they were not asked to give any figures. In one of the subpapers of an economic professor at Munster University, to the general surprise, some figures showed up. These were the figures and this economic professor took calculations which he had made on the Chernobyl accident, transferred them to Germany and regarding the fact that of course the population density in Germany is much higher than in this part of the Soviet Union, he came to the extremely high figure of about 6,600 billion dollars. So these figures are disputed very much, but I have not found any serious researcher who backs this professor and this particular method in Germany. The question is open of course, this is just one figure and I certainly cannot compare, as far as I understood, it with the obviously thorough calculations you have tried to make.

J. Yasinsky

If I could just add a little bit also. I am not familiar with either of the two studies, but we have over the last several years in Westinghouse done various comparative calculations involving externalities and one of the reasons why, in considering for nuclear power non-routine impacts, you don't see order of magnitude differences, is because you consider the non-routine event that can occur times the probability of that occurrence and you look at your analysis considering your number of plants and the years over which they are going to operate. That means non-routine operations have a relatively small externality cost. On the other hand there have been some analysis done, one by Pace University here in the United States, which generate very large externality costs for nuclear power. What they do is an analysis for an individual plant and they assume the Chernobyl-type event. That is not realistic. So again, as I indicated in my opening remarks this afternoon, it is certain in my mind that we are going to see the use of externalities, but the real challenge, and I think we are demonstrating here this afternoon, is which externalities do we consider and how do we calculate them in such a way that we end up with an objective comparison? I think that we have just demonstrated that we have a way to go on that.

B. Eyre

With regard to our own study, we did not do any detailed analysis as part of our work. We took published data, but I should emphasize that our severe accident externality was for a world-class design modern PWR with proper containment and so on. There is all the difference in the world between that and a Chernobyl-type reactor without any containment. We then averaged the externalities over a very large area.

N. Bains

Mr. Yasinsky, I would like to supplement the information that you just gave about the Westinghouse study. We actually used that study for Nine-Mile Point-1. I worked with Jim Motley and we used that data to actually verify the viability or the correctness of Pace study. We had submitted an economic analysis for the unit to the New York State Public Service Commission and working with Jim Motley we came up with those estimates and submitted those. The Commission wanted us to give a written justification for not using Pace estimate. So I ended doing a lot of work in that regard, and while I was doing that, I also had an opportunity to read a summary of the Prognos Study that has been discussed here already.

I would to share with you what I understood from Prognos Study. It was pretty much along the lines which you just said: that Prognos essentially assumed that Chernobyl had happened in Germany, disregarding the vast differences in design, containment structure, the reactor physics differences; everything was disregarded. Because of the higher population density, they assumed the same release to the atmosphere and multiplied it by the population, (so that) now you have a very large millions person-rem, dose. They then assumed a certain person-rem for each disease or fatality and assigned a dollar value. That is how they came up with over 6000 billion dollars.

I think since then other work has been done that has discredited that number altogether. Not only that but also the Pace number which was about 4 cents per kilowatt/ hour for nuclear field cycling including normal operation and so forth. So even though dust hasn't settled on that yet, it looks like there is a lot more work that needs to be done. Fortunately for us, the Commission accepted our position and accepted our economic analysis and the plant is still running. This year we had about a 72 per cent capacity factor and we did a 55-day outage on target under budget. It goes to show how these studies can be used not only for future resource allocation, but to shut down the current running units as well, would be a misapplication of that.

I have some other questions for the speakers. George Gunther, you also mentioned in your speech that you had, as part of the integrated resource planning, prepared an estimate for the nuclear externalities. I would be curious to find out what that value is in cents per kilowatt/hour, and if that was accepted by the regulatory agencies, and what other studies you considered while doing you own estimates?

G. Gunther

I'm sorry, I can't give you any of the numbers associated with that. As I indicated they were simply done as sensitivity analysis and were purely internal kind of exercises.

N. Bains

I see. They were presented to the Commission or regulatory agencies?

G. Gunter

No.

N. Bains

The next question is for Mr. Eyre. I was very pleased to see your numbers on the externalities. They came pretty close to what we had come to, and I would like to find out again, what other studies did you consider for those estimates?

B. Eyre

Briefly, I think that the numbers that we put in for externality were really illustrative and were not based on an analysis of their own. But I think an important source of evidence or information in our case is David Pearce's work at London University which I think you know is being published pretty widely.

N. Bains

My next question is for Mr. Strauss (concerning the) Prognos (Study). I am just curious if any rebuttals have been presented to the Prognos Study, or is that study accepted as is by the nuclear industry and especially the government regulatory agency?

L. Strauss

This has been a study for the Federal Ministry of Economics. When I wrote to them and asked them what is your comment on the figures given by the author of the sub-paper: either you agree with those figures, then you should do everything to shut down all nuclear power stations as quickly as possible, or if you don't agree, you should say that and to the public. The Ministry answered (that) we get such a great number of studies that it was just a study among many others. But I must say that the Prognos is an excellent study, but not all the subpapers. That is the difference. This Prognos study in a very clear way shows the structure of the problem and the questions to be tackled. It gives an excellent overview on the international literature on the issue, so I defend it. But the regulatory authorities haven't taken notice of it, as far as I know, and the industry, well, they gave one comment on it in a publication, but there is no official point of view to that. It is just a study and this is freedom of science.

Session 3/Séance 3

Chairman/Président
R. Carle, Electricité de France

Meeting/Managing the Demand for Electricity
E. Linn Draper, Jr., American Electric Power Co., Inc.,

**Sustainable Development and Choices of
Electricity Generating Technologies**
Y. Akiyama, The Kansai Electric Power Co., Inc. (Japan)

**Electric Power for a Sustainable Development -
The Italian Scenario**
S. Barabaschi, ANSALDO

Low Environmental Impact: Swiss Power Generation
K. Küffer, NOK

Discussion

MEETING/MANAGING THE DEMAND FOR ELECTRICITY

E. Linn Draper, Jr.
Chairman of the Board
American Electric Power Co., Inc.

Abstract

Electric utilities in the United States will meet their customers' future energy needs increasingly through non-traditional means. Many utilities will contract with non-utility generators for new capacity to meet load growth. Demand side management and load management to shift load, clip peaks and fill valleys are strategies utilities will use to reduce the need for new generating capacity. Natural gas will enjoy a short-term advantage as a fuel for electric generation due to environmental concerns and costs associated with coal. Design work is under way for standard design nuclear units. Optimists predict that orders will be placed for new nuclear units by the middle of this decade. At American Electric Power, an integrated resource plan features 44 demand side measures (22 residential, 13 commercial, 9 industrial) anticipated to reduce generating capacity requirements by about 900 megawatts by 2012.

SATISFAIRE ET GÉRER LA DEMANDE D'ÉLECTRICITE

Résumé

Aux Etats-Unis, les compagnies d'électricité auront de plus en plus souvent recours à des modes de production non traditionnels pour répondre à la demande d'énergie des consommateurs. Beaucoup passeront avec des autoproducteurs des contrats pour se procurer le supplément de puissance installée nécessaire pour répondre à l'accroissement de la demande. La gestion par la demande et la gestion de la charge pour décaler celle-ci, écrêter les pointes et atténuer les creux pourra être la stratégie retenue par les compagnies d'électricité pour réduire les besoins de nouvelle puissance installé. A court terme, le gaz naturel semblera avantageux pour produire de l'électricité compte tenu des problèmes d'environnement et des coûts liés au charbon. On travaille actuellement à la conception de tranches nucléaires standardisées. Les plus optimistes annoncent des commandes de nouvelles tranches nucléaires vers le milieu des années 90. Chez American Electric Power, un plan de gestion intégrée des ressources prévoit 44 mesures de gestion par la demande (22 en direction du secteur résidentiel, 13 du secteur commercial et 9 de l'industrie), qui permettront de réduire les besoins de puissance installée d'environ 900 megawatts d'ici 2012.

American Electric Power is in the enviable position of not requiring new **base load** generation until 2005, or more than 10 years down the road, according to our latest forecast. I say enviable because no utility wants to build, period. I could recount the whole bloody history of utility construction woes since the early Seventies. Suffice it to say, as illustration, that AEP is still involved in litigation over a plant, the Wm. H. Zimmer Generating Station, that was started in 1972 as a nuclear plant, converted to coal firing and placed in commercial operation in early 1991.

But build we must, at least, someone has to build the capacity we will need in the near future. Just a few words about the United States as a whole as seen by the U.S. Council for Energy Awareness. I believe this will put in perspective what I have to say about the way AEP has approached its generation planning.

In July USCEA published an **Energy Update** entitled "Electricity Supply & Demand into the 21st Century -- Who's Planning What, 1992-2002." Electricity is seen as a *resource*, of which the nation will need 148,400 megawatts between 1992 and 2002. How will this need be met? A significant portion -- 70,000 MW or just under 50% -- will come from non-traditional sources, namely, non-utility generators and demand side management measures. If this forecast is correct, utilities will be required to supply, with their own new units, only about half of the projected growth in load. How the world has turned....I believe that the two concurrent developments evident in this USCEA forecast -- the breaking-up of the vertical integration of electric utilities and the use of conservation as a bona fide resource in planning -- are the two most significant **externalities** that have affected utility expansion plans.

We are reading more and more about dispersed or distributed generation. That is, smaller generating units closer to customers. It's an alternative to the big, central station power plant and related in part to the non-utility generation trend. This dispersed capacity is of modular design and construction and can be tailored to one or more customers in a cluster or small load center. Because these units would be small, there is certainly less financial risk. Emerging technologies, such as photovoltaics, fuel cells, solar, wind — which have had virtually no impact to date — might be ideal for these applications in some locations.

But, across the country as a whole, many, many natural gas turbines are going to be built in the near future, for the reasons just stated and two others — the impact of the Clean Air Act amendments of 1990 on the cost of building coal-fired plants and the current limbo of nuclear power.

Regarding external contingencies, we don't know what to expect from that most amorphous of environmental issues — global climate change. Whether Congress will resurrect the idea of a tax on carbon-based fuels or restrict "greenhouse gases" is anyone's guess. I can tell you what we have done.

President Clinton has pledged that the United States will reduce greenhouse gas emissions to 1990 levels by the year 2000. Electric utilities, largely through the Edison Electric Institute (EEI), are exploring environmental measures, energy efficiency, conservation and other means to meet the president's goal. On August 12, AEP announced that it will join the U.S. Department of Energy in establishing voluntary programs to limit emissions of greenhouse gases. Included are fuel switching, upgrading power plant efficiency, forest management, tree planting, renewable energy sources, methane recovery, cogeneration of electricity and the capture and recovery of chlorofluorocarbons. We will work long and hard to foster this spirit of cooperation.

I don't believe that the current tilting toward natural gas as a fuel for electric generation will be a long-term, nationwide solution. Price and deliverability will militate against a long-term dependence on gas. Much of what I have said thus far bodes well for nuclear, but first some background. Twenty years ago, 29 nuclear units supplied 5% of the electricity used in the U.S. Today, 109 commercial nuclear units supply 22% of the electricity. During the same 20 years, oil's share of electric generation plummeted to 3% from 17%. The electric utility industry has done an admirable job weaning itself from oil. On the face of these facts — oil rightfully out of the picture, coal with current and looming environmental problems despite its abundance and gas a partial answer at best — it would appear that nuclear is poised for a giant step forward. So, it would appear....but, of course, we are still dealing with the consequences of Three Mile Island in 1979.

There are hopeful signs. The Energy Policy Act of 1992 "streamlined" the nuclear licensing process for advanced standard-design plants. Last March the Advanced Reactor Corporation, an organization of 16 utilities, including AEP, supporting development of new nuclear plant designs, signed a contract with Westinghouse Electric Corporation for sharing the cost of engineering work on a 600-megawatt design. This past June ARC signed a comparable contract with General Electric for a standardized 1,350 MW design. Some progress has been made at the Yucca Mountain site toward the permanent disposal of radioactive waste.

Optimistic statements have been made that orders for new nuclear units will be placed in the mid-1990s. Maybe. The certification process for three standardized light-water reactor designs is currently under way at the Nuclear Regulatory Commission. In this unforgiving economic climate, nuclear has miles to go before it is competitive. After the design and licensing hurdles are overcome, there will be staffing and O&M.

Now, as to American Electric Power and its view of the future, at least 20 years hence. AEP has just recently updated its **Integrated Resource Plan** or IRP. This is not just a load forecast or chronology of new generating units, though it is surely that, but a broader view of how supply-side and demand-side resources can be brought to bear on the future electricity needs of our customers.

The first step in any IRP is a **base-case** load forecast and an estimate of the amount of resources needed to serve the load reliably. Once the requirements are established, supply-side and demand-side options can be identified and screened. Then, the resource expansions are developed and evaluated, giving equal treatment to both options. The evaluation involves analytical simulation techniques with a generous dose of common sense.

AEP goes about its load forecasting from the **bottom up**. That is, forecasts are developed for each customer class in 16 economic regions across the seven states of the AEP System -- Michigan, Indiana, Ohio, Kentucky, West Virginia, Virginia and Tennessee. Energy forecasts are determined through end-use simulation and econometric time-series analysis. The Hourly Electric Demand Forecasting Model, which simulates typical hourly load shapes by customer class and jurisdiction, is used for peak demands.

Economic conditions, population trends, energy prices and weather are the major influences on energy and peak demand requirements nationally and in the AEP service area. According to DRI/McGraw-Hill, U.S. economic growth will be moderate with a 2.3% annual increase in real GNP from 1992 to 2012. Population growth will be slow -- 0.2% per year -- in AEP's area during the same period. Manufacturing output will be mixed with slow growth in some major industries, namely primary metals.

The forecast assumed a rise of 3.3% in average real price of electricity in 1992, followed by a decline of 1.6% per year through 1999. This price forecast is consistent with our plan for compliance with Clean Air Act amendments of 1990. Weather conditions are assumed to remain constant. So much for the major factors that went into the forecast.

AEP's total internal energy requirements are projected to increase by 1.2% a year, from 104,865 gigawatthours to 133,003 GWH in 2012.

How much reserve margin will we plan for in serving this load? We assume a system unit availability rate of 77%, based on experience. If we have 30 capacity-deficient days per year, the reserve margin would be about 18%.

Given the high cost of new capacity, we — like most other utilities — have been vigorously upgrading generating units in order to get more than 50 years of service out of them. If we retire **3,415 MW** of generating capacity, add nothing over the next 20 years and engage in no DSM activities, the AEP reserve margin would be negative 12% (-12%) by the winter of 2012-13. (By the way, that 3,415 MW includes the 60 MW derating of our Gavin Plant to account for flue gas scrubbers.) In an either/or scenario, we would have to add 7,500 MW of capacity by that time. Or we would have to reduce peak demand by about 6,300 MW through DSM to enable existing capacity (less retirements) to serve the 2012-13 demand.

In terms of hardware, AEP looked at 10 options: combustion turbines, gas-fired combined cycle units, pulverized coal with scrubbers, pressurized fluidized bed combustion clean coal technology, pumped storage hydroelectric, lead acid battery energy storage, wind turbine farm, solar photovoltaic, fuel cells and nuclear advanced light water reactor. Costs and operating data were developed either internally or derived from outside sources, primarily the Electric Power Research Institute.

These supply-side options were screened quantitatively by comparing their levelized total annual cost (that is, fixed charges plus variable operating costs) versus capacity factor relationships, based on an assumed 33-year facility lifetime and a discount rate of 11.15%. The objective was to eliminate the most costly alternatives from further study and reduce the options to manageable size. The relationships among the options were graphed. (Exhibit I on the following page.) Each technology is represented by a straight line showing the relationship between its total levelized annual cost per kilowatt and annual capacity factor.

Initial screening eliminated storage battery, wind, solar, nuclear, fuel cell and 450-MW pulverized coal. These six were too costly or were too uncertain in application.

AEP like many utilities has a present capacity mix that is primarily **base load** -- primarily coal or nuclear -- with little peaking or cycling capability. Hence, the next addition would appear to be peaking capacity, simple-cycle combustion turbines. Pumped storage hydro would be a suitable peaking-cycling option. As time goes by, with the retirement of base load units, the addition of new base load can be anticipated — PFBC, coal-fired with scrubbers and combined-cycle units. So, the finalists in our resource expansion analysis are simple-cycle combustion turbine, pumped storage hydro and combined-cycle units. All three have the lowest overall cost. New gas-fired capacity was limited to 10% or 3,000 MW of total capability to provide diversity of technologies. To cover future base-load capacity requirements, PFBC and 900-MW pulverized coal were retained as potential alternatives, being the lowest-cost options among base load units.

On the demand side, we cast a big net to gather as many DSM measures as possible. DSM measures fall into one of four categories: conservation, load shifting, valley filling and peak clipping. The initial list included 179 potential DSM

Exhibit I

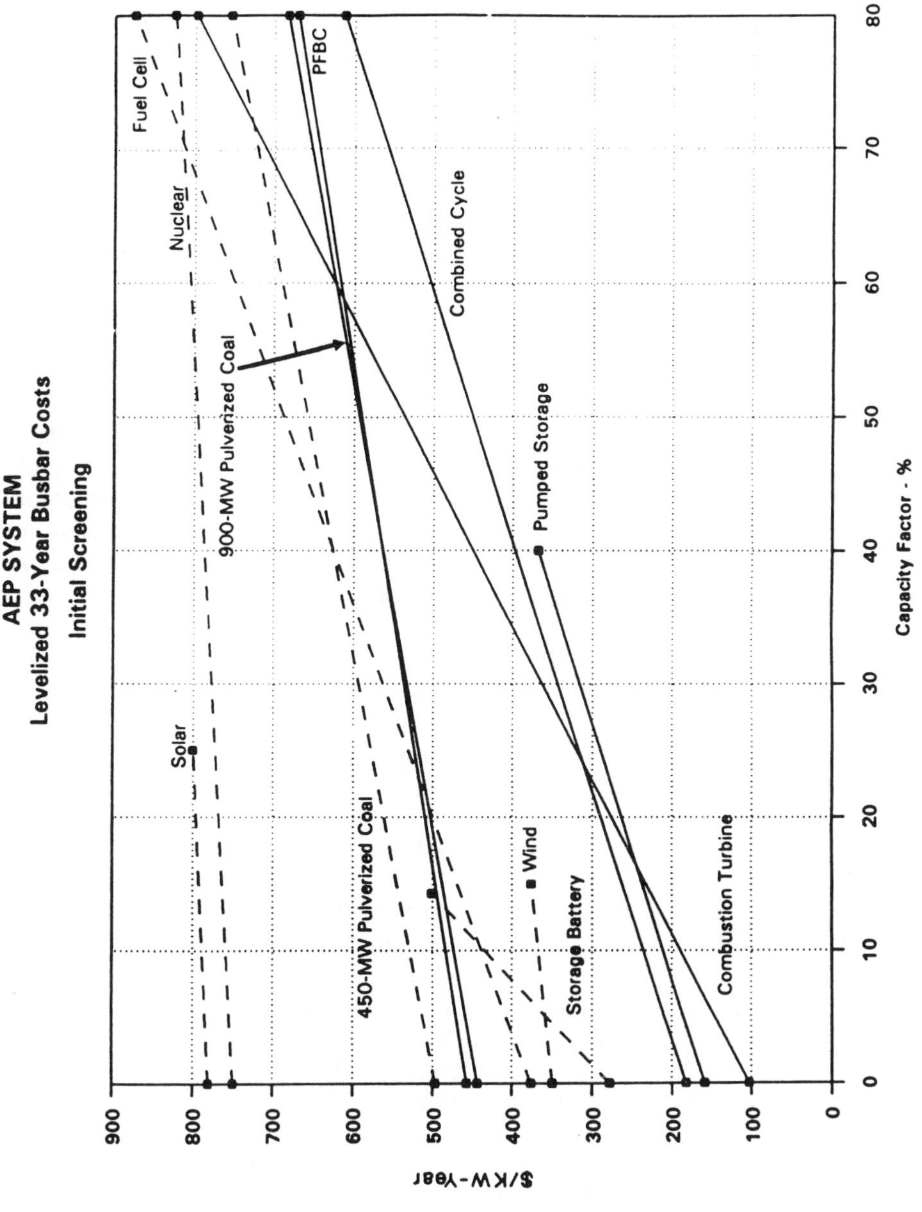
AEP SYSTEM
Levelized 33-Year Busbar Costs
Initial Screening
Fuel Cell
Nuclear
PFBC
Combined Cycle
900-MW Pulverized Coal
Pumped Storage
Solar
450-MW Pulverized Coal
Wind
Storage Battery
Combustion Turbine
900
800
700
600
500
400
300
200
100
0
$/KW-Year
0
10
20
30
40
50
60
70
80
Capacity Factor - %

measures — 73 residential, 50 commercial and 56 industrial. For each preliminary load impact and incremental cost estimates for participant and AEP were compiled. (Exhibits II and III on the following pages.)

Screening consisted of three stages. Technical infeasibility, low market potential, lack of cost or load impact data and high cost cut the list to 89 — 38 residential, 27 commercial and 24 industrial.

The second stage involved a structured, qualitative ranking procedure performed by a task force of several departments. Each DSM measure was given a relative ranking. The top 80% in each customer class made the grade for further analysis. So, the list was cut to 73 — 31 residential, 23 commercial and 19 industrial).

The third stage involved a detailed cost/benefit analysis of each of the 73, including applying standard DSM economic tests. Utility costs include DSM program costs — equipment, installation, operation and maintenance, Administrative and promotion expenses, increased supply costs — energy, capacity and T&D costs, and lost revenue, which must be recovered. Benefits include avoided supply costs, revenue gains and sulfur dioxide emission credits. Participant costs may include incremental equipment expenses (installation and maintenance) and bill increases. Benefits include avoided equipment and maintenance expenses, bill reductions and incentives.

The Total Resource Cost or TRC and Ratepayer Impact Measure or RIM have gained general regulatory acceptance across the U.S. These were the primary tests applied by AEP. Also, we used the Utility Cost and Participant Cost tests. TRC was used as the economic screening criterion for conservation and load shifting DSM measures. RIM was used to evaluate valley filling measures. Full-scale implementation was assumed to begin in 1993, except for TranstexT and the electronic lamp which are expected to be available in 1995.

Forty-four DSM measures — 22 residential, 13 commercial and 9 industrial — passed the TRC test and are currently being considered for implementation.

The 22 **residential** programs fall into seven groups — refrigeration, space cooling, space heating, lighting, water heating, new construction and rates. Many of the programs will take the form of **incentives**. We will give rebates for higher-efficiency refrigerators; remove old refrigerators, freezers and room air conditioners; give rebates for higher-efficiency room ACs; offer direct load control of central AC; do energy audits and low-cost weatherization; offer low-interest loans for insulation and storm windows and doors; offer free audits and insulation installation for low-income families; offer incentives for residents to replace incandescent light bulbs with higher-efficiency lighting; offer incentives for the

Exhibit II   DSM PROGRAM SCREENING

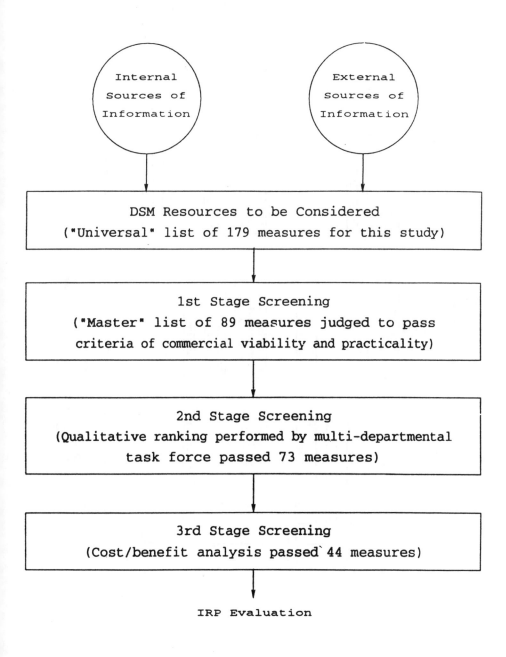

Exhibit III

BASIC BENEFIT AND COST COMPONENTS
OF
ECONOMIC TESTS FOR DSM PROGRAMS

BENEFITS

	Participant Test	Total Resource Cost Test	RIM Test	Utility Cost Test
	Bill Reduction	Utility Avoided Generation and Other Capital Cost	Utility Avoided Generation and Other Capital Cost	Utility Avoided Generation and Other Capital Cost
	Incentives		Revenue Gain	
	Avoided Participant Cost	Avoided Participant Cost		

COSTS

	Participant Test	Total Resource Cost Test	RIM Test	Utility Cost Test
		Utility Increased Generation and Other Capital Cost	Utility Increased Generation and Other Capital Cost	Utility Increased Generation and Other Capital Cost
		Utility DSM Program Cost	Utility DSM Program Cost	Utility DSM Program Cost
	Participant Cost	Participant Cost	Incentives	Incentives
			Revenue Loss	

Basic identities in net benefits or NPV (Net benefit = benefits minus costs)

(1) RIM = UC + Net Revenue Impacts

(2) TRC = RIM + PC

     RIM = TRC if Participants Receive No Net Benefits From The Program

(3) UC = TRC + (Incremental Participant Cost - Incentives)

Note: Above identities assume same discount rate for all economic tests and no free riders in the program.

202

purchase of high-efficiency electric water heaters; offer direct load control of water heaters; offer the lease or purchase of off-peak, storage water heaters; offer technical assistance and financial incentives to builders who install higher levels of insulation and high-efficiency electric heating and cooling and water heating; and, finally, we will offer off-peak rates and variable spot pricing to shift load.

Commercial DSM measures include HVAC, storage cooling, water heating, lighting, refrigeration, new construction and rates. We will offer incentives for high-efficiency motors and pumps and economizer controls and for storage cooling in office and retail buildings; install water heater wraps in restaurants; offer incentives to upgrade lighting in offices, stores and schools; offer incentives to grocery stores to replace compressors and motors and to install floating head pressure controls; provide technical assistance to builders; and, offer time-of-use rates for office buildings and retail stores.

Industrial DSM plans are aimed at motors and lighting. We will offer incentives for motor replacement with higher-efficiency or adjustable-speed-drive motors, depending on horsepower rating. And there will be incentives for upgrading fluorescent lamps and ballasts and for replacing mercury vapor lighting with metal halide or high-pressure sodium.

That's a quick trip through the gamut of DSM measures that AEP will tackle. Several of the residential measures, we are already doing in some jurisdictions, namely second refrigerator removal, audits, low-cost and low-income weatherization measures and storage water heating.

*

The first step in integrating supply side and demand side options is to develop a supply-side-only program. This resource expansion took into account results of PROVIEW, a capacity expansion optimization model. The DSM measures were evaluated in terms of their **aggregate** impact. The bottom line of the integrated resource expansion is that the AEP System will require 6,500 MW of new generating capacity by the winter of 2012-13, with the first peaking unit forecast for 1999. Beginning in 2005, as I mentioned earlier, a mixture of base load and peaking capacity would be added to the System.

The integrated plan reduces the Total Resource Cost by about $970 million and reduces capacity requirements by 900 MW.

*

The foregoing is an overview of how one utility — a holding company with seven operating companies in seven states — is dealing with resource planning. In varying degrees each electric utility has been going through the same process. The variances generally involve the assumed impact of DSM measures and how the generating capacity will be built. That is, will the utility build it or will a non-utility company build and operate?

*   *   *

SUSTAINABLE DEVELOPMENT AND CHOICES OF ELECTRICITY GENERATING TECHNOLOGIES

Yoshihisa AKIYAMA
President,
The Kansai Electric Power Co., Inc.

Abstract

A trilemma between energy availability, ecological consideration and economic development will certainly come out as a most demanding issue to be prudently solved in Asia where many developing nations continue rapid economic expansion. Advanced industrialized nations should carefully select appropriate electricity generation technology options for themselves by pursuing more opportunities for sustained development and also by paying a serious attention to the economic take-off potentials of the developing nations and encouraging their self-support efforts to respective stages of their economic development. A most important point here is our sincere recognition that the natural resources and the environment are the assets which the future generations entrust to us. These generations call for our safeguard or guarantee of the natural resources and the environment for the future. We should, therefore, create a new nuclear energy age as the most appropriate option for the global environment preservation.

DEVELOPPEMENT DURABLE ET CHOIX
DES TECHNOLOGIES DE PRODUCTION D'ÉLECTRICITE

Résumé

En Asie, beaucoup de pays en développement connaissent un développement économique soutenu et le choix entre disponibilité en énergie, écologie et développement économique risque de poser bientôt un problème crucial qu'il importera de résoudre avec la plus grande prudence. Les pays industrialisés avancés devraient se doter de la meilleure option technologique pour produire leur propre électricité dans une optique davantage axée sur les possibilités de développement durable et de décollage économique des pays en développement, et encourager les efforts de ces derniers, aux diverses étapes de leur développement, en vue de leur indépendance économique. Les ressources naturelles et l'environnement sont des richesses que nous ont confiées les générations futures : c'est là un point capital que nous devons à présent reconnaître. Ces générations nous appellent à sauvegarder ou à préserver les ressources naturelles et l'environnement pour demain. Il nous reviens donc d'amorcer un nouvel âge de l'énergie nucléaire, qui semble l'option la plus apte à préserver l'environnement de notre planète.

I. Introduction -- The Role of Asia

We now find ourselves caught in a trilemma of energy, ecology, and economy.

Technological innovation, market economy, and democracy have combined to allow advanced industrial economies to achieve their current levels of prosperity since the end of World War II. World energy consumption has risen by a factor of 4.5 in the process, 90 percent of which has been met by fossil fuels, resulting in over 20 billion tons of carbon dioxide emissions per annum.

The world population is 5.3 billion now, and it is estimated to grow rapidly in future mainly in the developing world, and can reach 10 billion people by the year 2050. Such a population growth could seriously deplete our supply of fossil fuels and cause serious problems in the global environment, threatening the very foundations of civilized society as we now know it.

It is feared that this crisis is likely to manifest itself most seriously in the Asian region. Many of the Asian nations have finally taken off after many years of poverty and stagnation when their GDP growth per capita remained practically at nil, and they are to achieve rapid economic growth after Japan and NIES countries.

Asia now accounts for 59 percent of the world population, with its high rate of population growth likely to continue in the foreseeable future, while its rate of intra-regional energy self-sufficiency is 43 percent comparatively lower than any other regions of the world. Therefore, it is not long before energy resources shortage might plague in the region because of its predicted higher economic expansion in the coming years. Then it is indispensable to promote electrification if the Asian nations are to move into modern societies and achieve desirable levels of living standards. However, if these nations continue their economic expansion, the total world emissions of carbon dioxide could double the present rate of emissions in the not distant future.

Against this background, electric power source alternatives in Asia are bound to exert far-reaching influences on global efforts to cope with the trilemma of energy, ecology, and economy.

II. Asian Factors

1. The Current Asian Situation -- The Hub of Global Growth

Asia now has a total population of 3.1 billion, or 59 percent of the world total, with China, India, and Indonesia accounting for 40 percent of the world population by themselves. On the other hand, GNP in Asia is 20 percent of the world total, while its per capita energy consumption is a mere one-tenth that of the advanced industrial countries.

In recent years, East Asia, Southeast Asia in particular, has grown remarkably with its real rate of annual growth at 8.4 percent during the ten years of 1980s. This is about as twice as much of the average growth rate of 4.3 percent in the same ten year period as posted in developing world other than East Asia. There are several factors to characterize higher economic growth in the Asian nations, although they may vary somehow from one nation to another. One common factor or pattern is that they have been successful in fostering export-oriented industrial structures.

These nations invited direct investment from Japan and other foreign advanced nations to stimulate production, create employment and strengthen their export capacity (mainly to the market in the USA). Such an economic policy helped stabilize political situations in these nations, which, in turn, could induce more foreign direct investment and more business activities in a more favorable manner for rapid economic expansion.

The region's macro-economic performance has been quite good, too. An analysis of the inter-industry relations table compiled by the Ministry of International Trade and Industry shows that in case the final inter-regional demands increase in respective regions of the world the Asian region (except Japan) has the highest co-efficiency of induced production to the other regions of the world, thus serving as the center of growth of the world economy.

2. Asia's Problems--Serious Shortage of Electric Power

Asia, of course, is not without problems of its own.

Population growth rates are very high to begin with. There has been an increase of 1.4 billion in the past 30 years, with an estimated 1.6 billion to be added in the next 30 years. 60 percent of this additional population will come from three nations of China, India and Indonesia. Demographic controls are urgently needed in the years ahead.

In this coming 30 years, India's population will increase from the present 800 million to 1.4 billion people, nearly the same level of the present Chinese population. China has adopted a one-child rule and Indonesia a two-child rule. However, little practical effect is predicted. An Indonesian survey has shown that in 1980 when their electrification rate (a ratio of a number of electrically serviced cities, towns and villages to the total numbers of cities, towns and villages in the nation) was 6 percent, the Indonesian birth rate was 37 babies per each

1,000 persons of the population, while the birth rate came down to 27 per 1,000 persons of the population in 1990 when the electrification rate jumped to 35 percent. The survey concludes that there is a very high inverse correlation between the electrification rate and the birth rate.

Another problem has to do with the poor foundations of energy supplies and difficulties in securing adequate supplies. Rates of energy self-sufficiency by region are 91 percent in NAFTA and 50 percent in the EC. Asia has a rate of as low as 43 percent, while its dependence upon the Middle East oil supply is substantially high. China and Indonesia account for 70 percent of exploitable oil reserves in the Asia. However, the ratio of reserves to production has been rapidly decreasing from 24.6 years ten yeas ago in China down to 22 .2 years now, and from 19.4 years ten years ago in Indonesia down to 10.5 years at present.

This depletion is accelerating with every passing years. At present, China is still a net oil exporting nation selling about 350 thousand bbl/day . However, given current high rates of economic growth, they will likely become net importers of oil possibly by the turn of the century, buying 500 to 1,000 thousand bbl/day from international market.

Thirdly, electricity needs in Asian countries in recent years have been rising faster than their rates of economic growth, causing an acute shortage of electric power across the region. The situation is compounded by poor funding and difficulty of siting, the latter being due in large measure to the "not-in-my-backyard" opposition by residents, as seen in Korea and Formosa. Accordingly, more nations are beginning to take electricity demand control measures by promoting wise and efficient use of energy and electricity and to improve power production efficiency. However, their programs have not seen good results because of shortage of technologies and funds.

3. Harmonizing Environment and Economy -- Eliminating Poverty before Preserving Environment

In Asia "promoting development and eliminating poverty" takes priority over "harmonizing development and the environment." Air pollution by SOx and NOx emissions has reached serious proportions and the people are becoming increasingly concerned about the environmental issues. With 80 percent of domestic needs currently met by poor quality firewood or charcoal, electrification will contribute significantly to environmental improvement. Electrification requires huge funds, which would not be available without economic growth.

You should be seriously concerned with what environmental policies and measures will be followed in Asia, particularly in China, India, and Indonesia. This is because they will give a significant impact on the world picture.

Supposing that the national income per capita in these three Asian countries may rise, which is quite likely in the early part of the next 21st century, to US $1,900 (which is equal to the present level of Thailand and Malaysia as combined), the current rate of carbon dioxide emissions across the world will be doubled. It is of great consequence for the world and its future to identify the best mix of electric power generating technologies that will be acceptable in terms, not only for energy and the economy, but also for tropical forests and other environmental concerns and safety.

III.   Technological Alternatives--Internalizing External Costs

1. Japan's Options in the Post-War Years -- Reconciling Growth and Environment
[1945 - 1973]

Fifty years ago, Japan started to rehabilitate itself from the abyss of hunger and destruction.

In 1951, six years after the end of World War II, electric utility industry in Japan was privatized and reorganized into 9 utility companies. At that time, they generated 41.3 billion kilo-watt hours, and per capita electricity consumption was 389kWh per annum, which was 80 percent of that of China today. In those years, the Government of Japan maintained a low commodity price policy during and after the war time and the electricity prices were forced to be at a cost level lower than the reproduction cost. Therefore, as utilities added more supply capacity, their financial deficits had snowballed and they could hardly secure necessary funds to finance more capacity additions. Electricity prices were set thus at lower levels than the values of other energy sources, so citizens, offices and industries tended to use more electricity. While revitalized domestic production meant rapidly growing demands for electric power, power source development was slow to catch up, and power cuts and blackouts were not at all unusual in daily life. In order to solve the problem of chronic shortage, electricity rates were raised and foreign funds were introduced to keep the financial position in balance. Half the construction cost of any large scale power source development project, however, had to depend on government funds, with the government making substantial fiscal allocations to basic industries such as electric power generation, iron and steel, cement, and coal mining.

Electric enterprises promoted power source development focusing on hydro- and coal-fired power plants. Improved fossil-fuel power generating technologies raised thermal efficiency from 17.2 percent in 1951 to 30.5 percent in 1960. The economics of fossil-fuel power generation were thus greatly improved.

In contrast, economically exploitable hydro-power sites decreased and got more difficult to find so that the hydro-dominated power generation pattern had shifted to the fossil-fuel dominated one in which the base electricity load was met by large capacity fossil fuel plants and the peak load by large capacity storage-type hydro stations. In addition, depressed oil prices which prevailed in those years encouraged a shift from coal to heavy oil as the major source of thermal power.

On the other hand, the Basic Nuclear Energy Law was enacted in 1955 in order to promote research and development of nuclear energy and use this form of energy for peaceful purposes, and various nuclear-related programs were carried out. As a result, a 357-megawatt Tsuruga Unit 1 was commissioned into operation by the Japan Atomic Power Company in March 1970, this being the first boiling water (light water) reactor in Japan. It was followed by a 340-megawatt Mihama Unit 1 by the Kansai Electric Power Company in November 1970. This was the first pressurized water (light water) reactor in the nation. Both of these first units were completed by introducing the U.S. technologies. In April 1970, a 350-megawatt Minami Yokohama Unit 1 of the Tokyo Electric Power Company started operation as the world's first LNG firing power plant.

In July 1970 Kisenyama pumped-storage hydro power station of the Kansai Electric Power Company began to send electricity. The rated capacity of 466 megawatts is the world's largest single unit capacity for this type of hydro stations. By the same year, per capita electricity consumption had reached 2,608kWh per annum (which is equal to that of the NIES nations today).

[1973 -   ]

In 1973, a full-scale flue-gas desulphurization system went into operation for the first time in Japan, and the oil crisis that erupted in this same year proved to be a major turning point in the power generating situation in Japan.

Special emphasis has since been placed on promoting efficient and wise use of energy and environmental protection, with the result that SOx and NOx emissions (by volume) per each kilowatt-hour generated by fossil fuels in Japan remain very low at about 1/18th and 1/7th respectively of the European and American levels.(See Table I)

Table I. EMISSIONS PER KWH GENERATED BY THERMAL POWER PLANTS IN 1988

(g/kWh)

	SO_x	NO_x
Average of 5 Developed Countries*	6. 7	3. 3
Japan	0. 3 8	0. 4 6
Kansai Electric Power	0. 1 9	0. 2 4

(Note) 5 developed countries include U.S., West Germany,
United Kingdom, France and Canada.
(Source)The Federation of Electric Power Co. of Japan

Such environmental costs have already been internalized in the electricity prices in Japan, and as a result, the Japanese electricity rates and charges got higher about 30 percent than those billed in several industrialized nations. In the meanwhile, various measures and investments were made to improve efficiency of household electrical appliances and promote efficient and wise use of energy at industrial plants and factories so that the cost burdens on electricity consumers may be reduced. You might be interested in knowing that electricity use by the standard household refrigerator has reduced by 66 percent, and the kilowatt-hour use to produce each ton of electric furnace steel is now decreased by 20 percent. More wide-spread use of less energy consuming appliances and accelerated investments in environmental considerations have generally contributed to the economic growth in a positive manner.

The most important option in the process has been the positive exploitation of nuclear capacity, which now adds up to 34,419 thousand kilowatts or 19 percent of total capacity in Japan and generates 222.3 billion kWh or 28 percent of the nation's total generation. As a result, carbon dioxide emissions per GDP(tons/1 trillion yen) was nearly halved from 1.26 in 1973 to 0.68 in 1991.

Japan has, thus, successfully shifted power source alternatives in the last 50 years to reconcile growth and the environment.

2. Asia's Power Source Options

Let us take a look at Asian countries. We have to consider several factors as we look at possible options for power generating technologies for the Asian nations.

(a)   Most of the nations depend upon the supply of energy sources and particularly of oil from outside of the region, notably from the Middle East area. Such dependence is increasing year after year, and in this context the stable availability of energy resources is extremely important to them. Energy situations in Russia may give a large impact in this region.

(b)   In order to sustain economic growth in a balanced manner with the environmental preservation, it will be required to improve the energy use efficiency by 30 to 50 percent in this region. However, the market force mechanism is not as yet established to function well, and these nations are still with the shortage of funds, technologies and human resources. You should pay a careful consideration to these issues.

(c)   In addressing the environmental issues from a global perspective, the advanced industrialized nations are required to take the initiative to effect more efficient and wise use of energy, promote use of energy sources to replace fossil fuels, and transfer energy technologies in steps to the developing nations.

(d)   Until the time when the developing nations can reach a stage of economic development to take off by themselves, the industrialized world should make a wide spectrum of international cooperation ranging from fund allocation, technological transfer and human resources training to know-how or expertise about the market force mechanism. In this case a careful consideration will have to be made in order to tailor such cooperation to respective local circumstances.

As an example, the joint environmental research program between China and Japan includes an important option of the projects to give up directly burning coal for non-industrial purposes and replace coal with briquets. This fuel conversion will help raise heat rate value and reduce exhaust gas. Mixing lime is also expected to improve the desulphurization effect. This particular option was not initially included in the proposed areas of Japanese cooperation, but it represents a stage that Japan has once cleared in the process of postwar development. It is a case that points to the necessity and the feasibility of multi-faceted patterns of cooperation to fit different conditions in different countries. A review of the course of development of advanced countries shows that energy consumption per GDP gets lower over time, so that late starters do benefit from technological progress in early starters.

With advanced countries ready to transfer their power generating technologies, including capital and education, to developing countries, it is hoped that the latter will find it possible to internalize environmental and other external costs at a lower price.

3. Assessing Different Technological Alternatives

In order to accommodate a trilemma between energy availability, environmental protection and economic development, the advanced industrial nations should develop the technologies for wise and efficient use of energy and gradually shift their life style to that of less energy consuming and less resources consuming patterns. This is a fundamental requirement as we see that these industrialized nations use three quarters of the world energy supply, while only one quarter of the world population lives and enjoys life in these nations.

Based on this assumption, we would like to evaluate several power generation technologies available, because the importance of power generation technologies to address a "trilemma" issue is becoming increasingly larger as the electrification rate is growing not only in the industrialized nations but particularly in the developing world.

As we evaluate these power generation technologies, we can look at them from various aspects, but we consider them from the following specific aspects; static evaluation, comprehensive evaluation and dynamic evaluation.

(a) Static evaluation

This is one of the most representative evaluation methods. In terms of the energy resources reserves, the ratio of the reserves to production is smallest with oil and then comes natural gas. Accordingly, the high priority options should be energy sources which can replace oil. In view of SOx emissions control, natural gas which contains no sulphur content is most desirable as well as nuclear energy which puts no burden on the environment.

Power production technology by burning natural gas in the combined cycle gas turbine units can promise as high as 47 percent heat rate. (The average heat rate of the fossil-fuel power stations in Japan is 37.1 percent.) Therefore, the utilities in many countries are moving toward building and operating more natural gas firing plants. However, the combined cycle gas turbine units emit a fairly large quantity of NOx, and the low NOx emissions power production technology should be urgently developed and established.

According to the prediction by The Institute of Energy Economics, Japan, natural gas supply in the world market will become short by 50 to 200 million tons by the year 2010, unless the gas price increases remarkably from the current level.

In terms of the ratio of reserves to production, coal supply can last for 232 years and look like very much promising. But on top of air pollution by its SOx and NOx emissions, coal-firing plants produce carbon dioxide emissions about 30 percent as much as oil-firing plants, and they raise another difficult issue. In order to alleviate this problem, the development of clean coal technologies is going on, and if they prove feasible, they can effectively apply to power stations in the developing world.

As far as uranium is concerned, the evaluation of nuclear power production may possibly vary significantly depending upon whether the world nuclear industry can take and follow steadily the spent fuel reprocessing and fast breeder reactor strategy for effective use of uranium resources and upon whether the industry can be successful in securing technological safety and non nuclear proliferation measures. However, the power generation technologies to give solutions to a "trilemma" issue cannot be considered other than nuclear energy.

Upon making the static evaluation like this, we have to look carefully at the uneven regional distribution of natural resources in the world. The uneven regional distribution of natural resources had so often become a major cause for various wars or conflicts, and in order to eliminate any more of such repeated tragedies, serious attention to and consideration of the uneven distribution of energy resources must be made.

(b) Comprehensive evaluation

As and when we evaluate the options of power generation technologies to address the global environmental problem, the simple analysis of the greenhouse effect gas directly discharged from power production process can hardly suffice the evaluation purpose.

Energy production process generally consists of exploiting, processing, transporting and refining fuel resources, and power generation process includes manufacturing of equipment and components of power stations, construction of stations and treatment and disposal of varied wastes. These processes are naturally subject to evaluation, and the exhaust quantity of greenhouse effect gas coming out of each of these processes will have to be studied to consider the power generation technologies.

According to the estimates prepared by the Central Research Institute of Electric Power Industry of Japan, the greenhouse effect gas coming out of each of the processes is 269 grams per each kilowatt-hour generated at coal-firing stations. This is at the highest end. The gas quantity is 6 grams per each kilowatt-hours at hydro power stations and 7 grams per each unit generated at nuclear power plants.

What interests us is that solar energy may emit seven times as much greenhouse effect gas as nuclear energy and that this energy fails to prove cost effectiveness, although solar energy is generally attracting popular attentions as an energy source friendly to the earth. (See Table II)

(g-c/kWh)

		CO_2 Total		Methane	Total	Investigated Power Plants	
		Fuel	except Fuel			Capacity	Reliability Factor
Fossil	[Coal]	2 4 6	1 0	1 3	2 6 9	1. 000MW	7 5 %
	[Oil]	1 8 8	8	4	2 0 0		
	[LNG]	1 2 9	3 3	1 6	1 7 8		
Nuclear		—	7	0	7		
Hydro		—	6	0	6	10MW	4 5 %
Geothermal		—	1 1	1	1 2	10MW	6 0 %
Wind		—	1 9	1	2 0	100KW	3 5 %
Solar		—	5 2	3	5 5	1MW	1 5 %

(Source) Central Research Institute of Electric Power Industry Report. May 1992

(c) Dynamic evaluation

Japan advocates and publicizes its "The New Earth 21" Plan in the international community. This plan is based on a concept that our common planet Earth has been deteriorated during the past two hundred years after the Industrial Revolution, and it basically aims at regenerating the Earth during the century to come.

For this purpose, the plan calls for improved accumulation of scientific knowledge, technological development to introduce and promote wise and efficient use of energy and clean energy development, expanded application of carbon dioxide absorbing measures, development of next generation energy sources et cetra. By continuously implementing one energy measure to another as soon as feasible and practical, the plan is to substantially reduce the emissions of the greenhouse effect gas from every energy-related process.

Such ultra-long term evaluation will be eventually required.

Some argue that until the time when the developing nations can put them on the track for sustainable economic development by cutting off a vicious cycle of 3Ps (poverty, pollution and population), these nations should introduce and apply the power generation technologies which are not necessarily efficient enough so that the electrification rate in these nations can be improved. We should give some evaluation to such an argument.

Inefficient technologies may tend to deteriorate the quality of the environment for the time being. However, the improvement in the electrification rate could bring about decreased birth rate, population control, availability of surplus funds by accelerated industrialization to finance environmental control, internalization of environmental cost through market force mechanism or principle, and promotion of wise and efficient use of energy and so many other turns or cycles for favorable developments.

It may sound like a "which came first, the chicken or the egg?" question. However, isn't it important to start and run a dynamic model for regenerating the Earth from a long term perspective?

(4) Outlook for Respective Power Generation Technologies

Hydro Power

Much is expected of hydro power development in Asia as electrification is advancing rapidly in its rural areas. In terms of cost effectiveness, large capacity reservoir type hydro power development proves attractive. However, it may pose difficult problems as well, which include the rejection by the residents who are to be relocated from the water submerged sites, deforestation and necessity of building longer distance transmission lines.

We look at country by country. In Indonesia, there is a high potential for electricity demand growth. However, the nation consists of a great number of isolated islands and the river water flow varies widely in rainy season and dry season. Therefore, their priority is shifting from building large capacity hydro capacity rather to expanding coal-firing plants. Small capacity hydro power stations will hardly prove beneficial in the future, and it is suggested to build more medium-sized hydro facilities by so doing to reduce the requirements, even though small, of fossil fuels for power production. Medium-sized hydro plants may not cause many residents to be relocated.

Thailand has many plains which are not considered suitable for hydro power development. In Vietnam of which land stretches long from the south to the north, the transmission lines may pose a difficult problem.

China has the areas with abundant, exploitable hydro resources in its western part of the land. However, most of these hydro resources are left unexploited, because the transmission lines should be long enough to reach these remote areas. The Government of China is to carry out large capacity hydro power plants like San Xia dam as national projects. However, this project will force many local residents to be relocated and their rejection has caused the project behind the planned schedule.

There are, on the other hand, successful examples of using the reservoirs for growing fish or as tourist attracting sites. They will be requested to commit to building multi-purpose medium-sized hydro power projects.

Fossil fuel Power

We take a look at China, for example. China now has the installed plant capacity of 140 million kilowatts, of which as high as 70 percent depends upon coal capacity. Average heat rate of coal-firing stations is 28.7 percent. (Japan's average heat rate is 37.1 percent.)

Coal consumption is 428 grams per kilowatt-hour generated, as compared with 316 grams at coal-firing stations in Japan. Station energy use is 7.9 percent as against 3.7 percent in Japan. Sulphur content in the coal they use is 1.7 percent. (Coal burnt by Japanese utilities contains 0.1 percent of sulphur content.) Thus, there is a lot of room for improvements in coal plant performance in China. They have stringent environmental control standards and codes. However, they have so far installed only two desulphurization devices for practical operation. As a result, SOx emissions from all the coal-firing stations in China amount to a total of 4.2 million tons a year, which is 30 times as many as in Japan.

Current energy consumption in China is the third largest in the world, next to the United States and Russia, but they plan to add another 100 million kilowatts, mostly coal-firing capacity, by the turn of the century. Coal consumption for power production will, therefore, increased to 710 million tons of oil equivalent by the year 2000 from 490 million tons in 1988. It will go up further to 1 billion tons or two times as many as 1988 rate by the year 2010.

The currently deteriorated quality of China's environment is extremely serious as illustrated by a fact that their falling dust is about eight times as much as the annual average value in Japan and that suspended particulate matters are about 12 times as much as the annual average in Japan. Accordingly, the environmental control measures are urgently needed, which will in turn require a huge amount of money.

An initial realistic option is the semi-dry sprayed fire system, currently being tested at Huangda. This is a mid-level technology device whose efficiency is at a level of 70 to 80 percent, but is one of the realistic low cost options, which can be built at about 60 percent of the construction cost and be operated at about 80 percent of the running cost of the state-of-the-art facilities. Another attractive promising clean-coal technology is the dry electronic beam method, which produces fertilizers like ammonium sulfate and ammonium nitrate as by-products, and allows to pass external costs to the prices of by-products. Japan is now aiming at achieving 47 percent of thermal efficiency by the natural-gas burning combined cycle system. There is a possibility of transferring this technology to natural gas producing areas.

Nuclear Power

As electricity is highly efficient and easy to manage in environmental considerations, it tends more and more to be the preferred source of energy across the world, and nuclear power represents a major alternative having the most contribution to make.

Japan has formulated and implemented the "Action Plan To Arrest Global Warming" looking to the year 2010. This plan is a national program as prepared to practically develop and implement what is included in the Earth Regeneration Plan or "One Hundred Year Plan" which is a comprehensive, long term vision for global warming prevention.

This action plan covers the years from 1991 to the year 2010, and aims at stabilizing per capita carbon dioxide emissions approximately at 1990 level for the years after 2000, and also at reducing methane and other green house effect gas to a minimum possible.

Measures to implement include deliberately considered actions in specified groups so as to build urban or city areas structures, establish traffic systems, create energy supply structures and develop life style which are all less carbon dioxide emitting.

As envisaged macro economically in this plan, energy conservation is to be maximized to 36 percent per GDP. This value is exactly same as the record of 36 percent per GDP which Japan achieved to date since 1973 when the first oil crisis took place by investing a lot of money in less energy consuming equipment and plant and by individual citizen's determination for energy savings. As energy utilization efficiency has now been improved to an extremely high level , we will be eventually required to exercise more intensified efforts than before.

When we look at electricity demands in the period of the plan, the value of electricity demand elasticity (a ratio of electricity demand growth to economic growth) is estimated at 0.68 in the first ten years and at 0.5 in the second ten years of the plan. When we note an increasing trend of the electrification rate in the most recent years, the plan calls for extremely stringent measures and programs for electricity savings.

As far as solar energy, wind power, other new energy sources and carbon dioxide fixation technologies are concerned, as large as possible applications are considered given most innovative technology development available. When it comes to the electric utility industry, the plan requests to introduce new energy sources at a level of 5.7 million kilowatts by the year 2010. If we are to achieve this target value by that year, we will have inevitably to step up efforts for more advanced technological innovations.

In addition, one of every two houses will have to be built as "solar energy producing and consuming house", and latent waste heat in the rivers and streams in the city areas should be utilized to meet as high as 10 percent of non-industrial heat requirement.

Even if we are to resort to the maximum measures presently conceivable, we have to build additional nuclear capacity of 72.5 million kilowatts by the year 2010 in order to stabilize carbon dioxide emissions at 1990 level. We are now strongly appealing this to each of the citizens.

In this context, the Japanese electric utility industry is working on a program to pursue the optimum and best generation mix to increase nuclear to 40 percent of total installed plant capacity and to 60 percent of the nation's total kilowatt-hour production.

Utilities put aside reserve funds to cover high level radioactive waste costs, decommissioning costs and other back-end costs and internalize them into the revenue requirements. Electricity production cost is calculated at 13 yen per kilowatt-hour generated by hydro power, at 10 to 11 yen by fossil fuels, and at 9 yen by nuclear. Therefore, nuclear energy proves cost competitive.

APWRs and ABWRs as currently planned for construction are designed at 90 percent of utilization factor, which is improved much from the present level of about 70 percent of the existing reactors. It will certainly promise improved cost effectiveness for future nuclear energy.

The most critical problem about nuclear energy is to win understanding and confidence by local citizens of plant siting. Nuclear power generation in Japan is highly and extremely reliable with the rate of forced outage at as low as 0.1 per reactor per year.

Everything possible will be done to further ensure safety in such areas as aging degradation, extremely low frequency accidents, high-level radioactive wastes, reprocessing technologies, and international nuclear non-proliferation. In order to dispel the nagging sense of anxiety and uneasy feeling about nuclear on the part of the residents, emphasis is also placed on measures to improve social perception of nuclear safety in general.

Kansai Electric Power Company set up the Institute for Nuclear Safety System and this institute considers the nuclear safety both from a social study aspect and from a technological system perspective, and aims at establishing nuclear technology in a balanced harmony with the human society.

According to a model simulation of the human sense of anxiety changing over time, which was developed by Professor Kinoshita of Kyoto University, if any single major accident happens to take place anywhere in the world in the next ten years, the sense of anxiety about nuclear power would quite likely to go beyond the threshold value and make people never forget it for good. Therefore, the responsibility for securing nuclear safety should be commonly shared in the world community. (See Figure 1)

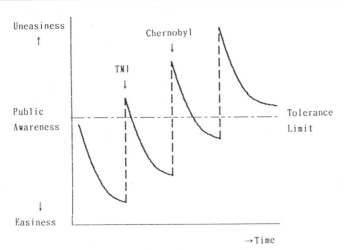

Figure 1. Serious Nuclear Accidents and Irreversible Public Awareness

In Asia at present, nuclear power plants are in operation in Japan, Korea, Taiwan, and India. Another plant is in a trial operation and two more are under construction in China. Indonesia is planning to construct 12 reactors over the next 15 years. Joint nuclear feasibility study is under way between Indonesia and Japan. This study covers not only reactor type selection strategies and site surveys, but also education of nuclear engineers and personnel and other problems in the context of a comprehensive national development program.

In addition to respective individual programs for nuclear cooperation, it is important to secure nuclear safety commonly and jointly through mutual inspection by an international organization like WANO.

New energy sources

Japan has set forth a target value for developing new sources of energy to provide a total of 5.7 million kilowatts by the year 2010. This capacity equals to 2 percent of total generating capacity of Japan. The target value includes 5.5 million kilowatts by fuel cells, 100 thousand kilowatts by solar power generation and another 100 thousand kilowatts by wind mill power.

It is planned to develop 1.05 million kilowatts of phosphoric acid type fuel cells by the year 2000. Designed efficiency of 41 to 43 percent could possibly be achieved with this type of fuel cells. Experiments indicate that the cells could run as long as 13,000 hours and suspend operation from time to time by accidents of various reasons. These running hours are still short of the designed hours of 40,000 hours, and the operation reliability should be necessarily established. There is a trade-off between reliability improvement and cost reduction and compact design of cell units, and the evaluation has difficult aspects.

Technological development is also going on with a new type of molten carbonate acid fuel cells.

Electric utilities in Japan plan to introduce a total of 2,400 kilowatts of solar photovoltaic power by 1995. Because of very limited land resources in Japan , it is difficult to adopt the concentrated placement of the units. Dispersed on installation, for an instance, on the house or building roofs may possibly be expanded in the future. As far as power generation efficiency is concerned, it is 18 percent or more with multi-crystal type cells and 12 percent or more with amorphous cells. Both types of cells have already achieved the target efficiency as set forth, and another 5 to 10 percent improvement in efficiency could be expected.

In the initial year of cell operation, there is noticed degradation of about 10 percent. However, it does not pose any significant difficulty in promoting wide applications.

The problem is cost. Total cost including photovoltaic cells and inverter units could exceed 1.2 million yen per kilowatts installed, and even if we expect technological progress, it is extremely difficult to reduce the cost down to 200,000 yen per kilowatt installed which is set forth as a target value to achieve by the year 2000.

Operating reliability of wind mill power units has been proved sufficiently for practical application. In this mode of power generation, utilization factor will largely influence the cost effectiveness. Given the installation cost of 300,000 yen per kilowatts installed, we should look for sites where the average wind velocity of 6.5 meters per second can be available throughout the year in order to have the utilization factor at 22.7 percent or more. Such sites are very limited in Japan.

There are a number of locations in Asia where solar power, wind mill power and geothermal power appear to be more promising than in Japan. In China, for example, they have ample plain land of about 3,366,000 square kilometers suitable for solar photovoltaic power generation. However, even if they install a 50,000 kilowatts solar capacity per each 1 square kilometers, total power generation from this source could add up to 2,960 million kilowatt-hours a year, which would supply 0.2 percent of the total electric energy requirements in China by the year 2000.

Potential wind mill power locations in China exist along the coast line as long as 100 kilometers or 10 percent of total coast line length of China. These locations can accommodate wind power capacity of about 880,000 kilowatts.

However, this capacity can send out about 1,750 million kilowatt- hours a year, which is a very small fraction of 0.12 percent of the total electric requirements in China.

Carbon dioxide fixation technology

The Research Institute for Innovation Technology for the Earth was set up in the Kansai Science and Cultural City in Japan. This institute, in close collaboration with private organizations and the Government, proceeds with a comprehensive research program on carbon dioxide fixation technology and many other technologies related to the global environmental protection.

5. Optimum or best generation mix for Japanese utilities

Each of the Japanese electric utilities has formulated its own best generation mix, by taking into a careful, comprehensive consideration the stable availability of fuels from diversified supply sources, cost effectiveness, and global environmental considerations so that a "trilemma" issue between energy, ecology and economy may come to solution.

The Kansai Electric Power company completed the first commercial nuclear power station in Japan in 1970 and has since then positively expanded and will continue to expand nuclear capacity to meet the base loads on its power system. Its optimum generation mix is to have nuclear energy at 40 percent of total installed plant capacity and at 70 percent of total power generation. (See Table III)

Table I. OPTIMUM MIX OF POWER SOURCES OF KANSAI ELECTRIC POWER

(%)

	k W	k W h
Nuclear	4 0	7 0
Fossil-Fired	4 2	2 1
[Coal]	9	7
[Oil]	2 1	7
[LNG]	1 2	7
Hydro	1 8	9
[Conventional]	7	6
[Pumped-storage]	1 1	3
Total	1 0 0	1 0 0

In order to diversify fossil fuel supply sources, the company promotes building more LNG-firing and coal-firing power stations, by so doing to gradually reduce the dependence upon oil.

As we go more for nuclear, it is indispensable for us to have the real understanding of the necessity to establish a closed nuclear fuel cycle and in particular the necessity of plutonium recycle use in the reactors. Such understanding will have to come both within and from outside of the country.

Electric energy output by using plutonium in the fast breeder reactors is about 60 times as much as that available simply by fueling U235 (of which R/P ratio is 68 years) to the light water reactors. Energy output which can be obtained from one ton of plutonium equals to about 1,750 thousand tons of oil equivalent.

In Japan, the Government and the nuclear industry make it a basic point to "produce and hold no surplus plutonium", and take every possible effort to secure nuclear safety and maintain nuclear non- proliferation as well as positively disclosing related information and data in order to win the understanding from the international community.

It might appear to many people that Japan takes a prominent place in the world in the area of plutonium resource utilization, and there is a growing serious concern in some sectors of the international community to allege that Japan should maintain an intention to use nuclear for military purposes.

However, Japan has been all the time careful enough to disclose related information to the public, and she definitely recognizes this nation should be responsible for clearly explaining and convincing the rest of the world that nuclear power development and plutonium resource utilization can contribute much to solving the natural resources problems and the environmental issues of the world. It does not necessarily benefit Japan alone.

More than half of the century will have to be spent before the fast breeder reactor technology is firmly established. It will be too late to launch and initiate the development effort from the moment when the world finds it should readily need the FBRs.

IV. Conclusion -- A New Approach to "Dialogues with the Future"

Selecting power generating technologies in response to the "trilemma" mentioned at the beginning is a very difficult task, to say the least.

While nuclear power development, which should be the key to success, is stalled in advanced countries, in developing countries, where eliminating poverty takes priority over environmental concerns, fiscal and technological limitations narrow the range of effective options. Can the "trilemma" really be resolved by modern rationalism thinking alone? The cold war confrontation has now come to an end, and we are entering into an age when various world problems should be solved and shared by many nations in a versatile manner. In this context, the global environmental problem is nothing but a critical test for Pax Consortia system, because every nation and every citizen on the earth is responsible for this serious issue. Three years ago, opinion leaders in the Kansai district got together and established a "Global Environment Forum-Kansai," and drew up "Five Basic Principles for Action":

1. The Principle of Asset Management over Generations;
2. The Principle of Cooperative Development between Developed and Developing Worlds;
3. The Principle of All Citizens being Responsible for the Environment;
4. The Principle of Science and Technology with a High Priority on the Global Environment; and
5. The Principle of Creating a Metabolic Life Style.

While each and every one of the five principles are no doubt of great importance, it is the first principle of "asset management over generations" that is most directly relevant to the problem of selecting power source alternatives and needs to be addressed under a long-term perspective.

We need to embrace the view that "We owe our future generations ample resources and a healthy environment." We need to make decisions from the point of view of our descendants. I would like to propose that a "Dialogues with the Future Commission" to be set in the OECD.

The "Dialogue with the Future" Commission should include the members who represent the present generations and the persons who play roles on behalf of potential representatives of our younger future generations. And this commission is requested to discuss the long-term issues associated with energy availability, global environment, and nuclear power generation technologies from an entirely new set of concepts and make drastic recommendations and advice to the governments in the world. This year is the 40th anniversary of the historic "Atoms for Peace" Declaration by the late U.S. President Eisenhower.

In the most recent months, peaceful use of nuclear energy appears to be retarded and get stagnant in one way or another. However, I am much inclined to gather that our future generations are sending an honest message or signal strongly calling for a new nuclear energy age capable of coping in a most appropriate manner with depleting energy resources and worsening global environmental issues.

High population growth

World population

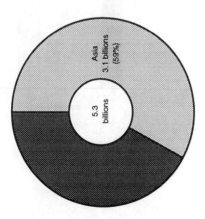

Asian population

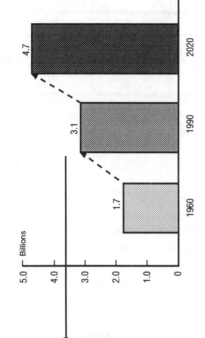

Birth rate and mortality rate (1991)
Per 1 000 persons

	Birth rate	Mortality rate
INDIA	30 persons	10 persons
CHINA	21 persons	7 persons
INDONESIA	27 persons	9 persons

SELECTED ENERGY INDICATORS BY REGION IN 1991

Low Energy Self-Sufficiency Rate

	EC	NAFTA	ASIA
ENERGY SELF-SUFFICIENCY RATE	50.4	91.1	43.4
OIL SELF-SUFFIENCY RATE	21.2	72.5	27.5
RATE OF RELIANCE ON MIDDLE EAST OIL	31.6	11.3	65.8

Ratio of oil reserves to production

Depletion of oil reserves

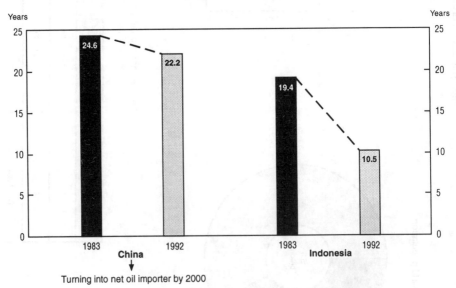

China

Turning into net oil importer by 2000

228

GDP per capita (1990)

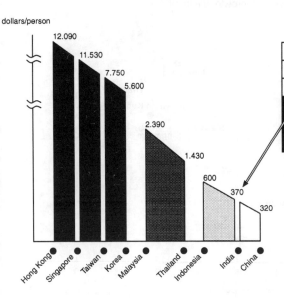

dollars/person

| | 12.090 | 11.530 | 7.750 | 5.600 | 2.390 | 1.430 | 600 | 370 | 320 |

Hong Kong • Singapore • Taiwan • Korea • Malaysia • Thailand • Indonesia • India • China •

Real GDP growth rate in 1980S
Ten years

	Annual average
Developed countries	3.0%
Developing countries	4.3%
East Asia, Southeast Asia	8.4%
Southwest Asia	5.5%
Latin America	1.6%

Asia and its impact upon global environment
Increase in CO₂ emissions of 3 Asian countries

Total of China, India, and Indonesia

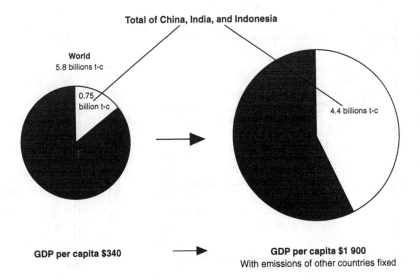

World
5.8 billions t-c

0.75 billion t-c

4.4 billions t-c

GDP per capita $340 → **GDP per capita $1 900**
With emissions of other countries fixed

Rapidly increasing electricity demand
Annual average growth rate in 1980's

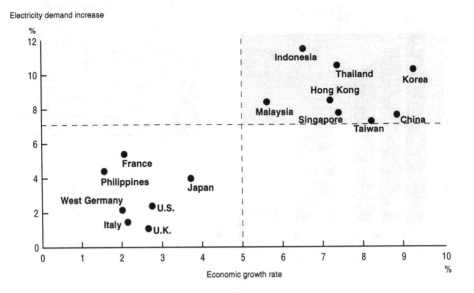

Electricity demand increase

Electrification rate and birth rate

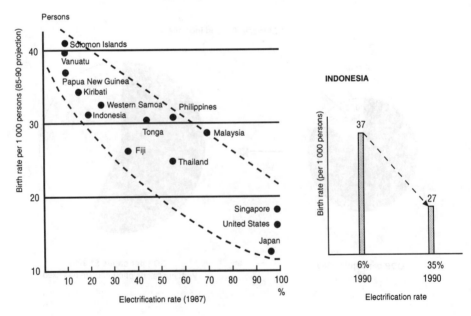

ELECTRIC POWER FOR A SUSTAINABLE DEVELOPMENT

THE ITALIAN SCENARIO

Sergio Barabaschi
ANSALDO

1. INTRODUCTION

If we examine present world distribution of per capita energy consumption, we see that there is quite a close correlation between consumption and the economic development of the various countries.

However, there is not a simple correlation between the consumption and local decay of natural environment (which can be noticed both in some consumer cultures and in poorly developed societies).

The world consumes an even higher energy amount. Consumption growth concerns above all developing countries ("DC's"), both because population increases concentrate in such countries and because per capita energy consumption rises, there as a consequence of better standards of living and expanding business activities.

A decrease in energy intensity can be noticed in industrialized countries: the large infrastructures requiring lots of material and energy have already been implemented. This event will still need a long time before occurring in DC's, which are located in the growth segment of energy intensity curve. On the other hand, the decrease in energy intensity does not necessarily entail less consumption in industrialized countries.

If we consider electric power consumption instead of global energy consumption, we detect a trend of even greater increase, and this is due to the fact that almost everywhere electric power penetration, viz. the portion of energy requirements met by electric power, is growing.

Electric power often allows to save energy because it: increases production processes quality (less rejects) and products life (less wastes), brings flexibility into production processes as well as means of automation, automatic control and electrotechnologies, enables a more rational energy exploitation (such as in the case of microwave applications or electric power based iron-making), and in general it affords the possibility to use energy when and where it is actually required.

Moreover electric power is a basic component of the present technological changes in the scenario of our society, such as the coming of the information and communication age, and it is also necessary to cope with problems such as materials recycle or water shortage.

Another great advantage of electric power consists in the fact that it allows to diversify primary energy sources, e.g. diversification in the prevailing hydrocarbons exploitation through the resort to nuclear energy and an expansion of coal burning.

Anyhow, current economical and environmental problems play a much more essential role than worries about availability of energy supply. It was recently observed that some of the environmental impacts of energy generation and exploitation are not localized close to releases, but show instead regional or even worldwide features.

Electric power generation is often accused of causing environmental damages (and it certainly is responsible too for a part of them), but in particular this is true only as for SO_2, which gives the most important (per cent) contribution. In general, it is by far easier to cut down the pollutants of a large centralized plant such as a thermoelectric power station rather than those of small spread-out users such as cars or gas range flames.

CO_2 implies instead quite a different consideration: the total contribution of electric power generation to carbon dioxide rate is less than one third of the total figure. It is both possible and necessary to curtail pollutant releases, CO_2 included, in electric power generation. A viable method consists in efficiency increases, through dual and combined cycles coupling gas and steam turbines with natural gas exploitation and nuclear energy.

In the long term, new technologies may become available: e.g., fuel cells, which will match a very high efficiency (prospectively, more than 60%) with very good environment impact features and photovoltaic for demands that are located far away from electric transmission systems.

Finally, there are several systems for electric power generation which drastically reduce CO_2 release since they are based on "no combustion" processes (hydroelectric, geothermal, solar, wind, nuclear type).

Nuclear fission energy, nowadays not exploited in italy, contributes for a 17% share to world electric power generation, and up to 33% share to EEC electric power generation.

A further promotion of such a kind of energy requires: organizational and institutional changes, with special reference to permits, capitalization of the huge experience acquired with western operating plants, standardization and implementation of a system near to intrinsic safety. As a whole, these conditions should make it possible to achieve a simple and fast procedure for approval, both of construction and operation of new plants in just one step.

Today renewable sources provide, on world scale, a contribution to commercial energy requirements in the order of 7%, mostly stemming from large hydroelectric plants. This contribution can be regarded as being much greater, if we also include the energy sources which do not go through commercial networks, and particularly wood as well as vegetal and animal wastes in the third world countries. However, a significant share of such contribution originates from an incorrect exploitation of forests and vegetation, leading to deforestation and desert expansion, and cannot therefore be referred to renewable sources.

Renewable sources are especially suitable for a decentralized development pattern spreading throughout a territory. This is particularly important in many developing countries, which still lack large energy infrastructures, above all electric networks.

Future energy system shall have a degree of diversification and flexibility by far greater than the present one, so as to cope with the two different kinds of requirements of developed and developing countries (better exploitation efficiency in the first case, decentralized plants in the second instance). On a global scale, it will deal with primary energy and electric energy in a sounder way.

A continuous expansion of electric power will also go on. Among others, electric power is a practicable and flexible vector which is limited in distance today. Tomorrow these distances could however be increased, on the one hand by high temperature supraconductivity development, and, on the other hand, by increases in electric vehicles.

Moreover, an important role will be played by electric power based technologies, both of signal and power type: for network automation, load control, standby power cutdown, modern storage systems and small plants remote control.

Before concluding these introductory consideration on the global energy issues, let me comment the most stimulating paper by D. Gray and others of the Mitre Corporation, which analyzes the potential world development in the next century and its impacts on energy demand, resources and environment.

Basing their analysis from the energy per capita data from the World Bank, the authors divide world population according to present per capita energy consumption and study the evolution of the population of each group and of their energy demand.

233

Instead of splitting the population in five groups as I did in my figure 2, they identify nine groupings. Basing its assessment on several interesting considerations, the paper reaches some conclusions that I feel useful to remind.

As on the figure 3 of this paper that reproduces the figure 8 of the mentioned Mitre report, the paper shows how total energy demand will reach about 1100 exajoules, with a 3.3 fold increase over current energy demand world-wide. This large increase derives almost entirely from the population groups that are presently below pro-capita energy consumption of 2 toe per year.

I quote from page 27 of this Report the main conclusions:

Of the total energy used in 2100, electricity can be the dominant form. As pointed out by Starr, "electrification will be the major path for economic growth world-wide". It is quite feasible to assume that electric power could provide as much as 80% of world energy demand. This would assume that a considerable percentage of transportation energy would also be electric. If 80% of world energy was electric, then 880 exajoules of primary energy would be required to be generated. This would correspond to about 13,000 electric power stations world-wide with a capacity of 1,000 MWe each.

Environmental implications

. Sulphur dioxide emissions can readily be eliminated by using advanced control technologies. Short-term solutions, like retrofitting old technologies with scrubbers, do not make for a sound environmental policy in the long term.

. Annual carbon dioxide emissions may not increase very much over present values for the next 50 years if end use efficiency improvements can be obtained and a modest increase in nuclear and renewables can be sustained.

Energy resource implications

. All energy resources, coal, oil, and gas, as well as nuclear and renewables, will be needed in the coming century to fuel world development.

. The potential contribution from biomass appears limited. Apart from the very real problems of competing land use for food and water resources, it is not clear if using certain biomass systems are actually net energy positive or negative. For example, Amoco has estimated 2.2 times as much energy is needed to grow corn and convert it in ethanol in the U.S. than the energy contained in the ethanol.

. There does not appear to be a real energy resource problem for at least the next 50 years, but after that, in the absence of a surprise, even if more oil and gas are found, the world had better be ready with acceptable nuclear and solar based technologies.

. High use of nuclear energy will require spent fuel reprocessing.

R&D implications

. No single energy technology can solve the upcoming energy problems that will face the world. Energy R&D cross-cutting between these separate technologies will be necessary to provide solutions, i.e. combine fossil/nuclear/renewable technologies into an integrated system approach.

. Solar energy, particularly photovoltaic (PV) and electric energy storage R&D, will become very important.

. PV use in small scale applications (residential, village) in developing countries will reduce requirement for large, central fossil generating facilities, and for long range power transmission.

2. THE ITALIAN SCENARIO

I have now to be more specific and to provide some data on the italian situation. For this let me draw from the most interesting report of Europrog (April 1992) and from the recommendations of the famous report of IEA on energy policies. (References 1 and 2)

The major developments in Italian Energy Policies during 1990 and 1991 concerned the implementation of the National Energy Plan (PEN) which was originally prepared in 1988.

The basic premise of the PEN is that Italy needs to decrease dependence on imported energy, particularly oil, while reducing the environmental impact of energy production, transport and use, especially the emission of CO_2. Much attention has focused on balancing electricity supply and demand amid opposition to nuclear power and to the siting of coal-fired power stations. The increased demand for gas in electricity generation and other uses is also an important issue in the PEN and its implementation.

The most important data on the Italian situation are reported in the figure 4 to 10.

Net imports of energy increased slightly in 1990. Domestic production met 16.5% of TPES (Total Primary Energy Supply). Italy's dependence on oil (59.1% Of TPES) and imports (97.2% of oil supply) remains among the highest in the IEA.

In 1990, oil demand decreased 1.8% To 91.5 . Oil's contribution to TPES fell from 60.7% in 1989 to 58.2% in 1990, though this share remains one of the highest in the IEA. The use of oil for electricity generation continued to fall, the share decreasing from 49.6% to 48.1%.

Natural gas supply reached 39 in 1990, an increase of 6% compared with 1989. This increase was mostly met by imports, which rose 8.1% to 25.3 . Suppliers were the former USSR (46%), Algeria (35%) and the Netherlands (19%). Domestic production rose 1.4%. The share of gas in electricity generation rose from 16.6% in 1989 to 18.3% in 1990.

Coal contributed 14.6 Mtoe to Italy's primary energy balance in 1990, increasing its share to 9.5% compared with 9% in 1989 and 7.4% in 1979. This increase was absorbed by the electricity generating sector and the industrial sector, essentially the steel and cement industry. Coal use in the residential/commercial sector remained stable at 0.1 Mtoe. Almost all this coal was imported, mainly from the United States (52%) and South Africa (24%). Remaining imports were distributed among Australia, Poland, Germany, the former USSR, China and Colombia.

In 1990, about 8.4 Mtoe of renewable forms of energy was produced in Italy, roughly 5.4% of TPES. About 92% came from hydropower and almost all the remainder from geothermal energy, wood and other forms of biomass. Substantial increases in the contribution of renewables are planned by 2000 and government support is being provided to help achieve this aim.

Regarding electricity, in 1991 the total demand of electricity in italy reached 235 billion kWh, with a per capita demand of 4170 kWh per person, one of the lowest in OECD countries. (OECD average was about 8000 kWh). Electric energy penetration was 35 per cent and is expected to rise to 38 per cent in the year 2000.

The programme for increases in generation capacity highlights the growing role of repowering combined cycle units, as well as multifuel power stations. Following a referendum on nuclear power in 1988, it was decided that nuclear power would be developed only when inherently safe reactors became available. As a result, ENEL lost 3280 MW from nuclear power plants in operation or under construction, and a further 2000 MW of nuclear capacity due to come on-stream in 1997. Multifuel capacity should be providing about 20000 MW of capacity by 2000, overtaking the contribution of oil-fired stations. Coal use is expected to increase.

With demand increasing for natural gas, both nationally and internationally, as an environmentally favourable and convenient resource, sufficient quantities of gas may not be available at the right price to enable all plans for new gas-fuelled plants to be realised. Moreover, while procedures for securing approval for the siting of power stations have been streamlined, local opposition to the new coal-fired facilities must cast doubt on the practicability of reaching targets for coal-based generation.

The role of natural gas in electricity generation is expected to expand significantly over the next decade.

The situation in the electricity generation sector remains difficult. In 1990, it was necessary to import 15% of electricity consumed and to rely heavily on oil-fired capacity. Plans to reduce dependence on these sources focus on three main areas: a substantial increase in gas-fuelled generation; continued coal use, albeit in multifuelled power stations; and a greater contribution from independent generation, particularly from renewable energy sources and co-generation. However, these policies entail certain risks.

Independent production, excluding municipal generation, represented 13% of total electricity produced in Italy in 1990. The amount of electricity produced reached 26.3 twh and production capacity was 7700 MW. Independent production covered 18% of industrial electricity demand. Law 9/91 introduced in January 1991 provides independent producers with a financial incentive to sell electricity, but it does not deregulate independent production, as ENEL remains the sole purchaser and prices are set by the government. The main rationale for this law is to encourage improved energy efficiency and increase generation capacity. The effect on independent production is likely to be significant, as such production has become a commercial venture and industries have announced plans to increase their electricity generation significantly.

District heating has expanded considerably, with heated volume more than trebling between 1981 and 1989. In 1991, the total heated volume reached 44.3 million cubic metres. Co-generation associated with heating rose by 15% in 1990, reaching 725 GWh.

Notwithstanding Italy's low electric energy intensity (1990 intensity was only 0.250 kWh per each US dollar of GNP compared with an OECD average of 0.470), improved energy efficiency remains a key objective of the revised PEN to reduce both the nation's import dependence and the environmental impact of energy activities.

Energy taxation remains high and though this probably contributed to the low energy intensity of the economy, its unevenness among the various energy sectors may need to be adjusted.

The transport sector is largely absent from the current strategy to promote energy savings. ENEA has been studying the effect of improved traffic management and the use of alternative fuels, particularly for urban road transport. Nevertheless, the fuel consumption benefits of improvements in the fuel efficiency of passenger cars have been offset by large increases in traffic. If these trends continue, as seems likely, the growing oil consumption of the road transport sector will increase.

The development of electric generation capacity up to the year 2000 (figure 10) has been defined: 1,500 MW will be provided by 5 combined cycle units of 300 MW each; 1300 MW will derive from repowering of existing steam units; 200 MW will be provided by conventional gas turbine units.

On the site previously assigned to the Montalto di Castro nuclear power station, a multifuel (natural gas-oil) power station is now under construction. It consists of 4 steam generating units of 627 MW each, repowered with gas turbines for a further installed capacity of 800 MW.

A "Medium Term Programme" for 5 000 MW consists of: 340 MW combined-cycle plant fed by a coal-gasification system and using the Sulcis national coal; 1 500 MW will be due to repowering (800 MW) or conversion into combined-cycle of old steam units (700 MW); 1 800 MW of coal, water, fuel oil, natural gas plants (300 MW units); 1 200 MW of new combined-cycles plants; two gas turbine units in Sardinia (200 MW).

In July 1990 ENEL has reached a preliminary agreement with some large industrial self-producers which should allow for a substantial increase of the electricity generated on behalf of ENEL, through repowering or installation of combined-cycle units in existing industrial factories.

The programme of new plants in operation up to 2000 includes also following efforts: completion of "multifuel plants" under construction totalling 6 400 MW; new hydro plants totalling 2 600 MW (of which 1 750 is pumped storage); new geothermal plants for about 800 MW and a corresponding tripling of the current production (from 3.2 TWh in 1990 to 9.1 TWh in 2000); gas turbines totalling 400 MW (200 MW of which still to be authorised); still unsited multifuel plants in Sardinia (900 MW) and Sicily (1 200 MW of which 300 MW of repowering plants); out of a wider programme of using new sources, two wind-farms (2x10 MW) and photovoltaic plants (3 MW) are under construction.

Furthermore, to face the uncertainties affecting both the "reference" demand and the timely completion of all the planned plants, ENEL planned a 3 000 MW of additional capacity (1 200 MW of repowering plants, 1 800 MW of combined-cycle plants) whose contribution in 2000 has been taken into account only for the "high" scenario.

As for decommissioning, in Italy the growing difficulties in finding new sites and the need of improving the overall efficiency of the existing equipments, insuring at the same time very high environmental standards, has lead to plan for an extensive revamping of the existing fuel-oil units (increase of their operating life of at least 15 years).

If the above programme is implemented on schedule, in 2000 it will provide a maximum net capacity of 80 600 MW.

The expected new commissionings and the guaranteed capacity from abroad under already defined contracts will lead, at the year 2000, to a capacity surplus of about 3.8 GW in the "reference" scenario of a foreseen electric demand, for the year 2000 of 315 billion kWh.

In term of environment control in Italy, the environmental legislation established strict technical rules for assessing environmental compatibility of plants, by specifying the percentages of abatement of pollutant emissions to be reached in the coming years. Italy signed the Helsinki protocol of abatement of SO_2 emissions and the Sophia protocol of abatement of NO_x emissions. Furthermore, the national legislation was adapted to the EC directives and in some cases it provided for much stricter limits.

To comply with this, ENEL planned a wide range of actions on new and existing plants. In particular, measures were scheduled on plants (flue gas desulphurization, denitrifacition, low NO_x burners, high-efficiency electrostatic precipitators) and on modes of operation (use of natural gas and oil with low and extra-low sulphur content.

Completion of the environmental actions on schedule will radically curb pollutants emissions. In spite of 41% increase in thermal generation of electricity in the "reference" scenario (53% in the "high" scenario), sulphur oxides will decrease by 65% (61% in the "high" scenario) and nitrogen ones by 38% (33% in the "high" scenario) in 2000 vs. 1990.

Finally, a still unresolved problem is the build up of CO_2 in the atmosphere. Energy savings and the anticipated greater use of natural will certainly contribute to hold down CO_2 emissions. Nevertheless, since the emission-free nuclear plants have not been, so far, considered for electricity generation in italy, CO_2 emissions will increase by 30 percent from 1990 to 2000.

3. CONCLUSIONS

To complete this panorama on the energy situation in Italy and in particular on electric energy, let me report the general recommendations made in the IEA Report on Italy energy policies.

I quote from page 202:

". Further strengthen co-ordination between relevant national administrations (in particular the ministeries of industry, of environment, of transport and of scientific research) and agencies, and state-owned utilities to ensure that effective energy policies integrating environmental considerations are developed and implemented;

. Carry out a full assessment of the policy measures needed to meet environmental committments and develop a comprehensive strategy (with full interministerial co-ordination) to limit greenhouse gas emissions that would take into account the cost-effectiveness of measures likely to help limit CO_2 emission levels, including those applied to the transport sector;

. Continue to pursue pricing policies that lead to greater equivalence between prices and costs and that help stimulate new supplies economically;

. Ensure that the incidence of energy taxation for different products and end-uses remains consistent with the achievement of energy policy and environmental goals."

To complete this presentation, let me add some personal remarks.

Italy has practically no domestic fuel and therefore must rely on importing to satisfy its needs. This represents a great risk both in term of cost and also of fuel availability.

We should therefore look again very seriously at the nuclear option. This option can be considered only in the frame of a new european nuclear generation programme.

I am confident that in the near future the nuclear production of electricity may find a new positive momentum both in Europe and also in the USA. It is undoubtedly the best option also for the so called "Sustainable Development".

In Europe we have had in the last months a series of brainstorming meeting that produced a report which I consider most interesting.

Let me quote the main Comments and the Recommendations:

" The group of experts, convened by the Commission, reviewed the energy situation in general and reaffirmed that the protection of the environment must be a main driving force in energy policy. The CEC should contribute to place different energy sources in their right perspective, taking account of technical as well as of economical aspects.

The community has presently the most important park of nuclear installations in the world which is supplying about one third of its electricity. The group emphasised the maturity of the community's nuclear industry, the efficiency of its safety authorities and the excellent safety records of operating nuclear plants and invited the Commission to search ways to maintain a competitive and competent european nuclear industry.

Among fossil energies only coal is available in quantities that can satisfy the longer term energy demand. Coal could compete with nuclear energy but the increasing concern about potential climatic changes due to emissions, in particular CO_2, might set limits to its future deployment.

Preserving the nuclear option implies a continuous technological development to demonstrate and improve its safety in the broadest sense, taking into account the continuously growing demand on safety in all technological areas. A coherent approach on a community level to outstanding problems could help to regain public acceptance of nuclear energy.

Future research should include areas such as a safety of current plants, radioactive waste management, radiation protection and safety features of new reactor design.

A new, large PWR, based on a evolutionary concept is presently developed by the industry of some European countries. It has safety characteristics which go far beyond current practice and which include the use of a containment concept which can withstand a severe reactor accident. These safety characteristics, which are unique, should be supplemented by an adequate and focussed Community Safety Research Programme.

There have been a number of developments in the field of nuclear energy in recent years, these include the delay in the need for the introduction of fast breeder reactors, the existence of large quantities of plutonium not needed for weapons, the situation relating to safety culture in Eastern and Central European Countries (PECO) and the CIS, the continuing difficulty in agreeing how to deal with long-lived nuclear waste and the proliferation of nuclear weapons."

In the light of these, the experts made the folllowing General Recommendations:

"The Community should:

. Reaffirm the place of nuclear energy in an overall energy strategy;

. Implement a strong and coherent R&D programme on nuclear fission safety with the objective to support further development and to reach a common understanding of the problems involved;

. Contribute to the preservation of know-how about reactor types, which might be needed for the longer term;

. Cooperate with Eastern and Central European Countries and the CIS in spocific R&D fields and assist them in solving particular problems.

Future Community programmes should be further integrated with programmes performed in the member states. Existing networks between research organizations, operators, industry and authorities should be reinforced and enlarged."

Energy issues are too important and too complicated to be solved by a single country and must therefore be faced by the widest international cooperation. Electric energy must be promoted for its intrinsic advantages both for the development of the industrialized countries and also for all the others which are still in the initial development phase.

BIBLIOGRAPHY

1) Euroelectric Report april 1992

2) ENEL Plans for 1991-93, October 1990

3) Energy Policies of IEA Countries - IEA Report May 1991

4) Plant Technologies by B. Musso - Paper presented at Pisa 2/12/1992

5) Global Development, Energy and Environment - by Sergio Barabaschi - Introductory lecture to the Conference. "50 Years from the Fermi Pile", Pisa - Italy, 2/3 December 1992

6) Energy Issues after the Rio Conference, by Sergio Barabaschi - Roma June 1992

7) Potential World Development and Impacts on Energy Demand, Resources and Environment, by S.W. Gouse, D. Gray, G.C. Tomlinson, D.L. Morrison - 1992 the Mitre Corporation

KEY ISSUES ON ENERGY

1) Local needs/Global needs (environm., fuel cost, technologies)

2) Total energy/Electrical energy (electrotech., cogeneration)

3) Generation of electrical energy (operating plants, new and future plants)

4) Transmission and Distribution (F.A.C.T.S., active filtering, storage)

5) Optimization of electrical energy use

6) New electrotechnologies

7) Environmental problems and future needs

PROJECTED ENERGY USE WITH ASSUMED EFFICIENCY IMPROVEMENTS FOR COUNTRIES GROUPED BY ENERGY USE PER CAPITA IN 1988

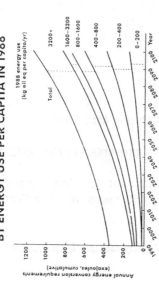

EVOLUTION OF ENERGY IMPORTS (1963–1988)

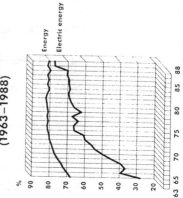

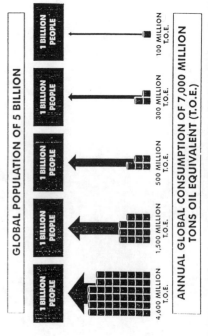

GLOBAL POPULATION OF 5 BILLION

ANNUAL GLOBAL CONSUMPTION OF 7,000 MILLION TONS OIL EQUIVALENT (T.O.E.)

244

GENERAL RECOMMENDATIONS
The Community should:

- reaffirm the place of nuclear energy in an overall energy strategy;

- implement a strong and coherent R&D programme on nuclear fission safety with the objective to support further development and to reach a common understanding of the problems involved;

- contribute to the preservation of know-how about reactor types, which might be needed for the longer term;

- cooperate with Eastern and Central European countries and the CIS in specific R&D fields and assist them in solving particular problems.

ADVANTAGES OF ELECTRIC ENERGY PROMOTE ITS EVER INCREASING PENETRATION

- *Converts easily into mechanical energy*
  (Less fatigue for man–kind)

- *Improves urban environment*
  (Electric transportation)

- *Increases industrial competitiveness*
  (Automation, robotics, electrotechnologies)

- *Helps saving total energy*
  (Microwaves, osmotic membranes, heat pumps)

- *Betters products quality and permits recycling*

AVAILABLE ENEL POWER (MWE) 1990

Geothermal (437)
Turbogas (1,691)
Oil fired (21,346)
Multi fuel plants (6,943)
Hydro (12,787)

AVAILABLE ENEL POWER (MWE) 2000

Turbogas (2,148)
Oil fired (18,850)
Multi–fuel plants (19,884)
Geothermal (1,118)
Hydro (15,422)
Repowering and combined cycle (7,407)

1991
PER CAPITA ELECTRIC ENERGY DEMAND
(kWh)

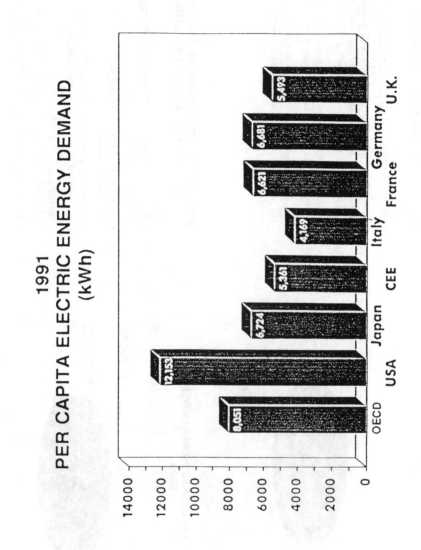

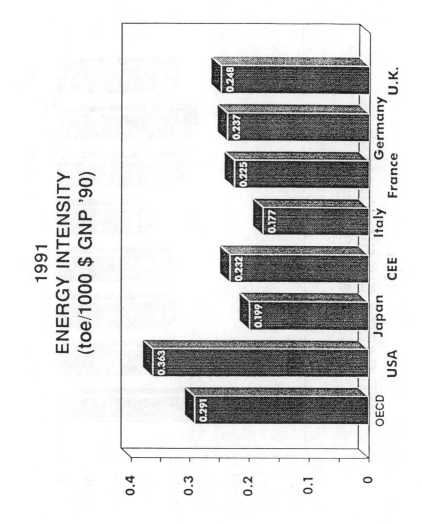

1991
ENERGY INTENSITY
(toe/1000 $ GNP '90)

OECD	0.291
USA	0.363
Japan	0.199
CEE	0.232
Italy	0.177
France	0.225
Germany	0.237
U.K.	0.248

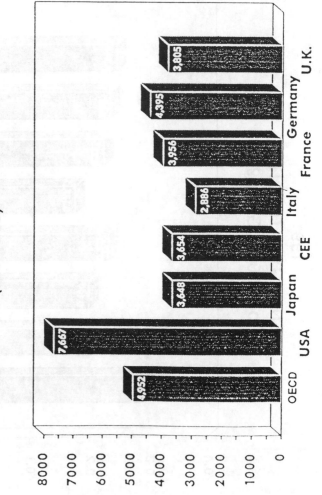

1991
PER CAPITA ENERGY DEMAND
(toe/1000)

OECD — 4,952
USA — 7,657
Japan — 3,648
CEE — 3,654
Italy — 2,886
France — 3,956
Germany — 4,395
U.K. — 3,805

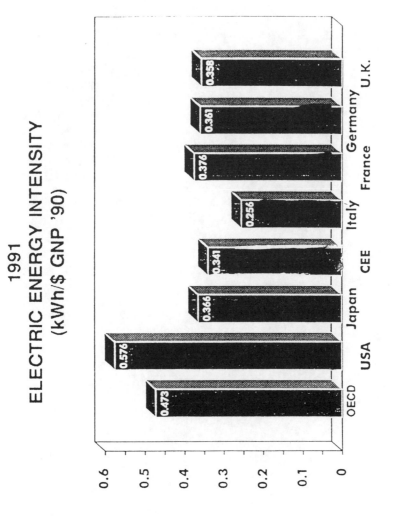

1991
ELECTRIC ENERGY INTENSITY
(kWh/$ GNP '90)

LOW ENVIRONMENTAL IMPACT: SWISS POWER GENERATION

dipl. Ing. ETH Kurt Küffer
Director NOK

Abstract

In the **years prior to 1990** the share of fossil fuels for power generation has not exceeded 5 percent of the total generation in Switzerland. The dominance of hydroelectric power and nuclear power led to the country's reputation for very clean means of electricity production compared to many other countries. **In the nineties** the increase of power demand in winter will be met mainly by imports, as a moratorium prevents licencing of new nuclear power plants untill the year 2000. In case of increased imports of fossil based electricity the external costs for power generation are rising significantly. - Although all options for electricity production are open **after the year 2000**, high population density, growing environmental concern and a landscape worth protecting require particularly clean power generating processes. A balanced combination of domestic generation on the basis of nuclear energy and gas with oil as storage reserve and import of electricity may be the solution for Switzerland to cover the future demand.

LA PRODUCTION D'ÉNERGIE EN SUISSE DANS LE RESPECT DE L'ENVIRONNEMENT

Résumé

Avant 1990, la part des combustibles fossiles représentait à peine cinq pour cent de la production totale d'énergie de la Suisse. Comparée à beaucoup d'autres pays, la Suisse avait la réputation d'utiliser des modes de production d'électricité très propres puisqu'elle privilégiait l'énergie hydroélectrique et nucléaire. **Dans les années 90,** l'augmentation de la demande d'énergie en hiver a été principalement couverte par des importations, un moratoire bloquant l'octroi d'autorisations pour de nouvelles installations nucléaires jusqu'en l'an 2000. Dans le cas d'importations accrues d'électricité produite à partir de combustibles fossiles, les coûts externes de la production d'électricité augmentent notablement. Bien que tous les modes de production d'électricité soient envisageables **à partir de l'an 2000,** la forte densité de population, les préoccupations croissantes d'environnement et le souci du paysage qui mérite d'être préservé font qu'il faudra choisir des modes de production particulièrement propres. Pour faire face à la future demande, la Suisse pourrait avoir recours à une solution conjuguant de façon équilibrée sa production nationale à partir de l'énergie nucléaire et du gaz avec l'utilisation de réserves de pétrole stocké et avec des importations d'électricité.

1. Swiss power generation

1.1 Historical development in rough strokes.

Compared to many other countries, Switzerland has earned a reputation for its very clean power generation. In the past 30 years the share of fossil fuels for power generation has not exceeded 5 % and has maintained an average of only 2 %. Before 1960 the circumstances were even clearer: hydro-power, virtually 100 %!

The dominance of hydroelectric generation is threefold:

- Switzerland is a mountainous country and rich of water;
- low interest rates over an extended period of time supported the construction of a park of hydroelectric power stations
- up to date no substantial fossil or nuclear fuel resources of economic value have been identified in Switzerland. The search for oil is still ongoing.

When it became clear in the sixties that the extension of traditional hydro-power alone could not meet the fast growing demand for electricity, nuclear energy was the choice for building additional power plants despite the lack of domestic nuclear fuel resources. The decisions to build the Beznau Nuclear Power Plants - two 350 Megawatt (MW_e) pressurised water reactors - as well as Mühleberg - a 320 MW_e boiling water reactor- were taken in 1965 and 1967 at a time when complete projects for oil-fired power plants were available. The following judgement tipped the balance in favour of the new technology:

- Nuclear power has a favourable cost structure. Given a plant running at base load - 80 percent load factor -, with an investment of $ 660,000 per MW_e, power generation costs of 19 mills per kilowatthour (mills/kWh) could be expected. The real generating costs turned out to be even more favourable in the first production years.

- Growing environmental concerns over the burning of fossil fuels resulted in strong opposition against fossil power plants.

- The dependence on foreign countries for the procurement of nuclear fuel and the problem of stockpiling was considered to be more easily dealt with. This is not surprising if one keeps in mind that in 1965 75 % of total primary energy requirements were covered by oil from abroad. The known oil reserves were at that time even more unilaterally located than today. In comparison stockpiles of uranium fuel for a supply up to two years were easy to build up.

- In terms of supply security, nuclear power ideally complements hydro-power. Nuclear power plants can be shut down for refuelling when electricity is in low demand in summer and hydroelectric production is at its peak.

1.2   100 years hydro-power - 25 years nuclear power

The strategy chosen for the build-up of power plants in Switzerland has been very successful. Run-of-river power stations and nuclear power plants generate base load energy. High pressure hydroelectric power stations with their reservoirs cover the peak demand. A few fossil fuel power plants serving as reserve power plants came and come into action very seldom (Picture 1).

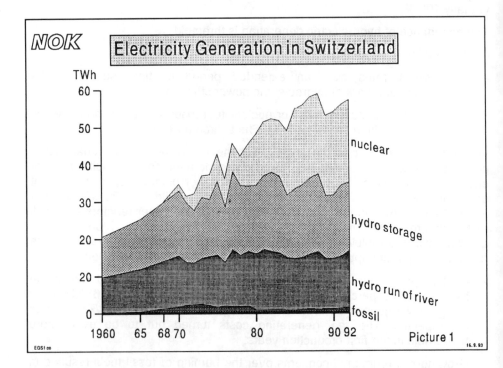

Picture 1

Plant performance of the Swiss nuclear power plants has been as good as or better than expected. During the 90 years of accumulated reactor operating experience in Switzerland, no plant worker or member of the general public has been exposed above legal limits. In addition radioactive releases to the environment have always been well within authorized limits.

The cumulative load factor of all Swiss nuclear power plants over their whole lifetime is in excess of 80 % and, compares favorably with the global average of below 70 %. The original expectation of 6500 to 7000 full power hours per year has been more than met (Picture 2). Development of power generating costs shows high continuity.

NOK

The electricity prices in constant money terms even have been reduced over the last decades.

1.3   Shadow on the future nuclear development

Fourteen years have passed since the pressurized water reactor in Gösgen (940 MW_e) and nine years since the last nuclear power plant - the boiling water reactor in Leibstadt (990 MW_e) - have been put into operation. For the past several years their production have been fully used to meet domestic electricity consumption steadily growing.

Two additional nuclear power plants to meet this demand could not be built for political reasons: Kaiseraugst (1000 MW_e) and Graben (1200 MW_e). Domestic hydro-power is already exploited almost to its limits and fossil fueled power plants had no chance for environmental and economic reasons. As a consequence Switzerland had to contract foreign electricity and "to build two nuclear power plants abroad".

In the years 1979, 1984 and 1990 Swiss voters rejected three times Federal initiatives to ban nuclear power.

The voters did however accept in 1990 an initiative, by a very small margin, which provides for a 10 year moratorium on licensing of new nuclear power plants. As a consequence no licensing process for new plants can be initiated up to the year 2000.

The past power generation program has earned Switzerland the reputation of being one of the cleanest power generators. The remaining years of the nineties will decide whether we succeed in further developing this route.

2. Meeting growing demand in the nineties:

2.1 Prospects of electricity demand

To figure out the electricity demand for the next decade, the leading Swiss utilities have assumed in their calculations an annual growth of 2 % in the Gross National Product up to the year 2005. This expectation is cautious-optimistic, but lies much under the possible growth rates stipulated by economists. It assumes that future appliances will use less energy and that it will be possible to mobilise higher awareness of the consumer for energy conservation through better information and motivation resulting in considerable savings. Nevertheless, a reasonable increase in consumption is expected specially in the industry and services sectors which will more than compensate for the savings already mentioned.

Strong advances in electrification in the services and residential sectors have led to higher electricity consumption. Switzerland has one of the highest computer densities (computers per capita) in the world and is likely to maintain this to remain competitive. As a result of scarce and expensive labour, mechanisation and automation in manufacturing are increasing.

Based on all the estimates from different consumer sectors and including anticipated savings through more efficient use, growth in electricity demand of about 2.2 % per year until 2005 is expected, if not further restructuration with phase out or relocation of industrial production to foreign countries is going to reduce the demand (Picture 3).

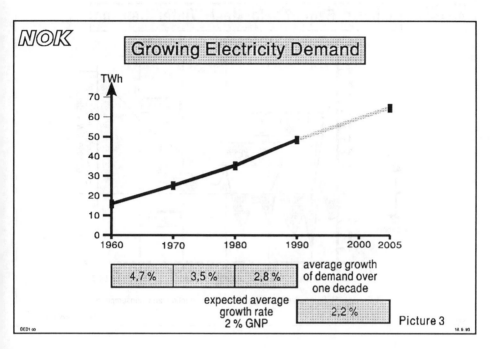

Picture 3

A smaller estimate of growth would be irresponsible. Currently a tough recession has temporarily stopped the growth of electricity demand. But past experience has shown that recessive phases are compensated later. During the last 50 years, consumption has increased in every decade independent of economic recessions.

2.2 Limited possibilities for electricity procurement

Electricity procurement must follow the seasonal nature of consumption. Consumption in winter is generally 15 % above consumption in summer, as a result of distinct seasonal differences and economic activities (e.g. tourism). About half of the demand in winter can at present be satisfied by hydroelectric power stations; the other half can no longer be supplied by Swiss nuclear power plants alone.

As explained earlier the licensing for new nuclear power plants cannot be started until at least the year 2000. For this reason and as a result of lower hydroelectric generation during the winter months, Switzerland has become a substantial net importer of electricity in winter. We forecast the current shortfall of 1...2 TWh to grow to 9 TWh by the year 2005 (Picture 4).

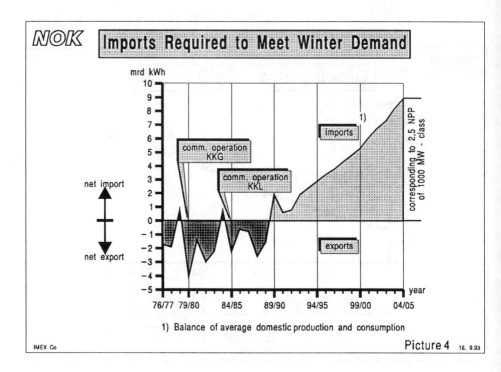

NOK — Imports Required to Meet Winter Demand

mrd kWh

net import / net export

comm. operation KKG

comm. operation KKL

imports 1)

exports

corresponding to 2,5 NPP of 1000 MW - class

year

76/77 79/80 84/85 89/90 94/95 99/00 04/05

1) Balance of average domestic production and consumption

IMEX Co

Picture 4 16. 9.93

Because of this reason, various Swiss utilities are holding long term contracts for electricity from nuclear power plants in France. The contractual capacity will reach 2650_e MW during the present decade. This means that Swiss utilities will have in France a production capacity similar to the current domestic nuclear capacity. These contracts will close the gap between domestic production and consumption during the time period up to the year 2005.

The electricity supply in summer is secure for a longer time as the generation of hydroelectric power stations is high and the demand rather low. Surplus energy is either exported or used in pumped storage power stations to support peak power demand.

2.3   Energy policy and activities in the nineties

Political events have cast their shadows on the energy scenery of Switzerland. An article dealing with energy policy has been included in the constitution with a referendum in 1990. The people have clearly expressed their wish to have sufficient, secure and diverse sources of economic energy, which are ecologically compatible. Further demands for energy conservation as well as for home based and renewable energies are also included. At the same time phase out of nuclear energy was rejected and a 10 year moratorium on licensing of new nuclear plants, as mentioned earlier, was accepted. As a result, the federal government has established a program called "Energy 2000".

This program intends to mobilise forces at all levels of government, business, as well as the private sector to concentrate on the common objectives given below:

° The total consumption of fossil energy and CO_2-emission should at least be stabilized by the year 2000 and reduced thereafter.

° The growth of electricity consumption should be reduced and the consumption stabilized from 2000 onwards. The main impact of the program should be in the achievement of savings, through more efficient and concious use of all types of energy and in particular, that of electricity.

° On the production side, the contribution of renewable energies should be increased considerably; home based electricity generation should be increased by 7.5 % in the next 10 years: 4 % should be achieved through the uprating of the 5 existing nuclear plants, 3 % through expansion of hydroelectric power generation, and 0.5 % through new renewable energy sources.

As far as possible, Swiss utilities are ready to support the above program in a constructive way.

Also last year a water protection law was accepted by public vote, which stipulates an increased amount of residual flow from hydroelectric power plants. This will reduce the production of electricity in the long run and thereby will accentuate the production shortage during winter months.

2.4   Cost situation

Existing hydroelectric power plants, the majority of which are decades old, produce electricity predominantly at very low production costs. These lie between 13 and 130 mills/kWh, the average being aproximatly 45 mills/kWh. The advantageous cost level is the result of a favourable cost structure of hydroelectric plants, where capital costs are much higher than operating costs. The long term advantage in the economy of production therefore lies in the fact that only the operating costs, that means only a small part of the total costs are affected by inflation.

Production costs are less favourable for new hydroelectric plants for the following reasons:

- The best sites are already used. New sites in general have less availability of water and head difference.

- Possible projects are slowed down because of complicated licencing procedures, environmental concerns, and increased expenses (e.g. residual water flow according to the new water protection law).

Therefore future projects generally tend to be more expensive. In addition, the risks associated with the costs due to increased environmental standards and licensing difficulties have made decisions to build new plants extremely difficult. The production costs of electricity in recently completed high pressure hydro plants were between 100 and 130 mills/kWh. Production costs for run of river projects are at 67 to 120 mills/kWh. After a depreciation period of a few decades, the production costs tend to decrease in real terms. From this point of view and because of ecological reasons (clean energy), such power plants should be built provided the licenses are granted.

The cost development at the five existing nuclear power stations is found to be satisfactory. Electricity production costs lie between 40 and 60 mills/kWh. This favourable level of cost is due to the high quality standards in plant desing, construction and operation as well as due to traditional conciousness of industrial labour regarding reliability and quality. The load factor of the five nuclear plants currently is above 83 %. The three oldest plants - Beznau I, 1969; Beznau II, 1971 and Mühleberg, 1971 - have invested heavily in upgrading and backfits. This has brought produciton costs to 53 mills/kWh which is still below the 60 mills/kWh for the newest plant Leibstadt (1984). The Gösgen plant (1979) with 40 mills/kWh has the most economical production.

The production costs for our nuclear power contracted in France amount to 40...50 mills/kWh.

3. Options after 2000 considering externalities

3.1 The future of nuclear energy in Switzerland

After the year 2000 the five Swiss nuclear power stations will still be in operation, based on their design lifetime of 40 years and their potential for life extension. Hence nuclear energy will still play a major role in the Swiss "energy mix".

New nuclear power plants, however, will be built only if

- nuclear power plants presently in operation have to be replaced and/or the **demand** is still increasing as predicted,

- the generating costs remain **competitive** within the European market

- the **licensing procedure** can be shortened (with the present procedure it would take 16...18 years from selection of site to connection to the grid),

- **public acceptance** is reestablished.

  Other key factors for the future of nuclear energy in Switzerland are maintenance and evidence of its high standards and excellent records concerning safety and environmental impact, resulting in low external costs. These considerations are addressed more detailed below.

3.2 Reactor safety

- How much safety is **necessary**?

  According to the Swiss Nuclear Law (paragraph 10) the utilities must take all safety measures **necessary** following the **state of the art** based on operating experience to protect human life and properties protected by law (including the environment).

- How much safety is **reasonably achievable**?

  The Nuclear Law requests measures to protect human life and properties protected by law only as far as they are both **neccesary** and **justifiable** (from a cost/benefit stand point; paragraphs 5 and 7). In other words nuclear power is given a chance both to be safe and to remain competitive.

However it is very hard to quantify this legal requirement into specifications to be followed by engineers and adaptable to operating power plants. This process is a newer drying up source for technical and political discussions and controversies.

Based on operating experience and plant specific, probabilistic risk assessment studies, measures for enhanced safety have been taken.

For the backfitting of the older power plants Beznau and Mühleberg, investments up to the level of the original investment costs of the power plants were made. The major areas of upgrading were improved protection against external events as well as improvements in emergency core cooling and removal of residual heat from the reactor. The licensing review process for the upgraded plants is completed in the case of Mühleberg and for the Beznau Plants well advanced.

The newer plants Gösgen and Leibstadt, were built according to tight standards similar to those used for licensing in Germany. Therefore only little backfitting was necessary.

All nuclear power plants will soon have containment venting systems.

As a consequence, probabilistic risk assessment studies have shown that for all Swiss nuclear power plants a core melt should occur less than once in 100,000 reactor-years per reactor. This is a factor of 10 better than the safety goal of the International Nuclear Safety Advisory Group (INSAG) of the International Atomic Energy Agency (IAEA) for existing reactors.

3.3   Safety goals for the next generation of nuclear power plants

The IAEA has proposed safety goals for future reactors such that serious nuclear damage must occur less frequently than once in 100,000 reactor-years. The frequency of a severe accident with both core damage and large radioactive release to the environment must be at least ten times smaller.

These safety goals are already achieved for the existing plants in Switzerland. If the 500 commercialy operating reactors were to just meet these IAEA-safety-goals stipulated for the next generation of plants, a severe nuclear accident without substantial impact on the surrounding neigbourhood could occur every 200 years worldwide; a severe accident with substantial release of radio-activity could occur every 2000 years. Compared to other risks we are exposed to, this nuclear risk should be acceptable. However a number of older nuclear power plants and most of the reactors of the former COMECON-countries can not be brought up to this standard.

Nevertheless, tighter safety goals are being discussed internationally and in Switzerland, to:

- reduce the impact of human error,

- increase reliance on passive safety principles and safety systems and

- **limit the consequences of severe accidents to the extent that emergency measures are no longer necessary outside the nuclear power plant.**

This last goal calls not only for a new level of reactor safety, which takes us a step closer to the ideal of absolute reactor safety, but also for a political decision to abstain from existing emergency planing. In the view of the Swiss utilities, this is an unrealistic goal: public acceptance of nuclear power generation will not increase even with increased reactor safety. There will always remain a residual risk which can be misused for public disinformation.

However, we do support research mainly in the field of evolutionary advanced reactors aimed at the development of simpler and therefore safer concepts.

3.4 Future nuclear power plants and Swiss preferences

In line with the above considerations, the next nuclear power plant to be built in Switzerland after the year 2000 should be a plant of the **"evolutionary"** type. The Advanced Pressurized Water Reactor (APWR) by Westinghouse-Mitsubishi, the Advanced Boiling Water Reactor (ABWR) by General Electric, the System 80$^+$ by ABB-Combustion Engineering, the advanced Boiling Water Reactor 90 (BWR 90) by ABB-Atom or the European Reactor (EPR) by Framatome / Siemens are examples for such a plant.

These reactors have a number of common features:

- Capacity between 1200 and 1400 MWe,

- No need for a prototype,

- Similarities with the latest built reactors with known characteristics like the convoy-plants from Siemens or the N4-concept from Framatome. These reactors have, in particular, conventional safety systems without innovative passive elements.

One or more new plants of the **"evolutionary-passive"** type, like the AP-600 from Westinghouse or the Simplified Boiling Water Reactor from General Electric, could also meet Swiss requirements because of their smaller capacity of 600 MWe and because of their improved safety design basis (systematic use of passive safety systems and inherent-safety -oriented characteristics).

However the increased space needed to accomodate the passive safety systems as well as higher costs of these systems could be a problem for their implementation in Switzerland.

Public acceptance of any new nuclear power plant will largely depend on our ability to convince the public that nuclear power will be needed to meet not only the increased demand for electricity but also ecological goals. It must become general knowledge that conservation of energy plus exploitation of renewable energies are insufficient to replace the need for new nuclear power plants. Also we must demonstrate that we are capable of handling and managing all wastes generated during the exploitation of nuclear power including storage in final repositories. Further improvement of reactor safety is an issue of secondary priority when compared to these issues.

The following generation of reactors, sometimes called **"revolutionary"** or **"innovative"** reactors like the Process-Inherent-Ultimate Safety reactor (PIUS) of ABB-Atom, the English pressurised water Safe Integral Reactor (SIR), the

American sodium-cooled Integral Fast Reactor (IFR) or the High Temperature Reactor (HTR) approach this objective of almost total safety. Commercially these plants will only be available well after the year 2000 if at all and therefore are presently of rather small interest in Switzerland.

3.5 CO_2-balance of electricity generation

As for other technologies, various alternatives have also been compared for new generation of electricity by assessing the ecological consequences. Applying these methods, called eco-balancing, to the Swiss power generation, it is obvious, that the nuclear energy contribution to the CO_2 balance is negligible whilst its contribution to power generation is significant all year long and dominant in winter.

The two last conferences of the World Energy Council (WEC) in the years 1989 and 1992 have shown clearly that global energy consumption will rise by a factor of 2 or 3 during the next 50 years. The major concern with this increase is **not** the availability of energy resources but the consequences of increased releases of greenhouse gases to the atmosphere, namely CO_2.

In 2020 the energy consumption will be 40 % higher than in 1990. In the same time, population will have grown from 5.3 billion to 7.8 billion inhabitants of our planet and in parallel the energy consumption per capita will have risen especially in the developing countries.

Another fact also has to be mentioned. Worldwide the actual energy mix, 76 % fossil energy in 1990, will hardly change in the future: Over 70 % will continue to be covered by fossil fuels! As a consequence, the emission of CO_2 will rise by about 40 % for the next 30 years. That is why ecological balances become more important in order to assess more precisely the external effects. Once known and accepted by the public, politicians and the industry, these effects can be converted into dollars and cents and included in the costs of every step of industrial production.

Since electricity moves a modern society, its development is of crucial importance. Therefore the energy mix for generating electricity is of prime interest. In the case of Swiss electricity, 92 % of the total power generation is free of CO_2-emission: 49 % nuclear and, 43 % hydroelectric.

The contribution of 8 % to greenhouse gases results from power imports, which are properly taken into account in accordance with the energy-mix of the 12 countries associated in the Union for the Coordination of Production and Transmission of Electricity (UCPTE, Picture 5) with 46 % fossil, 38 % nuclear and 16 % hydroelectric power generation.

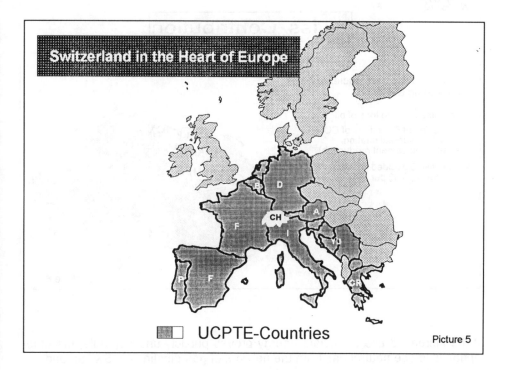

UCPTE-Countries

Picture 5

However, Switzerland is located in the middle of Europe and therefore developments in the neigboring countries need to be taken into account. In general fossil fuels have been chosen to meet the bulk of the increasing demand of electricity. France is a remarkable exception from this pattern: The forced substitution of oil by nuclear energy after the oil crisis in the seventies has resulted in a 30 percent reduction in total CO_2 emissions to the atmosphere in France during the past 20 years.

What could be a Swiss contribution? Using the existing 5 nuclear power plants sited in thickly settled areas for district heating as well, burning of about 2 million tons of oil or 6 million tons of CO_2 could be avoided. This amount is equivalent to 15% of all CO_2 emissions from residential heating, traffic and cement-production. Applying a disposal cost of $ 33 per t CO_2 (Seifritz 1992, ref. 1), this translates to about $ 200 million per year of avoided cost of disposal (Picture 6).

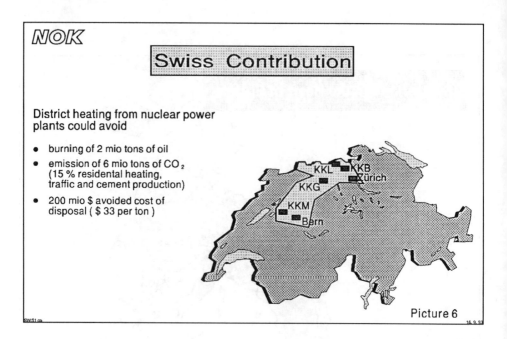

Swiss Contribution

District heating from nuclear power plants could avoid

- burning of 2 mio tons of oil
- emission of 6 mio tons of CO_2 (15 % residental heating, traffic and cement production)
- 200 mio $ avoided cost of disposal ($ 33 per ton)

KKL KKB Zürich KKG KKM Bern

Picture 6

If the increased use of nuclear energy proves publicly unacceptable, the other option is to use natural gas from the network of gas-pipelines in Switzerland.

Because CH_4 is a more effective greenhouse gas compared to CO_2, however, the total leak rate during extraction and distribution needs to be lower than 1% for natural gas to be ecologically viable.

3.6   Health aspects of the electricity generation

In trying to improve the precision of cost estimates for future production as a guide in choosing the right production technology, the concept of external costs becomes increasingly important. External costs may show up, for instance, as costs for health care. For that reason a comparison was made looking at the cost of health care as they result from generating power by burning coal and uranium, using the base data presented in the report "Medical Perspective on Nuclear Power" [ref. 2].   This is an interesting comparison as coal dominates the power generation in the US whilst uranium dominates the thermal power generation in Switzerland. The AMA report distinguishes between "illness and injury" and "mortality" as separate events. In order to calculate the frequency of those events as a consequence of power generation, the complete energy mix for both countries is taken into account.

The results are summarized in the following table 1.

The total events "illness and injuries" and "mortality" per Gigawatt-year (GWy) for different delimitation of the power plant combination are shown. The virtual part of the "imported" coal for Switzerland is calculated on the base of the coal part of 71,11 % on fossil fuels [Ref. 3]. From France no fossil generated electricity is imported.

	Production		Illness and injuries total events		Mortality total events	
	coal GWy	nuclear GWy	coal 337 events per GWy	nuclear 21,2 events per GWy	coal 21,6 events per GWy	nuclear 1,08 events per GWy
USA (1989)	172	64	57800	1352	3711	68
CH home based (1991)	0	2,5	0	52	0	3
CH with additional imported energy in 2005	0,1	4,2	34	89	2,2	4,5

Table 1 Total events "Illness and injuries" and "mortality"
        for both countries.

The second line of table 1 presents the results for the domestic Swiss production in 1991 only. The third line shows the approximate values, based on the energy-mix in Switzerland around the year 2005.

In the case of Switzerland with the additional imports it can be noted that the coal induced mortality is approximately half of that due the domestic and imported nuclear production.

The next table 2 reflects the relativ contribution of coal and nuclear fuel to the power generation mix in these countries.

	total electricity production GWy	Illness and injuries events per GWy of total production		Mortality events per GWy of total production	
		coal	nuclear	coal	nuclear
USA (1989)	338	172	4	11	0,2
CH (1990)	6,4	0	8,2	0	0,4
CH in 2005	8,4	3,9	10,5	0,25	0,54

Table 2 Specific events pro GWy of total production in each country.

Up to now, we tried to estimate the events caused by the different fuel cycles. The next step is to assign to those events a correct economic value, otherwise they can't be internalised. Two methods are known: **The human capital-method** is based on the average total loss of annual income. Therefore a mortality can be represented by a loss of 20 years working time. For an estimated annual income of $ 39,300 the cost of one mortality amount to about $ 0.8 million.

The approach more commonly used in the US is based on the willingness to pay higher salaries for work with a higher risk. Using this approach the cost of one mortality could be estimated to approximately $ 4.0 million. Using this number, the internalized cost of mortalities for nuclear generated power is 0.21 to 0.24 mills/kWh. The share of power generated by coal would result in a cost of 0.12 mills/kWh without taking into account the other external costs by hydro, oil and gaz generated electricity. These numbers clearly suggest that for Switzerland the potential import of coal generated electricity in the future will rise social costs in a significant way.

In summarising the results of these studies for Switzerland in about 2005, 42% of the social and ecological impact relates to the imports of coal related electricity and 58 % to domestic nuclear production, neglecting the costs of all other fractions of the energy mix. The small amount of imported coal related electricity is responsable for a third of the events.

In any case, we see with no doubt, that Switzerland is importing "events". This number is still very small, the total cost not reaching 1 mill/kWh.

In general we can say that the domestic Swiss electricity generation, while still dominated by hydraulic power generation and nuclear, has a direct social and environmental impact much smaller than impacts from other industry or societal activities, for example traffic!

4. Cost structure of new power plants in Switzerland

4.1 New power plants in Switzerland

The construction of new power plants in Switzerland has effectively come to a standstill in recent years. Therefore it is quite difficult to give an exact cost structure for new power plants. Nevertheless the following table tries to provide a "best guess" of a possible cost structure for future plants in Switzerland based on studies of our own and information from power plant projects abroad.

One area of uncertainty is the implementation of the new water protection law and the environmental protection law. Cost for new power plants could be affected appreciably if these laws were applied according to very high standards.

Type of Power Plant	Capital Costs		Fuel Costs	Operating Costs	Total Generating Costs
	$/kW	mills/kWh	mills/kWh	mills/kWh	mills/kWh
nuclear power plant (7200 hour of operation)	3330	48*	21*	13	82*
run-of-river power hydro plant station	5600	67...107	0	13	80...120
high pressure hydro plant with reservoir	4930	117...187	0	13	130...200
gas turbine power station (1000 hour of operation)	670	67	60	13	140
coal power station (4000 hour of operation)	1870	47	27	13	87
combined cycle plant (4000 hour of operation)	930	27	40	13	80

Table 3   Cost structure of new power plants in Switzerland

* including 16 mills/kWh for the back-end of the fuel cycle, decommissioning and final storage of nuclear waste.

4.2 External costs

These are assessed presently in a study of two Federal Departments, entitled "Externalities of Heat and Electricity Generation". The declared purpose of this study is to develop a tax system that would pay for the external costs of power generation otherwise born by society. The study points out the difference between the relatively well known social costs from **normal operation**, from pollution-induced health effects, and those, much less certain, from the **risk** of potential nuclear, gas or oil **accidents**.

The studies are not yet completed and highly controversial in all aspects. First estimates indicate the lift-up of electricity generating costs as follows:

Gas turbine power plant, oil fired   30 ... 70 mills
Gas turbine power plant, gaz fired   20 ... 40 mills
Nuclear power plant   0,2 ... 22 mills

The wide spred for nuclear power plants is due to quite different assumptions for the severest accident to be considered and the energy-mix used for enrichment services.

In principle, we expect a minor effect from any further inclusion of external costs on actual prices of Swiss nuclear electricity.

One important factor, namely the backend of the fuel cycle costs for plant decommissioning and for final storage of nuclear waste is already covered by taking into account provisions of up to 16 mills/kWh produced.

5. Conclusions

We have to keep all the options open for the future.

Switzerland is situated in the heart of Europe with a good transportation infrastructure in all directions. A close network of roads, railways, pipelines and highvoltage lines facilitates transport of primary energy as well as of electricity. Therefore, importing of fossil energy and electricity has be considered as a valid option for the time to come. High population density, high degree of industrialisation and a landscape which is worth protecting require power generating processes with high energy density, low land utilisation and minimal consequences for the environment.

Looking at the three types of fossil energy - coal, oil and gas - gas reaches the lowest sum of production and external costs.

Considering nuclear energy, Switzerland has acquired substantial experience in operation of nuclear power stations. We have to maintain the know-how in spite of the moratorium in order to build on it after the year 2000. Nuclear energy with its high energy density and low emission is the ideal form of energy for a densely populated country.

A major part of the external costs, the cost for the backend of the fuel cycle, for decommissioning and final storage of nuclear waste is accounted for already in the production cost. Future investment costs for nuclear power stations are quite uncertain, specially because a standard design has not yet evolved. Provided these costs lie within the range of 1500 $ per kW_{elec} as expected by modern reactor manufacturers, nothing will stand in the way to a renaissance of nuclear energy from the economic point of view.

Considering all the factors mentioned so far, it seems, at least qualitatively, that a balanced combination of domestic generation on the basis of nuclear energy and gas with oil as storage reserve and import of electricity may be the solution for Switzerland to cover the future demand.

References

1)   Seifritz, Walter (1992). **Zur Begriffsbestimmung CO$_2$-entsorgter fossiler Kraftwerke** BWK, Vol. 44 (1992), Nr. 6, June.

2)   CSA Report G (1-88) by the American Medical Association (AMA).

3)   Gattinger, Matthias et al. (1991). **Stromerzeugung und Strom-verbrauch - Verbundnetze.** Siemens Aktiengesellschaft, Berlin und München.

Session 3 - Discussion

R. Carle (Chairman)

Mr. Draper was talking about the demand side option and mentioned that you anticipated reduced generation capacity requirements by about 900 megawatts by 2012. I have seen and read a lot of papers with anticipations of that type and I wonder whether you have any experience with ex-post evaluation of figures that you had anticipated, were you really able in the past to reduce capacity in that manner or as far as you had expected when you started the programmes?

L. Draper

The answer to the question is no. We do not have experience and the reason for that is that the demand side management emphasis at AEP is a new emphasis. We are in the middle of the United States where the commissions for the most part have had less interest in conservation and demand side management, for several reasons perhaps, but a significant one being that the price of electricity in our area is relatively low compared to the price of electricity in the West Coast and in New England. Obviously, the higher the price, the more opportunity there is for alternatives to come in and make economic sense. So at this stage the demand side management enthusiasm in our area is more to avert the need for new construction than it is to reduce the current cost of electricity to the customer. That is different from other places.

We do not know whether 900 megawatts is really achievable in the next 2 decades or not. It depends a lot on how much you are willing to pay for it. The more you are willing to pay, the more you can get. We will have to see, as time goes by, how our commissions respond to proposals to implement new and more expensive programmes and, to the extent that they are not willing to pay, then we will retreat and the demand side management opportunity will be less. On the other hand, if they are willing to let us do those things and charge the customers and have the bills go up, we will do them and we may get more. So it will continually adjust. But we don't have that sort of data.

R. Carle

Yesterday morning we heard about some premium between the different types of energy sources. Ten per cent, for example, put on some more benign sources of energy for the environment. Have you something in mind or is it dollars?

273

L. Draper

Well at least at this stage on our supply side evaluations, a dollar is a dollar. We are not penalising sources because somebody thinks they have attributes that are desirable or undesirable. That again is largely a function of the area of the country in which the utility is located and the attitudes of its commissions. As I said earlier, we have seven different commissions and their attitudes are quite different, but I think it is fair to say none of them are as aggressive in promoting renewable energy sources and penalising traditional sources as the commission in say California or in New England, or in some of those other places. So we don't artificially skew the cost of the supply side options. We do try to include all the real costs of pollution control equipment, naturally.

J. Gray

Lynn, you referred to expecting significant changes over a period of twenty years. Would you speculate on the nature and character of those changes that are in your mind as you talk about and think about prospective changes in that period of time?

L. Draper

I guess the things I am thinking about are the sort of things that most people in the room would be thinking about, given what has occurred in the last couple of years with respect to inducements for non-traditional players in the utility field. I think that we are going to see, for a while at least, a lot of people who would like to generate electricity in a service area other than the traditional utility. Whether that independent power producer is really independent or an affiliate of another utility that believes it can make more money outside its own service area, we will see a number of projects of that sort. We now have in the Energy Act the ability or the requirement to move power in wholesale transactions; that is from a generator to another utility. I believe that there is a reasonable likelihood that this will spread to retail transactions. I think that that is unfortunate public policy, but I believe that it will happen. If it does, there will be significant winners and losers. The significant winner will be those systems that are low-cost producers of electric power that can go out and take customers from the high-cost producers. The significant losers will be the immobile customers of the high-cost producers, namely the residential and commercial customers of those high-cost producers. I think that that will inevitably lead to some consolidations in the utility business. We have seen a trickle of mergers and combinations in the last few years and I think that (this) is likely to accelerate as high-cost producers are no longer viable.

So I think that we are in for some considerable changes. I think that we will see some utilities continue to be vertically integrated, others that will choose to be primarily generators (and), many that will choose to be primarily distributors and

274

not be in the generation business in a significant way. We will see people spin off their generation assets and try to make more money as independent facilities rather than in supplying their own customers. While the details are hard to project, I think that we are going to see some turmoil.

R. Carle

I am tempted to ask you a last question which is a difficult one. We are, in Europe, building a European system of production. We try to harmonize, we try to optimise the global European production system. Which advice would you give to Europe? Are you ready to answer that?

L. Draper

My advice is worth what you pay for it! Not much!

I think that in many ways Europe is at the same stage we are, in the United States, with considerable uncertainty. There will be winners and losers. Those people who are low-cost producers are likely to be the winners and we must in some way find a way to protect the people who are likely to be losers. I would guess that in your circumstance you are a potentially winner. You have adequate capacity and low-cost. To the extent that you can take advantage of it, that is good. On the other hand, there are a number of places that are potential losers and their interests must be looked out for as well.

R. Carle

We know that the Japanese have tried to work on CO_2 trapping, on CO_2 combining and storing to stop emissions. That seems to me a difficult task and an expensive task. That is a way of internalising an externality. How much should we be able to pay to stop CO_2 emissions by this way?

Y. Akiyama

I believe that there are two kinds of technologies which can address the CO_2 emission problem. One is to try to trap CO_2 from the flue gas and fix them. The second one is to develop technology which would eliminate the emission of CO_2. Regarding the first category of CO_2 technology, which is trapping and fixation of CO_2, research and development work is being pursued at an organisation called RITE, the Research Institute of Innovative Technology for the Earth, (and) which is government and private sector corporative organisation. In our organisation, Kansai Electric, at the Nanko power generation plant, we also try to trap CO_2 from the flue gas and fix them. These are the sorts of research and development effort we are pursuing at this moment, but I suppose the cost could be very high and the technology has not been firmly established as of yet. The second category of technology is to reduce or eliminate CO_2 emissions as much as possible. Here in this area we are trying to introduce new technologies such as combined cycle and so forth in natural gas based power generation plant where we expect thermal efficiency up to 53 per cent which is considerably higher than the current thermal efficiency. So, thus we will be able to reduce emissions per kilowatt hour of power generated to a considerable extent. In addition, we should continue to pursue the development of higher efficiency equipment and energy conservation technology and so forth whose cost would be much less than the cost of development of the first category of technology.

J. Foster

Akiyamasan, of course China is a very big player in your part of the world and you gave us some inkling of what you expect in the future when you mentioned that you expected, between now and 2010, a doubling of their coal consumption. But since you are closer and much more aware of what's going there than the rest of us, I wonder if you could elaborate a bit on what you foresee with regard to the development of China over the next twenty or thirty years?

Y. Akiyama

China, of course, has the population problem as its biggest challenge, and now the government of China is taking a policy of one child per household, trying to arrest the population explosion in that country. But there certainly seems to be a limit to what they can do in pursuing this policy. So as I said in my presentation, although we do not see any direct relationship, we see some sort of relationship between the higher level of electrification and a lower birth rate. It means that a higher rate of electrification is leading to higher living standards, that the (higher) status of women in the society (is) perhaps leading to the lowering of the birthrate in that country. In addition, although some of the equipment could be less efficient than the leading technology, these could prove to be very useful in China in raising the electrification rate in that country, which I believe would be the key to the economic take-off of that country. However, the problem is that the volume of the fuel used for generation is very big and, therefore, we need to introduce lower-cost pollution arresting technology to that country because some of the equipment and devices we use in Japan, for example, are too costly for China to be able to introduce at this economic stage. So we would like to continue on that front.

In addition, the conversion of fuel from fire wood, for example, which is very inefficient, to more efficient fuels, such as coal and oil, will have a very important effect. We also have information exchange agreements with China to provide them with energy conservation technology. In addition, China shows considerably high interest in the development of nuclear power, particularly in trying to introduce nuclear to the southern part of that huge country. But for this the safe operation of the nuclear power plant is essential and we, Kansai Electric, have been providing support and advice by sending operators to the pilot operation of their nuclear power plant in China. At this moment China is exporting 350,000 barrels of oil to the international market, whereas it is expected to turn into a net importing country of oil in the year 2000, importing to the tune of 500,000 barrels. This means that in that time frame China will have an impact upon the oil market in the world to the tune of 1,000,000 barrels per day, which is a very significant impact upon the oil market. China, being a major player, I believe needs technical co-operation from Japan but also from the other countries in the world to address energy and other important problems.

R. Carle

I think you have shown us how Kansai Electric Power Company is involved in the preparation of the future, particularly regarding environment and I think we shall continue to look at your country with great interest. I think you will have more opportunity to ask questions to the four speakers at the end of the session.

K. Uematsu

I have no question, but I would like to make a brief comment on the proposal made by Mr. Akiyama. Thank you very much Mr. Akiyama for providing us with such a broad (picture) of the energy situation in Japan and the Asian countries. This provides very important information to us. My comment goes to your proposal that you made at the last conclusive remarks. You have been proposing the "dialogue with the future". That kind of a group should be commissioned within the OECD. Of course, I think that this is the proper time to have such a forum established within the OECD domain. However, this subject is much broader than the NEA's responsibility. Therefore I will take this proposal back to Paris and I will try to persuade the OECD to take up this subject item within the OECD's future activity, and I believe we shall stay in communication with Mr. Akiyama on this subject.

C. Willby

I should just like to ask if you could possibly clarify -- on one slide you suggested that the role of the CEC was to undertake comprehensive research and development programmes into nuclear research since some of the CEC programmes in the past in the research and development field have probably not been quite as, shall we say, efficient, as they might possibly have been. I wonder if you could possibly elaborate on why you saw this as a role for the CEC rather than as a role for member states who undertake a new nuclear power?

S. Barabaschi

Well, as you know, the CEC on this occasion was more a vehicle for transferring the information to the states more than as a possible source of funds. The framework programme on nuclear energy is very small compared to what we invest in Europe. The CEC only controls 3 per cent of the gross expenditure on nuclear energy in Europe.

C. Willby

So you are not really suggesting that this position should change substantially?

S. Barabaschi

Well we are suggesting that it may change in certain direction and we will insist that we make more comprehensive documents, but I expect this action more as a vehicle to restart proper discussions on nuclear energy in Europe than as a source of funds.

R. Carle

If I may add a word, it is clear that there is a general trend in Europe to try to harmonize nuclear safety, to find common views and to restore a general consensus about nuclear energy. Certainly, all initiatives are welcome in this direction, but it is clear that practically most will be done, I think, by individual members, by bilateral discussion and so on. In fact, the CEC is not funded to work on nuclear safety. Its domain is health physics, that's clear, but not safety, and the IAEA could certainly play a bigger role in this problem of safety.

May I ask you a question myself? You underlined, stressed the fact that there is a lot of progress regarding technology in general, electrical technology, electronics, and they can improve the way we use electricity. Do you think these can modify strongly the amount of energy needed by the developing countries in the future? Could they have a different way of development than we have had in the past, in Europe, for example?

S. Barabaschi

I don't know. It is very difficult to respond because we have so far (only) limited experience, but to give you an example, we have been building several solar driven water pumps in Africa. It is dramatic to see how difficult it is to design the electronic, how difficult it is to design a product that must be maintenance-free for ten years even if badly treated. So now we have developed together with S G Thompson, a very special pump, cheap, very rugged. In fact, it is an elaboration of a piece of the motor industry. This may be a case in which we may help, in certain particular situations, the reliability of these small plants for Africa.

J. Grawe

I think it is really admirable that such a small country as Switzerland has dared to go step by step on the nuclear path. I can only hope with you that nuclear will have a future in your country. It may be of interest to hear a little bit about your endeavours to solve and to handle the waste problem.

K. Küffer

Well, naturally, the solution to the waste problems is also one of the keys to nuclear power. Acceptance can only be re-established in a way to give us a majority if we can solve this problem. We have taken the following steps: we have now the site permit for an interim storage facility which will be adjacent to the Paul Scherrer Institute in Würenlingen. This is a centralised interim storage facility for all power plants. Then we have now selected a site for a low- and intermediate-level final storage facility in Wellenberg which is about 40 km east of Lucerne. It is in quite mountainous country. We have an agreement with the community that they are willing to co-operate with us for such a storage facility.

We will have to have a vote on this because the canton, which is the next political structure, has to give a concession. This concession is given by a public vote. This concession normally has to be paid for. This ballot will be, maybe, next year. We feel confident that the people of this canton will accept. Then we can start with the licensing procedure, and we hope that we can start to build this final storage place, and that we can start for the year 2000 which is necessary in order to overcome this moratorium.

For the high-active waste, we have two projects running: one in the granite underground, and one in sediments which are above this granite. We have made drillings, we have actually proven the safety within this geological material but we still have to get down to a specific site. We are working now on this problem. We will be ready within two years to tell the government where this could be done. this is what we have to fulfil. We don't have to start construction immediately because as you know we don't need a final storage for high-level waste until the year 2030 or 2035. We can wait with the interim storage for this final storage and we hope and feel it is necessary that final storage for high-level storage is done mutually in Europe. Not every country should do the same things for their own storage as we have a reprocessing situation with two plants in Europe. I am sure that we need only two final storage areas for high-level waste, and I hope we can arrange in Europe for such a situation in the few years.

R. Carle

If you ask French people, as we did that a few years ago, what is the main reason for the origin of the greenhouse effect, seventy per cent answer nuclear energy. Is it the same in your countries? How could we be so bad in public relations, and do you think that the work we are doing now to try to quantify these things, to have a scientific background about externalities, can help to reduce this gap in knowledge?

L. Draper

In the United States, I don't know the specific number of people who think that nuclear plants are the cause of greenhouse gases, but I do know that the US Council for Energy Awareness has worked very hard in their television commercials to indicate that nuclear plants are not the cause of greenhouse gases. In fact, the commercials are quite effective, sometimes to the dismay of people like myself who burn a lot of coal. They clearly show that nuclear plants have an advantage.

K. Küffer

Well, we have not asked this question in Switzerland, but we have another question, whether the answer was also quite interesting. We asked: "which energy form in ten years from now is going to give the largest amount of electricity"? The answer was solar power. So this is our problem.

S. Barabaschi

Well, I think that we have a large gap, as I said before, between the real public opinion and the perceived public opinion. We made an analysis of about 500 people, so it is a good group, just when we celebrated the 50th anniversary of the Fermi pile in Pisa last year. We found that the real position of the people is much more favourable to energy than thought by the policymakers.

Y. Akiyama

Let me try to share with you the example in Japan. We conducted a public opinion survey on nuclear energy and 70 per cent of the respondents said that they need nuclear power in Japan, whereas 34 per cent of the respondents expressed some sort of uncertainty about nuclear power. So the question and the challenge for us right now is to try to take various measures to allay such psychological uncertainty and anxiety on the part of the Japanese people. In this connection we learned a very important lesson quite recently. It was on the 9th of February 1991 at the Mihama Nuclear Power Plant of Kansai Electric. We had some incident where radioactivity was released from the plant at something like about 1/100 thousand of natural radiological background level. So it didn't have any impact, of course, on the health of the human body, but the problem at that time was that we couldn't give the precise figure of such radioactivity release outside the plant to the residents living in the neighbourhood. Certainly we could say that, technologically and technically, such a very low and very limited level of radioactivity could be negligible from the point of view of the human impact upon the health. But what people needed in that area at that time was the precise figure, number, on the radioactivity released. Therefore, it was decided to install the measurement device which can detect and measure very low levels of radioactivity. Therefore, in the future if a similar incident should happen, then we will be able to provide the people in the neighbourhood with a very precise and accurate data and information about the radioactive level activity released. Thus we feel that it is important to strike good harmony between big technology and the human psychology. I think that this would be the best way to allay fear and anxiety on the part of the people.

Unidentified

Mr. Küffer, given your country's success in implementing nuclear power by its 50 per cent share of your energy production, I am curious as to why the tide turned against nuclear power, where you have a referendum so you cannot build plants, and what you see as being the key to eventually building plants again in the early 21st century? Maybe a suggestion, we are talking about externalities here. We have seen some substantial numbers and facts today that actually say that not using the nuclear option is putting CO_2 gases and NOx in everyone's backyard. A public relations tack might be that in fact nuclear power is the better option. Perhaps we are not getting the message out well enough. I wonder specifically what you would be doing in Switzerland to relay that message?

K. Küffer

Well the case is quite clear. The nuclear power plant Kaiseraugst was quite close to a final decision when we went through all courts up to the Federal Court and it was really close to the final decision then and we had the Chernobyl accident . As we have seen on the Japanese graph, the dramatic drop of public acceptance.

Now, what is the key to come back to nuclear power? In Switzerland the majority is of the opinion that we don't need new power plants. That we have now demand side management, integrated resource planning, and a lot of these fancy words which nobody fully understands, but it's like a hope that with these techniques we will survive without new plants. So to economise, to use electricity more efficiently, this is actually the prime goal. This is also the prime goal of this programme "Energy 2000". If it fails, if we cannot stabilise the electricity consumption, then we have to something else. If we look (beyond) the year 2000, we know that we have definite lifetimes of these power plants, so we have to replace them. If we have eighteen years of time from the start (when we say here at this site we would like to build a plant), until we can commission it, eighteen years with our legislation which we have now, it is high time to start thinking about replacing the first generation power plants. So this is something that will come, but to economise and to use the new renewable sources like wind energy, like solar energy, this is actually the big hope. We have to prove that this cannot be a substantial part of our electricity production for the future until the public is again ready to talk about nuclear power. So this proof we have to do now.

Unidentified

My first question is for Mr. Küffer. You mentioned that you already have decided or picked an interim storage site for the spent fuel. I am curious to find out, have you actually moved any fuel from your registries to dry storage? If yes, what kind of storage cask have you used? Is that cost also included in the 16 mils that you mentioned in the decommissioning cost?

K. Küffer

First answer, yes we go to a dry storage in these containers and the second, yes it is included in the 16 mils per kilowatt hour.

Unidentified

I see. I am just curious to (know) what these casks are that you are using. Are they manufactured in Switzerland or who is the supplier for these casks?

K. Küffer

Now the Castor casks are manufactured in France and also in Germany, but not in Switzerland. Not yet, maybe someone will take a license up, but at the moment no.

Unidentified

My next question is for Mr. Draper. We were talking about your twenty year strategic plan for resource planning and you mentioned that 10 per cent of the fuel mix in that plan comes from natural gas. We were wondering what kind of presumptions had you made on the supply? Ten per cent of total projection in twenty years time could be substantial.

R. Carle

I would like to ask, "what assumption on the price"?

L. Draper

The 10 per cent in our case amounts to about 3,000 megawatts of capacity, but those are peaking units. So we are nowhere near that proportion of our total in terms of generation mix. We would expect to run those a few, or a few tens of hours a year. Our guess on natural gas prices is no better than anybody else's. We have seen in the last year in parts of the United States that use lots of gas, that the price has gone from one dollar a million BTUs to two dollars a million BTUs. It doubtless will go up and down dramatically as time goes on and that is the reason that we have limited our capacity to about 10 per cent gas. We just are not in the right part of the country to have a reliable supply at affordable prices.

R. Carle

Thank you very much. My duty is to close this session with a few remarks. I will try to be very brief.

I started this morning by telling you that I realised that all our problems were common problems, and behind different presentations we all faced the same situation. I could say the contrary, and I see better now the differences between the different countries and how the feelings about these externalities and risks may be different from one country to another. I don't know if I am right when saying that an externality is not exactly the same in California and on the East Coast of the States, but I am sure that it is not the same in China and in the United States or countries of Europe. I think we have to take care of that. We have to admit that, we cannot reduce some cultural approaches, but of course we must try to quantify this and try to better harmonize because we are all using the same planet and we are all taking fuel from the same basket.

I was struck by the discussion yesterday and this morning about risks. When an externality becomes a risk, the question is much more difficult because we have three terms of uncertainty about how people evaluate it. We have the value of the damage, we have seen the difficulties to define a nuclear accident or I would say a reasonable nuclear accident, what can happen, what cannot happen. We have the difficulty to define the probability, which is probably the best known factor, and we have to multiply the value by the probability; and most people, we must recognise, don't accept these multiplications as we do. By chance I heard on American T.V. someone speaking about food and drugs raising the risk of cancer. This women had a baby and she said very clearly I don't want any small piece of food, or drug which could cause harm to my baby - zero. She doesn't realise that these foods, these drugs, she already takes on many occasions have a finite probability of causing harm, but she doesn't know that. I mean people do not understand what a probabilities means. That is quite clear and we have to make a big effort to make these understandable by comparisons, by different practical ideas. So I would say that I think there is a big interest for the international organisations to take part in these discussions and to try to clarify this. We can make very big theories about risks, about probabilities. If we are not understood, if we are not credible, we certainly will not help to solve the problem and maybe the international organisations, I don't which one, but international organisation can certainly help to give more consensus and more credibility to all this business. I think that it was a good idea to organise this symposium, and we have to continue this work both at the national and international levels.

Session 4/Séance 4

Chairman/Président
R. Shelton, ORNL

Outlook for Costs by Energy Source
L.J. Williams, J. Fortune, G. Booras, EPRI (United States)

Outlook for Risks by Energy Sources
J. Grawe, VDEW, & University of Stuttgart (Germany)

**The Evaluation of External Costs from
Energy Sources - the EC-US Fuel Cycle Study**
A. Krupnick, Resources for the Future
A. Markandya, Metroeconomica & Harvard Institute for
    International Development
R. Lee, Oak Ridge National Laboratory
P. Valette, Commission of the European Communities

**Perspective on Energy Security and Other Non-Environmental
Externalities in Electricity Generation**
D.R. Bohi, Resources for the Future (United States)

**Prospects for Internalisation of Externalities
Where Do We Stand - What is Ahead ?**
P.M.S. Jones (United Kingdom)

Discussion

Closing Remarks

List of Participants/Liste des Participants

OUTLOOK FOR COSTS BY ENERGY SOURCE

Larry J. Williams
James Fortune
George Booras
Electric Power Research Institute

Abstract

The outlook for future costs of generating electricity will depend on a number of technical, economic, and regulatory forces. We examine these issues with special emphasis on technology and environmental factors. Of particular interest is the turmoil within the utility industry. The trend toward a more market driven industry is taking place in response to falling regulatory constraints. Environmental considerations will continue to shape energy system design. Some aspects of environmental externalities are briefly described. Then an example is constructed showing how the Massachusetts values for externalities could dominate resource allocation decisions. This simple example clearly shows the powerful influence that environmental externalities could have on power system design. The conclusion is that economic and environmental factors will continue driving generation technologies toward higher efficiencies and lower emissions. Because of continuing research, there will be a large menu of clean and efficient generating technologies available to meet the emerging needs.

NOTE: This paper reflects the views of the authors and not necessarily EPRI or the US electric utility industry.

LES COÛTS PAR SOURCE D'ÉNERGIE : PERSPECTIVES

Résumé

L'évolution des coûts de la production d'électricité dépendra de plusieurs facteurs techniques, économiques et réglementaires. L'auteur analyse ces facteurs en mettant l'accent sur les aspects technologiques et environnementaux. Les bouleversements affectant le secteur des compagnies d'électricité présentent un intérêt particulier. Aujourd'hui, l'industrie obéit de plus en plus aux forces du marché du fait de l'assouplissement des contraintes réglementaires. Les considérations d'environnement continueront à peser sur la conception des systèmes énergétiques. On trouvera une brève description de certains aspects des coûts externes d'environnement, puis un exemple montrant comment, dans le Massachusetts, l'évaluation des externalités pourrait prévaloir dans les décisions concernant l'allocation des ressources. Ce simple exemple montre clairement l'importance que peuvent revêtir les externalités environnementales dans le choix des systèmes énergétiques. En conclusion, les facteurs économiques et environnementaux continueront à orienter le progrès technologique vers des technologies de production plus efficaces et moins polluantes. La recherche ne cesse d'avancer, aussi disposera-t-on demain d'une large gamme de technologies de production d'énergie propres et efficientes pour répondre aux besoins qui se font jour.

I. Introduction

This paper will develop information useful for evaluating future cost trends for generation technology choices within the US electric utility industry. The major forces influencing costs are:

- environmental constraints and other regulatory requirements
- technology choice and future improvements
- fuel markets and other economic conditions

Each of these issues will be discussed in turn. Since forecasts of energy prices and interest rates are problematic, most of the attention will focus on technologies and environmental externalities. The paper will begin with an overview of the extraordinary pressures at work today that will result in significant and lasting structural change within the US utility industry.

II. Turmoil and Change Ahead for the US Utility Industry

For a variety of reasons electric utilities are presently experiencing more pressures than at any time in their history. The structural and other changes resulting from these forces will be significant and are likely to increase over the next decade. The long term future of the electric power industry will be permanently altered.

There are several significant issues that will affect electric power production throughout the world. Environmental concerns will play an ever increasing role in shaping world power systems. Among the dominant environmental issues that we are concerned about are:

- global climate change
- air pollution (air toxics, ozone, acid rain)
- electromagnetic fields (EMF)

In recent years, there has been growing concern that accumulation of man-made greenhouse gases in the earth's atmosphere will lead to undesirable changes in climate. This concern has led to a number of proposals, both in the United States and internationally, to reduce greenhouse gas emissions.

Global climate change presents many challenging problems from a scientific perspective and most research on climate change addresses specific scientific elements of the problem in great detail. For example, the extensive US research program on global climate change is almost entirely devoted to detailed studies of the scientific aspects of climate change, and little work is being done to develop and apply accepted methodologies for assessing the costs and benefits of alternative climate change management proposals.

289

Climate change management proposals could have extensive ramifications for both economic and environmental systems. Electric power production will surely be at the top of the list of impacted industries. Further research is needed to help clarify our understanding of costs and impacts, and to provide an acceptable framework for their comparison. This research could save billions of dollars in terms of unnecessary and onerous regulations.

The siting of electric power generating stations became much more difficult in the US in 1970. The National Environmental Policy Act became effective on January 1, 1970, followed later that year by the formation of the Environmental Protection Agency (EPA) and the passage of the first (federally enforced) Clean Air Act. The national program has had a large impact on air quality in the United States. Emissions of most major pollutants have declined substantially in the last two decades, and as a result, ambient air concentrations of major pollutants have fallen sharply. Furthermore, the 1990 Clean Air Act Amendments represent a major additional commitment to cleaner air, and will produce further dramatic improvements by the end of the decade. Title IV set a national cap on utility SO_2 emissions of 8.9 million tons after 2000, a reduction of 10 million tons from 1980 emission levels. The costs are large and rising. The direct costs of complying with environmental regulations in the US has been estimated at 120 billion US$ per year in 1990. The costs are still rising.

Since the 1970s there has been some concern that exposure to power-frequency electric and magnetic fields, or EMF, may cause adverse health effects in humans. Studies conducted in the 1980s led to the conclusion that electric fields are almost certainly not a problem, and attention is now being focused on magnetic fields. Epidemiological studies, which examine consequences of exposures in large populations of people, have yielded inconclusive results, though there are some consistent findings of associations between surrogate measures of EMF exposure and a roughly twofold increase in the incidence of some fairly rare cancers, especially childhood leukemia. Laboratory studies have neither revealed a mechanism for an EMF risk nor consistently shown similar effects in animals. Meanwhile, public perceptions – and public policy, in some instances – are well ahead of the science. Concern over possible EMF risk has already made it difficult to site new transmission facilities in the U.S. and some other countries in the developed world. If the existence of a risk is confirmed in the future, these difficulties will probably persist, and there will be increased pressure to develop and utilize low-field designs, including underground cable.

Another global issue that will affect electric utilities is the increased importance of international markets and the associated import and export of generation technologies and their associated services. Globalization has reduced the control and leverage that US utilities have over their equipment vendors. Furthermore, utilities will find increased competition developing not just from domestic sources, but from international groups as well.

This globalization of utility related markets will also bring opportunity. Developing countries and the developing markets will open up new business opportunities involving technology transfer from aggressive utilities pursuing diversification strategies. Carbon constraints and other environmental forces may finally lead to significant international trade in liquefied natural gas.

III. New Competitive Setting

Decreasing levels of regulation are transforming the arena in which utilities operate. The US electric utility industry is in the early stages of a process that has brought significant change to other US industries. These other industries include telecommunications, banking, railroads, airlines, and gas pipelines. Regulation in each of these industries has been responsible for some basic structural similarities. Under the regulatory umbrella there is no competition, bundled services, cross subsidized lines of business, over capacity, and relatively high costs. Deregulation brings in price competition, service unbundling, cost reductions, internal restructuring, and some early acquisitions. Once this process is underway, regulatory and emerging competitive forces lead to fundamental restructuring. This means intense price competition, product proliferation, broad-based competition, and more deregulation. Eventually this results in decreased profitability which in turn leads to cost restructuring, market restructuring, and industry restructuring. The final outcome is a market driven industry.

Utilities will be pursuing multiple business strategies with big winners and big losers. Consolidations, mergers, and possible bankruptcies will transform the industry. The "typical" electric utility company will cease to exist. Large unregulated generation companies may well dominate the independent power producer (IPP) industry. There will be transmission companies, distribution companies and energy service companies.

This new competitive environment will focus attention on costs. Economic pressure should intensify the development of more efficient generation technologies.

IV. New Generation Technologies

All forecasting is a risky business and technology forecasting shares that characteristic. Nevertheless several trends are underway and these trends can be expected to persist. There are strong economic and environmental pressures leading toward higher efficiencies (> 50 percent) and lower emissions. The units will have improved reliability and availability due to improved analytics along with higher quality databases. Project costs will be lower due to simpler modular designs which will allow faster construction. Finally, the O&M cost picture will improve due to the use of expert systems, better diagnostics, and resulting reductions in labor expenses.

In the area of nuclear generation, much has happened in the last decade. Utilities recognized that many serious institutional barriers (licensing procedures, state economic regulation, etc.) stood in the way of any new nuclear orders and set a program in motion at EPRI in the mid 1980s to make sure that if and when the institutional problems were sufficiently resolved to permit nuclear as an option, that new advanced nuclear technologies would be developed and available to meet their needs. Utility design requirements were developed for both large evolutionary and mid-size passive safety designs that would improve safety, simplify operation, and be amenable to a high degree of standardization in order to reduce design, construction and operating costs.

The institutional barriers were addressed by a new "Strategic Plan for Building New Nuclear Power Plants", prepared in 1990 by the Nuclear Power Oversight Committee, which integrated all the technical work at EPRI with the economics and nuclear waste work at Edison Electric Institute (EEI), current plant performance improvement programs under the Institute of Nuclear Power Operations (INPO), regulatory interface efforts at the Nuclear Management and Resources Council (NUMARC), public acceptance initiatives at the U.S. Council for Energy Awareness (USCEA), and government support activities at the American Nuclear Energy Council (ANEC). Some major successes have been achieved, such as the creation of an improved nuclear plant licensing procedure that requires all design and siting decisions to be made before the plant is built, not afterward as was the case in the 1960s and '70s. This new NRC regulation which added great stability to the process was reinforced by Congress and the President in the 1992 Energy Policy Act.

Other institutional barriers to new nuclear construction still loom large. Chief among these are a resolution to nuclear waste issues, and the various state economic regulatory issues. Primarily for these reasons and the fact that the new advanced reactor designs will not be certified and available for another 2-3 years, it is expected that no nuclear orders will occur until after 1995 or later. During that time, much additional work will be done to further reduce the cost of nuclear generation, primarily through very high degrees of standardization and centralization of engineering, procurement and spare parts, training, maintenance, and other O&M-related functions. For these reasons, nuclear cost projections are not included in this paper.

Next, we present some specific estimates about future fossil, and renewable energy generation costs. These cost estimates were prepared by EPRI staff and are mostly documented in EPRI's Technical Assessment Guide (TAG™) [1]. The TAG™ provides up-to-date information useful in preliminary resource planning activities for the US electric utility industry. The information is also useful for research planning and management by providing a consistent database of cost and performance estimates and environmental emissions for conventional and advanced power generation technologies, and a set of fuel price scenarios.

Figure 1 presents estimates of the total capital requirements for future fossil generating technologies. Total capital requirements in this context means money that is placed (capitalized) on the books of the utility on the service date. This figure includes total plant investment plus capitalized plant startup, fuel inventory, allowance for funds used during construction, and land costs. The light shaded portion of the bar near the top represents a range of "expected" capital costs, with the most likely value indicated at the center of the range. These ranges are speculative in that they are based on judgment more than on specific technical information.

The levelized constant dollar cost of electricity for each of these technologies is presented in Figure 2. Note that the costs for coal-fired plants assume that the plant burns 4.0 percent sulfur coal (e.g. Illinois #6). Other important assumptions are a 75 percent capacity factor and a 4.9 percent after tax discount rate. The fuel price assumptions (presented as a 30 year levelized cost, expressed in 93$'s based on start-up in year 2000) are:

- Coal cost = $1.43/MMBtu, with real escalation rate = 0.8%/year
- Natural gas cost = $4.13/MMBtu, with real escalation rate = 1.9%/year

Environmental Aspects of New Technologies

Environmental considerations will continue to shape future power generation systems. Consequently, the cost information on new generation technologies presented in Figures 1 and 2 is only part of the total picture. Figure 3 presents some revealing data on emissions and solid waste produced by current and near term fossil fueled technologies. Again, the coal based technologies assume bituminous coal with 4.0 percent sulfur-similar to Illinois Number 6. Comparing the capital costs for the coal technologies shown in Figure 1, it appears that lower emission coal technologies are available at higher capital cost. This may not be surprising, but it demonstrates how environmental considerations will tend to increase the costs of future power systems. Also note that R&D will provide future technological options with lower emissions and lower solid waste. Another well known fact that is apparent from these charts is that gas fired options are both clean and cheap. The major risk factor with gas will be future price and availability.

Figure 1

Total Capital Requirement for Future Fossil-Fueled Technologies

$/kW (1993$)

(Based on Bituminous Coal with 4.0 % Sulfur)

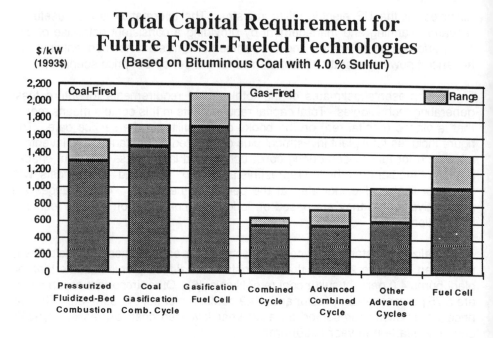

Figure 2

Cost of Electricity from Future Fossil Power Technologies

Levelized Constant Dollar Cost of Electricity, mills/kWh

(Based on Bituminous Coal with 4.0 % Sulfur)

(Based on average C.O.E. parameters for each technology)

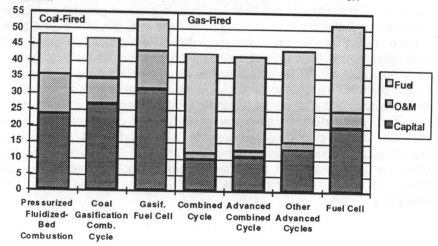

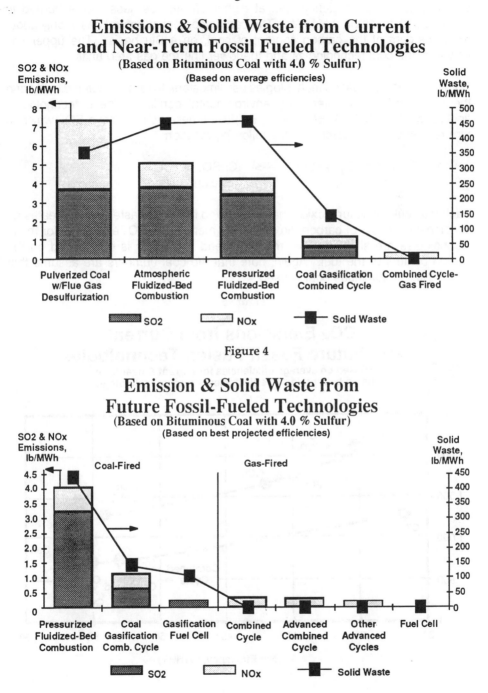

Figure 3

Emissions & Solid Waste from Current and Near-Term Fossil Fueled Technologies
(Based on Bituminous Coal with 4.0 % Sulfur)
(Based on average efficiencies)

Figure 4

Emission & Solid Waste from Future Fossil-Fueled Technologies
(Based on Bituminous Coal with 4.0 % Sulfur)
(Based on best projected efficiencies)

Another potentially important emission from fossil fueled power plants is the greenhouse gas carbon dioxide. Figure 5 shows CO_2 emissions from a variety of fossil-fueled generation technologies. In general, CO_2 emissions are inversely proportional to net efficiency; i.e.. at higher efficiencies, less fuel is burned to produce a given amount of MWh. The carbon content of natural gas is only about 55-60 percent of the carbon content of coal, on a Btu basis. The upper line connects the coal technologies and the lower line the gas fired units.

Some existing coal-based technologies use limestone to remove sulfur dioxide and thereby introduce an interesting environmental conflict. These technologies (including PC w/FGD, AFBC, and PFBC) use limestone to remove SO_2, which in turn releases CO_2 according to the following reaction:

$$1/2 \ O_2 + CaCO_3 + SO_2 \ -----> \quad CaSO_4 + CO_2$$
$$\text{(limestone)} \qquad\qquad \text{(calcium sulfate)}$$

Each molecule of sulfur dioxide that is removed from the waste stream creates an additional molecule of carbon dioxide. This means that CO_2 emissions for these technologies are slightly above the coal trend line which is only based on the carbon content of the fuel. Again, note that the coal numbers are assume that Illinois No. 6 bituminous coal is used.

Figure 5

CO2 Emissions from Current and Future Fossil-Fueled Technologies

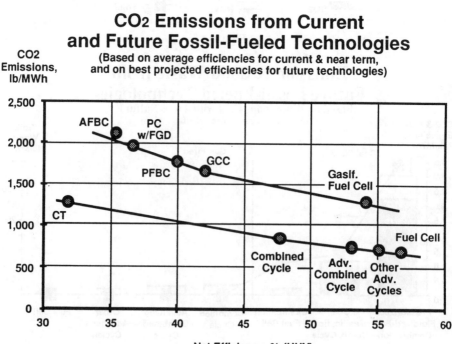

Increased attention to the environmental aspects of power generation has boosted interest in renewable energy options, as well as the potential for nuclear generation. Earlier research in the renewables area centered mainly on technology improvements. Today the emphasis has shifted to the marketplace-getting the technology into the field and nurturing markets in locations where existing renewable technology can be economically competitive. Efficiency improvements of the past decade have already opened up market niches for wind and solar technologies. The evolution of these markets should broaden the scope for all renewable applications: some believe that renewables could be contributing an increased share of future US energy supply, with biomass by far the largest source. Our estimates for the intermediate and long term costs of producing electricity from renewables sources are shown in Figure 6.

Figure 6

Intermediate and Long Term Outlook for Renewables

(Cost of Electricity in Favorable Locations)

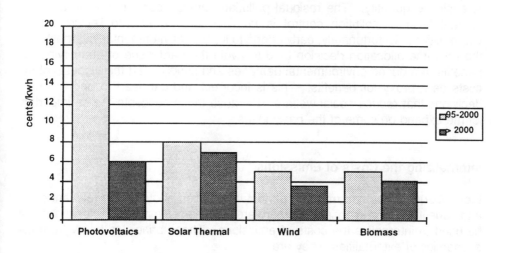

V. Cost Impact of Environmental Externalities

The widespread interest in the application of the environmental externality concept apparently arose out of the desire of regulators and others to "level the playing field" for Demand Side Management (DSM) activities and renewables. However, the rapid growth of DSM over the past 10 to 15 years has occurred without any boost from externalities. Nevertheless, by including the complete private and social costs of traditional electric generating technologies, it is thought that DSM and renewables will get an additional boost. Although current activity aims at the application of the environmental externality concept to the regulated utilities sector, it clearly has potential for much wider application.

Including environmental externalities in the alternative energy choice problem works best within a least cost integrated framework. An Integrated Resource Planning (IRP) framework can be used to compare the costs and benefits of traditional generation technologies with DSM, and renewables. In this setting all alternatives compete on an equal basis to arrive at the best combination of technologies. In practice, an important part of the externalities calculation is often missing. This missing part is the quantification of environmental impacts and the valuation of the damages resulting from alternative energy choices on a site specific basis.

Along with the capital, fuel, and O&M, the costs of generation technology should include the costs of abating pollution. A socially efficient amount of pollution control is obtained by equating the marginal cost of abatement control with the marginal benefit of reduced pollution. Uncertainty is an intrinsic part of the benefit side of the problem. The marginal costs for abatement for various technologies is easier to quantify. The residual pollution damage costs that exist after the optimal level of pollution control is in place are considered "environmental externalities". Unfortunately, early efforts to include environmental externalities in the resource allocation decision tried to avoid the hard issue of quantifying the benefits of reduced environmental damages and simply used the highest control costs as a proxy for benefits. This is incorrect and can lead to bad resource decisions that reduce social welfare. Because these concepts are so important let me expand on some of the basic ideas.

Internalizing the Costs of Emissions

Ideal World: If we had free and perfect information about all the relevant costs and benefits related to emissions, there are a series of simple principles that could be used to internalize the costs of emissions. These principles follow from the economics of externalities. They are:

- Any specific emissions target should be achieved at the lowest cost.

- Optimal emissions are achieved by balancing benefits and costs.

- Final goods prices should reflect the emissions control cost as well as the residual emissions damage cost.

- Emissions taxes or marketable permits can internalize costs.

Non-Ideal World: Implementing the above principles in the real world faces a number of serious problems. Among them are the following:

- The prices of <u>all</u> goods should reflect control and damage costs, not just goods regulated by Public Utility Commissions.

- Estimates of damage costs are expensive to obtain and have large uncertainties. Therefore it is hard to balance benefits and costs.

- Current emissions target levels are not achieved at lowest cost due to the extensive use of command and control environmental regulations.

For more on the economics of environmental externalities see Pigou [2], Coase [3], Baumol and Oates [4], Joskow [5], Burtraw and Krupnick [6], and Freeman et. al. [7].

Wide Range in Damage Estimates

Given the uncertainties and the non-transferability of results it is not surprising that a very large range for damage estimates are cited in various places [8, 9, and 10]. Figure 7 shows just a few of these numbers. Notice that the NO_x damage estimates vary by a factor of 100.

Figure 7

Wide Range in Cost Estimates

	Cost in $/Ton			
	Pace	*NYPSC*	*MA/Tellus*	*MA/Lave**
CO_2	14	1	22	2–10
NO_x	1,640	1,832	6,500	70–450
SO_2	4,060	832	1,500	60–990
TSP	2,380	333	4,000	100–1,500

*Best estimate–high estimate

299

The externality adder approach has been the most common way of "internalizing" the environmental impacts of residual emissions in the resource selection decision. However, other methods are also in use. These include rate of return and avoided cost incentives. Some states have used non-monetized methods: environmental impacts used as a "tie breaker", and various non monetary weighting and points systems.

Resource Selection and the Massachusetts Externality Values

Although the externality numbers that have been used are very uncertain, they are large and can make a significant difference for resource selection decisions. To demonstrate this in concrete terms, we will show how the environmental externality adders currently required by the Massachusetts Department of Public Utilities (DPU) can change the outcome of a generation technology decision. Docket DPU 89-239 requires the use of monetized values for environmental externalities for all filings involving resource cost-effectiveness tests. The monetized externality values that the DPU has selected for air emissions are based on control costs which the DPU considers to be a proxy for the damages. From an economic theory point of view, the control costs have no relationship to damages. However, an argument has been put forward that the control costs represent an "implicit" or "implied" set of damages on the part of legislators or regulators who formulate the control regulations. These values are reported by Stephen Bernow and Donald Marron of the Tellus Institute [10]. The values are SO_x at $1,500/ton, NO_x at $6,500/ton, and CO_2 at $22/ton.

Using the Massachusetts externality values, Figures 8, 9, and 10 use bar charts to display the cost of producing electricity with three supply side and two demand side options. The capital costs, along with fuel plus O&M are taken from EPRI's Technical Assessment Guide (TAG^{TM}) for a pulverized coal plant with flue gas desulfurization and a combined cycle gas turbine system. An existing coal plant, with no capital charges, is included as part of this chart in order to show how a coal plant with just variable costs could compete with the other options. Also included are two arbitrary Demand Side Management (DSM) options as contrast for the coal and gas supply side technologies. Figures 8, 9, and 10 present three cases in which only the gas price changes; from $1.50 to $2.50 to $5.00 per MMBtus. In all three cases the coal is assumed to be Pittsburgh number 8 seam coal with 2.1 percent sulfur and is priced at $1.31 per MMBtus (30 levelized cost in constant '93$). The supporting data for this example is somewhat different from the data supporting Figure 2.

Generation Technologies Compared:
Massachusetts Externality Values

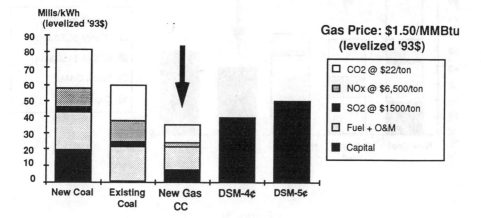

Generation Technologies Compared:
Massachusetts Externality Values

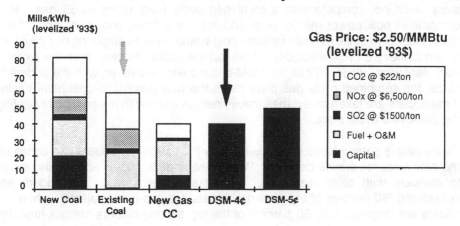

Generation Technologies Compared:
Massachusetts Externality Values

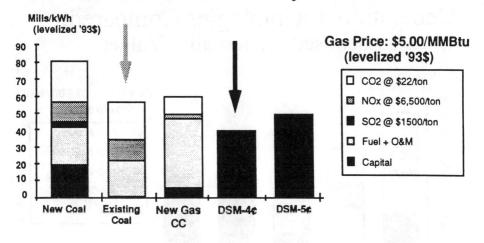

Figure 10

With gas priced at $1.50 the gas combined cycle plant is the clear winner. Even without the externality adders, the combined cycle plant produces electricity as cheaply as the coal plant with no capital charges. The situation changes when the gas price moves up to $2.50. If the externality adders are included, then DSM at 4¢ is the cheapest alternative. Again, if the adders are used, the existing coal plant could not compete with a combined cycle plant using $2.50 gas. In a competitive bulk power market (with adders), the existing coal plant would have to be shut down since even its variable cost would price its output higher than fully priced combined cycle electricity. If we set the adders to zero, then the existing coal plant wins. At $5.00 gas the DSM options win. However, with the adders in place, the combined cycle gas plant beats the new pulverized coal plant easily. If the adders are taken away, then the winner reverses with the coal plant beating the combined cycle plant.

These simple comparisons show clearly how the externality adders can dominate the new resource choice decision. With these adders, coal is apparently unable to compete with other resources. In Massachusetts, when the adders are considered, 90 percent of top scoring bidders are gas-fired projects; when the adders are dropped, only 50 percent of the top scoring bidders are gas-fired. [8]

Although not discussed here, recent technical work on externalities from nuclear generation shows that nuclear externalities will likely be much lower than fossil technologies. This is primarily because of the extremely low emissions from operating nuclear plants, the extremely low probabilities of nuclear accidents, particularly for the advanced designs, the existence of robust containments should an accident occur (e.g. TMI-2), and the fact that almost all potential nuclear externalities are already internalized in the plant cost (e.g., accident insurance, decommissioning, nuclear waste through the utility-funded Nuclear Waste Fund, etc.).

VI. Conclusions

In the next decade the utility industry will undergo significant changes which will place increased demands on generation technologies. Fossil technologies will continue to account for a sizable portion of new generation. Both economic and environmental considerations favor gas based technologies. Future gas supply considerations are a concern. Because of continuing research, there will be a large menu of clean and efficient generating technologies available to meet the emerging needs.

The environmental externalities issue offers the potential to policy makers and the public to improve environmental quality at the same time that the costs for obtaining that quality are reduced. That is the motivating force behind all the activity that is observed in the US and elsewhere. However, whether that promise can be realized is not yet clear. This is true due to the complexity of the issue and the many practical difficulties that enter once the simplicity of the textbook case is qualified by the costly and complex measurements required by the real world application of the simple underlying principles.

References

1. TAG™ Technical Assessment Guide Volume 1: Electricity Supply-1993 (Revision 7), TR-102275-V1R7 EPRI Member Edition, June 1993, Electric Power Research Institute, Palo Alto, CA.

2. Pigou, A.C. 1938. The Economics of Welfare (London, Macmillan and Co.).

3. Coase, R.H. (October, 1963) "The Problem of Social Cost", Journal of Law and Economics III.

4. Baumol, W.J. and Oates, W.E. 1988. The Theory of Environmental Policy, (Cambridge and New York, Cambridge University Press).

5. Joskow, P. L., 1992. "Weighing With Environmental Externalities: Let's Do It Right!", Electricity Journal 5 (4), p.53-67 ,May 1992.

6. Burtraw, D. and Krupnick, A.J. February 1992. "The Social Costs of Electricity: How Much of the Camel to Let into the Tent?" Discussion Paper, Resources for the Future.

7. Freeman III, A.M., Burtraw, D., Harrington, H., and Krupnick, A.J. 1992. "Accounting for Environmental Costs in Electric Utility Resource Supply Planning", Discussion Paper QE92-14, Resources for the Future.

8. Schmalensee, R.L. 1991. Testimony before the Massachusetts Department of Public Utilities, D.P.U. 91-131.

9. Lave, L. 1991. Testimony before the Massachusetts Department of Public Utilities, D.P.U. 91-131.

10. Bernow, Stephen S. and Donald B. Marron, Tellus Institute report "Valuation of Environmental Externalities for Energy Planning and Operations" (May 1990).

11. Barakat & Chamberlin, Contents Of Eprinet's Environmental Externalities Clearinghouse: State Regulatory News, July 22, 1992.

OUTLOOK FOR RISKS
BY ENERGY SOURCES

Professor Dr. Joachim Grawe
Vereinigung Deutscher Elektrizitätswerke
- VDEW - e. V., Frankfurt/Germany
and University of Stuttgart/Germany

Abstract

The paper discusses the notion of risk. It examines the risks of power generation from coal, oil, gas, nuclear and renewables (hydro, biomass, wind, solar). Its scope extends beyond the operation of the various plants and accidents. The fuel cycle and the production of the material for the respective generation facilities are regarded as well. Special attention is given to the "greenhouse effect". In the final chapters, the author tries a comparison and draws some general conclusions.

LES RISQUES AFFÉRENTS AUX DIVERSES SOURCES D'ÉNERGIE

Résumé

La présente communication traite de la notion de risque. L'auteur passe en revue les risques liés à la production d'énergie à partir du charbon, du pétrole, du gaz, du combustible nucléaire et des sources d'énergie renouvelables (hydroélectricité et énergies tirées de la biomasse, du vent, du soleil). L'analyse ne se cantonne pas uniquement à l'exploitation des diverses installations et aux accidents. Sont également examinés le cycle du combustible et la production du matériel nécessaire à chaque type d'installations. Une attention particulière est accordée à l'effet de serre. Dans les derniers chapitres, l'auteur tente de faire des comparaisons et tire certaines conclusions générales.

I. The Notion of Risk

Future energy systems will be determined by the concept of "sustainable growth". Therefore, technologies will be judged not only by their contribution to security and low cost of supply but even more by their environmental benignness and (relatively) low risks.

The word "risk" has a broad meaning. The Chinese use the same letter for risk and for chance. In this paper, "risk" shall be defined as the danger of damage and thus as the contrary of chance, which means the possibility of benefit (1).

The "classic" risk definition multiplies the frequency of damaging events with the extent of the damages, in extreme cases one of the factors being very high and the other very low. This scientific and technical approach is guided by the philosophy that has found its expression in the farmer curve: The tolerable probability of an event must be the lower the greater the damages are expected to be. In the sphere of law, the notion of proportionality applies for this.

The risks by energy sources may affect the life or the health of man, his goods, and his natural as well as his cultural environment. Lack of energy and thus the non-covering of the needs of people, the by far greatest risk, will not be discussed. The paper ist restricted to power generation and to the primary energies coal, oil, natural gas, nuclear, and renewables.

Accordingly, neither final energy carriers nor the demand-side are regarded. Two facts should be mentioned, however: At the application stage, electricity cannot be rivaled by any competitor as to its low environmental impacts; therefore, electrical storage room-heaters and, in particular, the heat-pump can very well compare with oil-fired central heating-systems in spite of the generation losses. This is at least true if the electricity to a certain extent comes from nuclear and/or hydro. Conservation, on the other hand, is by no means always environmentally favourable. Just three examples may illustrate that:

-   In energy-efficient homes the radon concentration is comparatively high (2).

-   So-called energy-saving lamps (compact fluorescent lamps) contain mercury, pose problems of waste disposal and are accused of harming their users by electro-magnetic radiation.

-   Poly-urethane, produced with CFC, is used in many insulation layers and in pipelines for district-heating based on co-generation.

Any technology converting energy into electricity or steam and power influences the environment or man in the one or other way. It goes without saying, too, that for every primary energy the whole chain starting from the extraction of raw energy - and, parallel to that, from the fabrication of the materials for the power plants - to the useful energy must be included in the assessment. All kinds of impacts have to be looked at.

Risk is multi-dimensional. Table 1 gives an impression of the resulting complexity. It makes any recording and comparison so complicated. There have to be solved, i. a., problems of availability and quality of data, of uncertainty (e. g. about thresholds for doses and about repair mechanisms of the human body and of nature), of boundary definition, and of comparability and commensurability of different types of harm (3).

But even if we assume all (objective) facts can be identified and evaluated by scientific research on a broadly accepted methodological basis the (subjective) risk perception by the people is often quite different (4). And it is this risk perception that matters in the end, with its typical over-estimations and under-estimations, heavily influenced by traumatic events, familiarity and other personal experience, and, last but not least, by the mass media.

II.   Risks of Coal

1.   Classic Noxes from Coal Burning

The burning of coal leads to emissions of the "classic" noxes sulphur dioxide (SO_2), nitrogen oxides (NO_x), dust and volatile organic compounds (VOC). The electricity supply industry has a high share in the respective total emission figures (5).

1.1   Risks Associated with Classic Noxes

SO_2 contributes to acid rain. This has led to overacidity of soil and lakes, e.g. in Scandinavia, with negative consequences for flora and fauna. Moreover, it causes severe damages to buildings and cultural monuments by corrosion. In higher doses, SO_2 can cause illnesses of the respiration organs such as asthma and chronic bronchitis and enhance cardiac and circulation diseases with fatal consequences mainly for older people. Several authors attribute 20 % of the deaths caused by lung-diseases to air pollutants, others even up to 50 % (6). Some parts of former Communist-dominated Central Europe, in particular the south-eastern region of the former German Democratic Republic, North Bohemia (Czech Republic) and Silesia (Poland) give illustrative examples of grave air pollution by SO2 and dust, caused mainly by power stations and individual heatings fired with hard or brown coal. Its human health hazards, especially for children, and impacts on nature are well documented.

NO_x are responsible for the formation of ozone. They have been identified as one of the main contributors to the phenomenon of the "Waldsterben" in Europe.

Dust serves as a transport media for heavy metals (cadmuim, lead, mercury) as well as for arsenic and other poisonous substances. If releases of nuclides from atomic power stations are held to be worth discussing it should not be forgotten that coal-fired plants emit about as much - or better: as little - radioactivity as those, namely around 0,01 mSv p. a.

VOC, finally, are said to be carcenogenous.

Depending on the height of the stacks the effluents of the classic noxes may either increase the local immission level or be carried by air over very long distances. Their effects are therefore not limited to national territories.

1.2 Risk Reduction by End-of-the-pipe Technologies

The emissions of classic noxes can be reduced considerably either by abatement (end-of-the-pipe) or by new generation technologies. In West Germany, the electricity supply companies have succeeded in bringing their emissions of SO_2 down to 10 % and those of NO_x to 25 % within the last ten years by installing flue-gas desulphurization and denoxification facilities in the framework of a DM 22 billion crash program (7). But even then there remain around 900 tons of SO_2 and of NO_x each for the generation of 1 TWh in retrofitted coal-fired units.

A Mitre study shows that in the US two thirds of the coal sulphur could be captured by scrubbers within ten years (8). By the clean Air Act of 1990 America's power stations are required to halve SO_2 emissions.

1.3 Reduction of Risks by New Power Station Concepts

Coal as a solid fuel is a fairly "dirty" energy carrier. The necessity to burn it to secure mankind's future energy supply on the one hand and the "must" of preserving the environment at the same time have stimulated the development of a series of so-called clean coal technologies. They cannot be described here (9). Worth mentioning are: advanced combustion processes with better steam parameters and thus efficiencies up to 44 % (hard coal) or 40 % (lignite), fluidized bed combustion (in particular its pressurized form-PFBC), and combined-cycle units with integrated coal gazification (IGCC). The last two technologies are expected to yield even somewhat higher overall efficiences.

It is evident that, in proportion to the lower coal input per kWh generated, the emissions can be reduced. The gazification has an additional positive effect as particulate matter and sulphur are extracted before burning the coal. The effects may be even over-proportional, as for noxes such as SO_2 the assumption of a threshold before severe damages occur seems to be well-founded.

Advanced combustion processes are state of the art and e. g. used for the modernization of the power station park in Eastern Germany. Demonstration plants for PFBC and IGCC, based on hard coal, are under construction in the Netherlands, Spain and Sweden. For brown coal, the "Kobra" project is being planned in Germany. These new technologies - or those of them that meet the expectations and prove to be economic - will be applied for the next generation of coal-fired stations after the turn of the century.

2. Risks from the Fuel Cycle

Coal supply with its front-end stages, in particular mining and transport, and the handling of the wastes cause additional risks.

2.1 Mining and Transport

Opencast mining always means changing the landscape and in some cases evacuation of people. Its negative effects can be repaired by full financial compensation and comprehensive recultivation after the exploitation of the deposit. There is no doubt that the highest risks for workers come from underground mining explosions. Table II shows pit-disasters in Europe with more than 50 deaths in the last 30 years. A number of comparable or even worse accidents happened in China, India, Japan, Simbabwe and South Africa during the same period.

But even the "normal operation" of coal production entails considerable mortality and morbidity risks for the miners by silicosis. In modern pits with efficient ventilation systems these can be reduced as well as explosion dangers.

All existing studies coincide in allotting a non-negligible share of the risks associated with coal to its transport. We have to do with enormous amounts. To generate 1 TWh about 320.000 t of hard coal are needed as compared to 220.000 t of fuel oil, 270.000 m^3 of natural gas, nearly 1 million t of biomass (wood) or 22 t of uranium. Reliable statistical data are available for railway and road accidents (10).

2.2 Waste Handling and Disposal

To burn coal means to produce cinders and slag. The residues from desulphurization (gypsum, calcium sulfite and sulfate sludges) and from denoxification have to be added to them. All these have to be handled, utilized if possible or disposed of.

Their quantities and their toxicity can pose serious problems, in particular, as far as they have to be brought to dumps and ponds. Heavy metals may be leached. For the ashes from fluidized beds we still lack a convincing solution.

An interesting case is the slagheap slide at Aberfan (S. Wales) in 1966 which caused 144 deaths, among them a school with more than 50 children. It was then the worst accident in the UK after World War II (11).

3.   Carbon Dioxide and the Risk of Global Warming

The greatest imminent danger to mankind and its natural environment as a whole probably arises from the "greenhouse effect". What exactly its consequences are for the climate - especially rise of temperature, regional shift of precipitations, or increasing frequency of extreme weather phenomena (e. g. ravaging storms) - and for the sea level cannot be discussed here. It may suffice to quote the Supplementary Report to the IPCC Scientific Assessment of 1992:

" • Emissions resulting from human activities are substantially increasing the atmospheric concentrations of the greenhouse gases: carbon dioxide, methane, chlorofluoro-carbons, and nitrous oxide;

• the evidence resulting from the modelling studies, from observations and the sensitive analyses indicate that the sensitivity of global mean surface temperature to doubling CO_2 is unlikely to lie outside the range of 1.5 to 4.5° C"(12).

Carbon dioxide is the most important greenhouse gas with a contribution of around 50%, nearly half of it coming from electricity generation. It is produced whenever and inasmuch a carbon containing energy carrier is burnt. As coal has the highest carbon content it causes specifically more CO_2 than other fossil fuels. The relations are about 100 : 82 : 72 : 48 for lignite: hard coal : oil : natural gas. From coal mining, gas wells and gas pipelines methane, another and even more hazardous contributor to global warming, escapes in smaller amounts. If this is made allowance for hard coal comes up to lignite, and the difference between oil and gas decreases. As to electricity generation, the figures attribuable to 1 kWh, of course, depend on the efficiency of the respective power stations. They range between 0,5 for combustion cycle gas turbines to 1.2 for brown coal steam turbines (13).

The amounts and the impacts of the CO_2 emissions will increase when and because more coal is burnt to meet the future energy needs. All studies agree that this is more or less inevitable on a world-wide scale. To some extent it will be compensated as the specific input per kWh goes down in relation to the achievement of higher efficiencies for future thermal power stations.

The effects of global warming cannot yet be judged in a well-grounded way. They are by all means enormous. By inundations, loss of fertile land (malnutrition and even famine) or spread of diseases they may even hit people directly and immediately.

III. Risks of Hydrocarbons

The liquid and gazeous hydrocarbons emit SO_2, NO_x and CO_2 as well, though to a lesser extent, with advantages for natural gas in relation to oil. Their environmental belances are thus gradually better than those of coal for the normal operation of power plants. There remain, however, remarkable hazards from the respective fuel cycles.

1. Oil

The main risks of oil emerge from accidents happening mostly in connection with fires, offshore drilling and tanker collisions or strandings. Between 1980 and 1990 seven major events of this type occured (Table III). There has been further pollution of the sea und of coasts by the leaking out of oil from damaged tankers since then. It is estimated that altogether 300.000 to 450.000 t of crude oil and derivates flow into the oceans every year. The "Amoco Cadiz" alone discharged 230.000 t off Britany in 1978. In the Mexican city of Guadalajara 227 people died from a gasoline explosion in 1992 (14). The ignition of the Kuwait oil-fields by Iraqui troops in the Gulf War, however, was a special case.

The British Health and Safety Executive carried out famous risk assessments for the petrochemical complex of Canvey Island in 1978 and 1981. It resulted in fairly high values (15). As in the nuclear industry the studies led to a substantial improvement of the safety level.

2. Gas

Methane (CH_4), the principal component of natural gas, is - as mentioned - itself an important greenhouse gas. The specific effect of one CH_4 molecule is at least twenty-fold higher than that of CO_2. To what extent natural gas leaks through the Russian production and transport systems is controversial.

There are several examples of desastrous gas explosions (including liquefied petrol gas) with dozens or hundreds of people killed within the last two decades. Among them are those on a caravan site near Tarragone in Spain (1978), in a Warsaw bank (1979), in a methane processing plant in Mexico (1984) and near Tscheljabinsk in Russia (1989) where a spark discharge between the wheels of a passing train and the rails set gas on fire that had been escaping from a parallell running pipeline. Here 607 passengers lost their lives (16).

311

Gas cryo-tankers during their stay at ports and facilities for liquefaction and re-gazification of methane at terminals present an additional high and - with the growing number of transports - increasing risk. Up to now no big accident has happened, however.

IV. Risks of Nuclear Energy

The risks of no other energy technology have been so thoroughly examined and so broadly discussed as those of nuclear. We know much more about them than in any other field. The publication of many assessment studies may have contributed to the fact that there exist obvious wide gaps between scientific-technical judgement of most experts and public opinion.

1. Radiation Exposure during Normal Operation

The discharge of radioactive materials from nuclear plants, particularly from power stations, is in most cases extremely low. In Germany, they have further gone down by at least one order of magnitude during the last ten years (except for Tritium and Carbon). The radiation exposure due to them amounts to 0.1 to 1 % of the natural irradation. It thus remain reliably in the range of the latter (17) varying between one and five mS in Central Europe. In some (inhabited) regions of Brazil, France, India, and Switzerland the natural irradation dose is by far higher.

Radiation effects of high doses (more than 200 mS) are well-known. As long as the linear no-threshold dose-effect model is maintained it has to be assumed that even the minimal influences from the normal operation of power plants can cause cancers as stochastic effects. Their contribution to cancer genesis has been estimated to less than 0,0001 % - with, by the way, nutrition and smoking taking the lead with 35 % and 30 % respectively (18). In-depth studies in various countries came to the conclusion that the irregular leukemia clusters cannot be explained by them (19).

The International Radiation Protection Commission uses as risk coefficient 500 fatal cancers, 100 non-fatal cancers and 130 genetic damages per one million people exposed to 10 mS.

As to the workers, the discussion is going on whether at the beginning of the "Nuclear Age" groups of them have been exposed to non-acceptable doses with life-shortening impacts. The monitoring of the surviving Hiroshima victims has led to lower limits for gamma rays some time ago. The ALARA-principle shall keep doses for the workers and the public low.

2. Risks Associated with Nuclear Accidents

The risk of releasing high quantities of radio-nuclides in a severe accident is the crucial problem for the use of nuclear energy. The question how to deal with low-probability high-consequence events is politically still open. The practice all over the world has more or less followed the principle: the higher the damage potential the less probable its realization should be. Many people, however, do not seem to accept any probability above zero at all. There is a general tendency, therefore, to try to exclude catastrophic consequences outside the plant site, for new reactors at least.

Risk assessment studies for the various existing nuclear power stations have resulted in prohabilities for core damage frequencies from 1 : 10.000 to about 1 : 100.000 per reactor and year in the Western World. If accident-management measures, i. e. accident prevention and accident mitigation by human intervention according to simulator-trained emergency procedures using untapped safety reserves (e. g. "bleed and feed"), are taken credit of these figures go down by approximately one order of magnitude. Older power stations often had to be retrofitted to reach this standard. Whether and in how far a core damage will have consequences outside the plant is open and depends i. a. on the integrity of the containment, the humidity in the reactor building, the location of an eventual leak, and the weather conditions.

In some countries "dynamic safety" is a guiding philosophy. Others took the TMI accident as the starting-point for requests by the supervisory authorities for fundamental retrofits. For the frequency of events that could not be mastered by the technical installations the power stations with the probably highest safety standards (the so-called "Konvoi" plants in Germany, commissioned in 1978/88) the value of 4 : 1 million is given. Again, accident management would decrease it to around 4 : 10 million reactors years(20).

This is a sharp contrast to the situation in countries with Soviet-type reactors. They lack sophisticated quality control, the necessary fire protection, sufficient redundancies of essential devices, modern control systems and a containment. The older PWR and in particular the RBMK reactors cannot be retrofitted to Western safety standards at reasonable cost. For some time the deficiencies might be made up for by a very careful and responsible way of operation.

The Chernobyl accident in 1986 was due to design inadequacies compounded by substantial operator errors and even gross negligence in starting and carrying-out experiments. It was classified in the top category ("major accident"), of the IAEA/NEA international nuclear event scale. Chernobyl can be identified with a worst case scenario. It is difficult to imagine a greater dispersion of radioactivity under real conditions.

Chernobyl's consequences cannot yet be fully assessed. 200.000 people had to be evacuated. There are doubts whether some of them might have stayed at their homes and others should have been or should be relocated. A very large area was contaminated and can neither be populated nor cultivated for years or decades. Among children, the number of thyroid gland cancers has multiplied in the most severely hit regions. Lack of information and knowledge, distrust of the authorities and fear obviously have led to massive psychic reactions of many people.

Much of the desastrous consequences must at least partly be attributed to the way the accident was handled by the local authorities and the Soviet government. Chernobyl thus showed that risks always have to be seen before the background of the "safety culture" of the society in question. This applies to all energy carriers, technologies and installations unless the latter are integral part of an international system as, e. g., gas pipelines or oil tankers.

3. Risk Reduction by Advanced Reactor Concepts (21)

As indicated, modern reactor development is guided by the objective to reduce as well the likelihood as the consequences of an accident and, if achievable, to definitely avoid catastrophes, including those caused by operator errors. For that purpose two approaches may be distinguished. One is to design a modified or different reactor type that reliably prevents a core-melting. The other is to take precautions that - in case it should occur - serious negative effects are limited to the plant site. Neither would the near-by living population then have to be evacuated nor would the soil be contaminated (Table IV).

To get there an "evolutionary" and a "revolutionary" concept are pursued. The latter tries to use so-called inherent safety characteristics such as the physical properties an natural laws (e. g. gravity) and passive systems (e. g. emergency cooling of the reactor tank by a water storage above it with re-condensation of the developing steam in the reservoir) to the highest possible extent with the aim to exclude the core-melting a priori. It tends to smaller unit sizes and lower energy densities which would result in longer "grace periods" where no intevention by the crew is needed. Promising examples are the advanced BWR in the USA and Germany. They may be called "partly revolutionary". In their normal operation they do not differ from existing light-water reactor plants. But the concepts to prevent resp. master a core-melting are new.

The fully "revolutionary" approach represents a new start. As examples, the Swedish PIUS and the HTR, particularly its modular type, again designed and - in the latter case - even tested by the Americans and the Germans, may be mentioned. They cannot rely on thousands of years of operation experience. Therefore it still has to be proved that the hardware will keep what the paperwork or demonstrations of precursor units (HTR) promise.

This experience, on the other hand, is a trump-card of the "evolutionists". It gives them the additional chance to appear earlier on the market.

Manufacturers in France, Germany, USA, and Japan follow this path, mostly for PWR, by modifying some construction and machinery elements, adding others (e. g. new types of coolers or a core-catcher), and reinforcing the containment to rule out an early leakage. The melting of the core would thus become a design basis accident.

If these andeavours are successful the risk of nuclear power stations would become infinitesimally small.

4. Risks from the Fuel Cycle

There still remains the fuel cycle. Its risks are again dominated by radioactivity.

4.1 Mining and Transport

What was said about strip mining of coal has to be applied here as well. But as the waste still contains rests of radiating material the stockpiles have to be put up and isolated carefully if the surrounded area is populated. Otherwise there can be a risk of accumulating unacceptable radiation doses. This is demonstrated by the pitchblende mines in Eastern Germany where the Russians kept producing uranium for military purposes more or less without any protective measures - one of the "inherited burdens" they left behind.

The transport risks of the fuel cycle front-end are negligible because the amounts are low and uranium is no intense emitter of rays. The same is true for used fuel elements under condition that transport containers are employed which meet the international safety standards. It may be different for plutonium as a product of reprocessing at least when transported as pure plutonium oxide. But here again very strict safety requirements exist.

4.2 Waste Handling and Disposal

This already leads to the much disputed waste handling issue.

A 1.000 MW nuclear power station annually produces - according to the burnup of the fuel - 25 to 30 tons of irradiated fuel elements containing non-fissionable uranium, plutonium and highly radioactive, heat-generating fission products as well as (ten times as much) low- and medium-active operating waste. This is an extremely small quantity as compared to household and even industrial waste (Table V). Depending on the way chosen to deal with the fuel elements the risks are more short-term or long-term. By reprocessing the plutonium would be recycled and utilized and thus important isotopes with long half-life periods would be removed from the waste. This at the same time minimizes the risk of plutonium misuse for military purposes. On the other hand, reprocessing itself with its liquid wastes (15 m^3 per reactor and year which can be concentrated), the transports connected with it and the use of plutonium bring about risks of their own (22).

The risks connected with nuclear waste can certainly be mastered. The effluents of reprocessing plants are controlled, held back by filters and diluted. For final underground storage several geological media, above all salt, granite, and clay, are being examined. It has been proven that such a deposit if built according to the state of the art keeps away the radioactive substances from the bio-cycle for at least 10.000 years. In this time these will have decayed to a very high degree. After around 1.000 years the dump can be compared to natural deposits of uranium ore with respect to radioactivity.

Sweden and Finland, the first countries to open ultimate disposal sites for non heat-generating waste, have chosen granite. For the fuel elements resp. the just as highly hazardous waste from reprocessing the question is open. Both anyway have to be kept in interim stores above ground for some decades. Tanks with liquid waste from the early (military) times of nuclear have sprung leaks and decontaminated the soil. Special vessels used nowadays have eliminated this risk.

5. Dismantling

Unlike conventional technologies the genuine risks of nuclear cannot be excluded as soon as the power station has been decommissioned. The last fuel has to be removed. Parts of the machinery and the buildings - altogether 6 % of the material - are contaminated or activated. Dismantling therefore has to be carried out in a way that neither the workers nor the public are exposed to radioactivity. Due to its decay this can be done with much less or even without any serious risk after having moth-balled the installation for some decades and when using robots for the demolition of the reactor vessel. Experience has been gained by the dismantling of some smaller reactors, in particular in the USA, and is being gained by several larger units in Germany.

V. Risks of Renewable Energies

1. General Remarks

Renewable energies are sometimes called "soft". Quite a few people hold them all around environmentally benign and free of risks. This judgement is too undifferentiated. The consequences of "the other energy crisis" (23), the rigorous felling of trees because of lack of firewood in various regions of the Developing

World, may serve as an illustration. There is certainly no energy technology without an impact on the environment.

Insufficient information and experience makes assessments difficult. Some effects will probably not be revealed before large-scale electricity generation from renewables exists. It is true, however, that the operation of solar, wind, or hydro power stations does not lead to emissions from combustion. Due to the low energy concentration of these sources and at least as long as only small units are installed high-consequence events with long-term large-area damages can be avoided as well. In addition, there is no fuel cycle (except for biomass to some extent). The predominant contribution comes from the high specific consumption of land and of material for the facilities leading to non-neglectable discharges at the manufacturing stage. Frequent transport, construction and mounting accidents result from the great number of devices. The risks for the workers generally exceed those for the public (24).

2. Hydropower

Run-of-river power stations influence, in particular, natural habitats, the surrounding landscape, and the ground water. Their risks are generally limited.

This is different for hydro plants with (vast) reservoirs. The latter require (fertile) land in river valleys and may imply the resettlement of large population groups. The Volta dam, e. g., flooded an area the size of the Lebanon. There have been examples where stagnant water has favoured the spreading of parasitic and waterborne diseases. Thus, the risk has materialized in damages. As always, these must be compared to the benefits the individual project or the technology yields.

In the history of hydropower utilization there have been numerous examples of dam ruptures all over the world up to our time. Some of them have resulted in thousands of casualties and in large areas lost by inundations and washing off of arable land (25). In the industrialized world, their number and consequences have decreased in the last two decades. Modern construction technologies will bring the probability of risks of large dams further down.

The possibility of accidents during the construction of the facility and of landslips remain, however.

3. Biomass

Among the multifold forms and applications of biomass the burning of wood, straw, and refuse in furnaces and the use of landfill gas as well as sewage gas in motors of block heating stations are of primary interest for electricity supply today

317

and probably to-morrow. The latter need not be discussed here. But the first group poses a variety of risks.

Emissions from biomass power stations, above all particulates, are specifically high because of the low energy content of biomass. Those from waste incineration contain potentially carcinogenic and toxic substances with dioxines among them. The installations need to have high burning temperatures and efficient filters to reduce them to acceptable levels. These substances and heavy metals can leach from ashes and cause groundwater pollution.

If rapidly growing plants, e. g. certain species of trees or C_4-plants like China reed, were grown for energetic purposes the effects of single-crop farming, fertilizers, and pesticides have to be regarded.

4. Wind Energy

It seems as if electricity generation from wind energy entails the least risks, altogether.

Noise is one problem that can be overcome or at least limited by carefully selecting the sites. The influence of wind converters on birds is being disputed.

There may be accidents from blade disintegration or collapses. Transport and setting-up contribute most strongly to the total risk balance (26).

5. Solar Energy

The two methods to produce electricity from solar radiation - using concentrating collectors on the one hand and photovoltaic cells on the other hand - have to be distinguished as to their risks for man and his environment. Among the solar thermal plants only the farm technology with trough collectors has entered the market by now. As long as thermal oil as heat-transfer fluid is used the risks of discharges leading to fires exist. For solar tower facilities cooled by sodium its inflammation may have considerable impacts as well (27).

Photovoltaics yield fairly modest net energy gains. Their by far greatest problems are connected with the manufacturing of the devices (28). It should be mentioned, however, that effects on the local climate and, in the case of large solar generators, changes of the albedo have to be faced. What this means is open.

Silicium, the classic primary product for solar cells, is an environmentally benign material. But its processing involves the use of solvents. The highly toxic fluosilicic acid arises. In a more general sense, toxic (e. g. phosphine) and hazardous (e. g. silane) gases play a role in the cell production. This is particularly true for some new types of cells based on compounds such as gallium arsenic or cadmium sulphide. These substances can hardly be found in nature. We do not yet know their effects on plants and animals.

Cells that have been installed may break or catch fire and by that set free toxic substances.

No definite solution has yet been found for the waste handling and disposal of solar cells and of production refuse. Dangers for the groundwater caused by chlorinated hydrocarbons cannot be excluded. Cadmium and arsenic may leach up to 100 % at deposits. By careful handling and by recycling these risks can be reduced. Of course, all this is a matter of quantity and will come up if and when energy supply demands a large contribution of photovoltaics ("big solar").

VI. Comparisons

Various attempts have been made to compare the risks of the individual energy sources. The "pioneer" study was presented by H. Inhaber in the seventies (29). Two books by Fritzsche and by Kallenbach and Thoene that sum up the discussion appeared recently (30). They are restricted to health risks but treat this subject in a comprehensive way.

Up to now, no entirely convincing method to compare different impacts has been found. The reasons cannot be explained here. They have to do with lack of data and analyses but as well with principal problems of assessment and the absence of a risk yardstick (31). There is general agreement, however, that only complete systems can be contrasted with each other which yield identical benefits, e. g. the generation of 1 GW per year with a defined availability of the power plants. This, if necessary, has to be guaranteed by back-up facilities.

Nearly all authors come to the conclusion that the use of coal and oil entails the highest risks for human health. Natural gas, nuclear and wind range at the end. Photovoltaics come in between. The greenhouse effect has not yet been fully taken into consideration.

Though the generally accepted results are poor there can be no doubt that we need to compare for the optimal allocation of resources. Weighing of alternatives and the choice of the "lesser evil" is inevitable in a given situation (32).

All consequences have to be looked at to our best knowledge. The reasoning that by giving solution B preference over solution A to avoid one risk we may run another is as trivial as seldom helpful. Of two evils choose the least is, on the contrary, traditional wisdom. But by evaluating the overall risks of a given technology it may be reasonable to deduct those avoided by it from the risks caused by it. e.g., the contribution of nuclear power to reduce the danger of climate change can be valued high (Table 6). Phasing-out nuclear would aggravate the CO_2-problem to an alarming extent and lead to additional emissions of SO_2, NO_x etc.

In the whole process, financial expenses, cannot be entirely neglected. The task remains to best prevent damages at lowest costs ("eco-efficiency" according to the Business Council for Sustainable Development) which is nothing else than the good old economic principle.

VII.   Summary and Conclusions

All energy sources have negative impacts. These can be reduced by improved technologies. This as well applies to all of them. In future, clean coal technologies, combined-cycle gas-fired power stations, advanced reactors and more decentralized hydro power plants may be more or less competitors for electricity generation. In the long run, solar applications may complement them.

Nevertheless, the risks cannot be brought down to zero. There is no free lunch. The fire has been tamed by mankind, but it still can destroy lifes and goods. The force of running water, the arising of radioactive substances by nuclear fission, and the burning of carbon to CO_2 remain. To arrive at a well balanced fuel mix by diversification may therefore be an excellent strategy to minimize the total risks.

Risks from energy utilization thus are inevitable for mankind. Their concentration has, no doubt, increased. Marchetti may even be right in saying that the higher the density of an energy source is the less material will be needed to use it (which means preservation of resources) but, at the same time, the higher its endangering potential will be (33). This potential, however, clearly as such is not the yardstick. Life has obviously not become and probably will not become more risky to the individual.

Identified risks are permanently reduced. The efficiency of energy technologies has been growing enormously, and the specific emission of toxic or hazardous substances has been decreasing accordingly.

For judging social adequacy of the risks connected with an energy source its benefits have to be incorporated in the assessment. Costs (including risks) need to be balanced against them. This presupposes that either those who are exposed to the risks or the whole society profits from the benefits. Risk analysis is just part of cost-benefit-analysis.

The assessment would not be complete if the risks of non-action were not taken into consideration, too. But this leads to speculation.

References

This paper is in many aspects and as to a great deal of data indebted to the general studies listed in ref. 1, 3, 4, 11ans 28.

(1)   J. Grawe, Wirkungen verschiedener Energieträger und -quellen auf die menschliche Gesundheit und die Umwelt, in: Geographie und Schule, Sonderheft: Energie und Umwelt, August 1992;

For the risk issue in general see also: CEC and others, Comparative Environmental and Health Effects of Different Energy Systems for Electricity Generation, Key Issues Paper No. 3, Senior Expert Symposium on Electricity and the Environment, Helsinki/Finland 13 - 17 May 1991; H. Khatib - M. Munasinghe, Electricity, the Environment and Sustainable World Development, 15th WEC Congress Montral 1989, Paper for the WEC Commission "Energy for Tomorrow's World", Plenary Session No 8; K. P. Masuhr - H. Wolff - J. Keppler, Identifizierung und Internalisierung der externen Kosten der Energieversorgung, Endbericht Prognos, Basel 1992

(2)   Swedish Ministry for Industry, After Chernobyl (Report Dsl 1986:11 by the Expert Group on Nuclear Safety and Environment), Stockholm 1986, p. 18

(3)   P. F. Ricci - M. D. Rowe, Health and Environmental Risk Assessment, New York 1985; R. Friedrich - A. Voss, External costs of electricity generation, in: Energy Policy 2/1993 p. 114 ff; E. Franck, Risikobewertung in Technik und Wissenschaft, in: G. Hosemann (ed.) Risiko in der Industriegesellschaft, Erlangen/Germany 1989 p. 43 ff.; J. G. Tyror - F. R. Allen - A. R. Taig, Risk Assessment and Comparative Choices, Paper No. 1.1.7, 14th WEC Congress, Montreal 1989

(4)   See e. g.: P. M. Sandmann, Hazard Versus Outrage: A. Conceptual Frame for Describing Public Perception of Risk, in: H. Jungermann - R. E. Kasperson - R. M. Wiedemann (ed.), Risk Communication, Proceedings of the International Workshop, October 17 - 21, 1988, Juelich/Germany p. 163 ff.; O. Renn, Risk Perspection and Risk Management, Paper No. 2.2.1, 14th WEC Congress, Montreal 1989

(5)   For specific data see: IEA, Electricity Supply in the OECD, Paris 1992, p. 58 ff.

(6) Helsinki Symposium (ref. 1) p. 36; Voss (ref. 2); H.-J. Wagner, Analyse der drei Varianten der Hohmeyer-Studie zu den externen Kosten der Stromerzeugung, Berlin 1993 (to be published in VWEW-Verlag Frankfurt a. M. shortly)

(7) M. Hildebrand, 10 Jahre Emissionsbilanzen der Stromversorger (1982 bis 1991), in: Elektrizitaetswirtschaft 1992 p. 1158 ff. [english version: "10 years of emission balances of the electricity suppliers (1982 until 1991)", p. 1169 ff.]

(8) S. W. Gouse - D. Gray - G. C. Tomlinson - D. L. Morrison, Potential World Development trough 2100: The Impacts on Energy Demand, Resources and the Environment, Paper No. 1.2.15, 15th WEC Congress, Montreal 1989,

(9) For detailed information see e. g.: K. Hassmann - W. Keller, Entwicklungstendenzen des Energieverbrauchs und Moeglichkeiten der umweltschonenden Stromerzeugung, in: Brennstoff - Waerme - Kraft 1989, 315;
K. Riedle, Auswirkungen der Klimadiskussion auf Kraftwerkskonzepte, in: Siemens AG, Kraftwerk und Umwelt (Energie-Treff '92), München 1992; Khatib - Munasinghe (ref. 1)

(10) Helsinki Symposium (ref. 1) p. 23

(11) Helsinki Symposium (ref. 1) p. 54; D. J. Ball - L. E. Roberts - A. J. Wilkinson, Societal Risks and Energy Sources, Paper No. 1.4.06, 15th WEC Congress, Madrid 1992

(12) J. T. Houghton - B. A. Callander - S. K. Varney (ed.), Climate Change 1992, The Supplementary Report to the IPCC Scientific Assessment, Cambridge (UK) 1992; see also: Protecting the Earth's Atmosphere - An International Challenge, Interim Report of the Enquête Commission "Protection of the Earth's Atmosphere" of the 11th German Bundestag, Bonn 1989, Table 10 (The Final Report is not available in English. Its findings do not differ from the Interim Report so far).

(13) S. R. Jones - D. C. Watson - W. I. Wilkinson, The Environmental Impact of Nuclear Power, Paper No. 1.1.03, 15th WEC Congress, Madrid 1992, Table 5; Helsinki Symposium (ref. 12) p. 44

(14) Focus (German weekly) 3/1993 p. 188 f. after "Oil Spill Intelligence Report" and other sources; Focus 7/1992 p. 67 ff.

(15) See Ball - Roberts - Wilkinson (ref. 11); Tyror - Allen - Taig (ref. 3)

(16)   See Grawe (ref. 1) p. 5 and 11; Helsinki Symposium (ref. 1) p. 54

(17)   A. Birkhofer - W. Thomas, Schadstoffemissionen aus Kernkraftwerken im Normalbetrieb und unfallbedingte Freisetzungen, in: atomwirtschaft (atw) 1990, 43 ff;

         A. Keller, Kernenergie in Europa und ihre radiologischen Folgen, in: atw 1993, 513 ff.

(18)   D. Henschler, Stochastische Risiken von Stoffen, in: A. Kuhlmann (ed.), 1. Weltkongress fuer Sicherheitswissenschaft, Cologne 1991, Volume 1 p. 431ff.

(19)   H. Wieczorek, Umweltmedizin in der Bundesrepublik Deutschland in: Umwelt Nr. 4/1993 p. 159 ff.

(20)   O. Gremm, Risiko-Studien, in W. Korff (ed.), Die Energiefrage - Entdeckung ihrer ethischen Dimension, Trier/germany 1992 (with data from a number of risk assessment studies) p. 196 ff.

(21)   W. Buerkle, Weiterentwicklung von Leichtwasserreaktioren, in: atw 1992, 404; J.-C. Leny, Franco-German Cooperation in Nuclear Development, Paper Presented at Wintertagung 1993 of Deutsches Atomforum, Bonn January 26/27, 1993; W. H. Young - J. D. Griffith - D. J. McGoff, Nuclear Energy and its Benefits to the Environment and the Economy, Paper 1.3.11, 15th WEC Congress Madrid 1992; R. C. Berglund - J. B. Redding, Advanced LWR Technology for Commercial Application, in: ENS, Towards the Next Generation of Light Water Reactors, Transactions of ENS TOPNUX '93, The Hague April 25-28, 1993, Volume I Invited Papers p. 225 ff; L. Noviello, Overview on Innovative Passive Reactor Types, ibd. p. 242 ff.

(22)   COGEMA, Nuclear Materials Reprocessing and Recycling - Evaluation and Outlook, Paris June 1993

(23)   Erik Eckholm, The Other Energy Crisis, Washington 1975

(24)   M. D. Rowe, Assessing Occupational Health and Safety Risks of Renewable Energy Technologies at the National Level, in: Ricci-Rowe (ref. 3) p. 237 ff; W. Kroeger, Risiken aus Energieumwandlung und -nutzung: Bestimmung, Darstellung, technische Strategien zur Risikominimierung bei Verwendung von regenerativen Energien, in: Kuhlmann (ref. 18) volume I p. 211 ff.; Muenchener Rueckversicherungs-Gesellschaft, Energiesysteme heute und morgen, Munich 1990

(25)   Kroeger (ref. 24) p. 212

(26)   A. Voss - A. Wiese, Gesamtbilanzierung einer Stromerzeugung aus
         Windenergie: Energie, Risiken und Kosten, Paper presented at Forum
         "Umweltvertraeglichkeit regenerativer Energietraeger - Risiko, Akzeptanz,
         Technikfolgenabschaetzung", Cologne 08/09-12-1993 (to be published
         shortly)

(27)   Franck (ref. 3) p. 55

(28)   H. J. Wagner - F. Pfisterer, Umweltaspekte photovoltaischer Systeme, in:
         Forschungsverbund Sonnenenergie, Themen 92/93, Photovoltaik 2,
         Cologne 1993, p. 21 ff.

(29)   H. Inhaber, Risk of Energy Production, 4th edition, Ottawa/Canada 1979

(30)   A. F. Fritsche, Gesundheitsrisiken von Energieversorgungssystemen,
         Cologne 1988; U. Kallenbach - E. Thoene, Gesundheitsrisiken der
         Stromerzeugung, Cologne 1989. Both publications give a lot of data on
         the consequences of the utilization of the various energy systems and
         therefore were consulted for the chapters II through V as well.
         See also: J.-M. Kaelin, Comparing the Risks of Nuclear and Other
         Energies, in: nucleus No. 7/1992 Berne/Switzerland September 1992; M.
         Bertin, Evolution des études sur la comparaison des risques des diverses
         énergies pour la production d'électricité, in: Revue générale nucléaire
         1992, 430 ff. J. Fiksel - A. Cox - P. F. Ricci, Health Risks of Energy
         Production: The Process Analysis Approach, in: Ricci - Rowe (ref. 3); M.
         Yokell - P. F. Ricci, Use of Input Analysis to Estimate Occupational Safety
         and Health Impacts of Electricity Generation Technologies, ibd. p. 208 ff;
         D. W. Pearce - R. K. Turner - T. O'Riordan, Energy and Social Health:
         Integrating Quantity and Quality in Energy Planning, in: WEC Journal Dec.
         1992, p. 76 ff.; Ball - Roberts - Wilkonson (ref. 11); Helsinki Symposium
         (ref. 1) p. 23 ff. and 43 ff.

         Most interesting are the attempts of life cycle analyses of energy options.
         See R. Frischknecht - P. Hofstetter - J. Knoepfel - P. Suter, Total Pollution
         Including "Grey" Pollution: Life Cycle Analysis for the Assessment of
         Energy Options, Paper No. 1.1.02, 15th WEC Congress, Madrid 1992

(31)   Helsinki Symposium (ref. 1) p. 19; Franck (ref. 3) p. 89; G. Hosemann,
         Gefahrenabwehr und Risikominderung als Aufgabe der Technik, in:
         Hosemann (ref. 3) p. 101 ff.

(32)   W. Korff (ref. 20) sees this as the fundamental principle of ethics

Table I: Classification of Risks

Source	Exposure	Means	Object	Effects
1) Fuels (supply) - extraction - transport	**1) Duration** - short - medium - long-term (permanent)	**1) Emissions** - harmful gases - chemical substances - heavy metals - radionuclides	**1) Man** - workers - public - life - health	**1) Extent** - local - regional - global
2) Conversion facilities - fabrication of materials - construction - (normal) operation - accidents - dismantling	**2) Mode** - single - accumulation - synergism	**2) Waste** - toxic - radioactive	**2) Nature** - air - water (ground, surface) - soil - plants - animals - biotopes - landscape - climate	**2) Duration** - short - medium - long-term
3) Waste Management - storage - recycling - disposal		**3) Consumption of land** **4) Noise**	**3) Culture** - buildings - monuments	**3) Time** - immediate - delayed
				4) Mechanism - deterministic - stochastic
				5) Degree - proven - assumed - hypothetical

Source: Grawe (1992), Helsinki Symposium (1991), Prognos (1992)

Table II: Pit - disasters in Europe 1962 - 1992

Date	Place	Country	Type of mine	Number of Deaths
07-02-62	Voelklingen	Germany	hard coal	299
27-02-62	Sarajevo	Yugoslavia	hard coal (?)	54
19-03-65	Ulusa	Turkey	hard coal (?)	68
07-06-65	Kakanj	Yugoslavia	brown coal	128
03-09-81	Zaluzi	Czekos-Slovakia	brown coal	65
07-03-83	Eregli	Turkey	hard coal (?)	102
01-06-88	Borken	Germany	brown coal	51
17-11-89	Aleksinac	Yugoslavia	hard coal (?)	90
08-02-90	?	Turkey	hard coal	68
26-08-90	Tuzla	Yugoslavia	brown coal (?)	180
03-03-92	Kozlu	Turkey	hard coal	500

Source: Stuttgarter Zeitung (German daily) of 05-03-1992

Table III: Major Oil Accidents in the Eighties

Year	Country	Type of Accident	Deaths
1980	Norway	Capsizing of platform "Alexander Kielland"	123
1982	Canada	Sinking of oil-rig "Ocean Ranger"	84
1982	Afghanistan	Crash of tank trucks in tunnel	about 1.100
1982	Venezuela	Explosion of oil tanks in power station	145
1984	Brazil	Explosion of pipeline	about 500
1988	UK (Scotland)	Explosion of oil-rig "Piper Alpha"	187
1989	Alaska	Average of tanker with loss of 42.000 t of oil	- *)

*) only sea and coast flora and fauna hit by one of the worse oil pests

Source: Grawe (ref. 1)

Table IV: Reactor Safety Requirements and Levels

I.   Fundamental requirements:

- Possibility of shutdown at any time

- Safe transfer of residual heat
  (long-term cooling)

- Retaining of radioactivity
  (integraty of containment)

II.   Safety levels

Reactors are conventionally designed for

level 1:   normal operation

level 2:   operational breakdowns

level 3:   design basis accidents

New
level 4:   Precautions for events beyond design basis
                accidents
                (core - melting)

Table V: Annual amounts of nuclear waste as compared to residues from coal-fired power plants, household and industrial wastes in Germany (West) [)]

	Type of waste	amount	relation
1.	high active nuclear waste	0,005	0,03
2.	total nuclear waste	0,15	1
3.	residues from coal-fired plants to be deposited	1	60,6
4.	total residues from coal-fired plants [")]	(16)	(1066,7)
5.	industrial waste	5	330
6.	household waste	30	2000

[)] Basis: Electricity produced from nuclear 130 TWh, from hard coal and lignite 188 TWh

[")] 15 million tons were recycled

Source: Klose, VDI-Bericht 807

Table VI: Supply-side options to avoid 10 MM t of CO_2

To avoid 10 MM t of CO_2 per year in Germany the following supply-side alternatives exist*):

- Increase the efficiency of 38 big coal-fired power stations (650 MW, 4000 h/a) from 37 % to 42 %

- Replace 9 of these by gas-fired power stations

- Instal a nuclear power station (1300 MW, 7500 h/a)

- Build 11300 big wind converters (500 KW, 2000 h/a)

- Fit out 11 MM houses with photovoltaic cells (10 m², 1000 h/a),

----

*)   It is assumed that otherwise the electricity would have to be generated in coal-fired plants with an efficiency of 37 %

THE EVALUATION OF EXTERNAL COSTS
FROM ENERGY SOURCES

THE EC-US FUEL CYCLE STUDY1/

A. Krupnick*, A. Markandya, R. Lee+, P. Valette++**

Abstract

The paper outlines the progress of the joint EC-US Fuel Cycle study. This study seeks to provide a methodological framework for precisely the evaluation of external costs over the complete fuel cycle, from fuel extraction to decommissioning, conservation technologies, solar and wind power.

L'ÉVALUATION DES COÛTS EXTERNES DES SOURCES D'ÉNERGIE

Résumé

Cette communication détaille l'état d'avancement du projet conjoint "EC-US Fuel Cycle". Cette étude cherche à donner un cadre méthodologique pour l'évaluation des coûts externes du cycle du combustible, de l'extraction jusqu'au démantèlement.

1/   This paper reports on the work of many of our colleagues who have been working together on this topic for nearly three years. We acknowledge their contributions, without which there would be nothing to report.

* Resources for the Future, ** Metroeconomica and Harvard Institute for International Development, + Energy Division, Oakridge National Laboratory, ++ Commission of the European Communities, Directorate General for Research (DGXII).

1. Introduction

1.1 What We Understand by Externalities

As the public becomes more and more resistant to investment in, and utilization of, energy sources that have significant detrimental environmental and social effects, the research community has been obliged to evaluate these effects more carefully, and policy makers are grappling with ways in which decisions about investment and operation of different energy sources can reflect their different environmental and other social impacts, such as employment and national security. Both in the Commission of the European Communities (EC) and the United States there is a movement towards 'social pricing', so that different energy sources bear the full costs of providing them.

To generate electric power, many activities are involved, ranging from fuel extraction to decommissioning of the power plant. These activities comprise a **fuel cycle**. Some discharges damage ecosystems, others affect human health or productive activities (agriculture, buildings etc.) and others damage recreational services. The emissions result in impacts, which in turn have economic costs associated with them. In addition to the impacts that are part of the environmental chain, there are others not associated with the environment that are important from a social point of view. Among these are issues of employment, energy security, and administrative cost of regulation.

Where the marginal impacts described above are not properly reflected in the prices charged to the final consumers of the energy sources, they are said to result in externalities. We leave open at this stage what is a 'proper reflection' of these impacts in the price. The joint study by the EC and the US attempts to develop methods to measure all the externalities (environmental and non-environmental) associated with electric power. The study aims to make a major contribution to the monetary valuation of the external costs of power (taking account of the different stages of the fuel cycle), and to disseminate the results so that policy-makers can use them to provide a more rational use of energy. The study was launched by the Department of Energy in the US and the Commission of the European Communities in November 1991, to "develop a comparative analytical methodology and to develop the best range of estimates of [damage] costs from secondary sources". This was to be done for eight fuel cycles (coal, nuclear, oil, gas, hydro, biomass, solar, and wind), as well as for four conservation options for electricity generation. This paper provides a justification of why such work is needed, what the main findings and lessons that have merged so far are, and where we see the effort going in the future.

1.2 Past Studies, and Why Research is Needed in this Area.

Growing worldwide concern about the environment has led to several studies which have estimated some of the externalities associated with electric power production and fuel cycles. Probably the most prominent of the studies are those by Hohmeyer (1988), Ottinger and others at Pace University (1990), Bernow et al. (1990) of the Tellus Institute, ECO Northwest (1987), Pearce et al. (1992), and the ongoing Hagler-Bailly study for the state of New York.

Hohmeyer's (1988) study is certainly one of the first important attempts to estimate externalities. He used a "top-down" approach. First, he identified other studies' estimates of the total damages (health costs, for example) attributed to air pollution. Next he estimated the fraction of the total emissions that are from electric power generation with fossil fuels (e.g., 28 percent). Then he multiplied this fraction times the health costs attributed to air pollution. The result is an 'estimate' of the health damages from fossil fuels. Hohmeyer's methodology enables him to assess "the big picture", but the methodology relies heavily on approximations and previous estimates of total damages. Furthermore it does not take account of the different stages of the fuel cycles, thus ignoring some important sources of external effects. Finally it cannot provide a tool for assessing site-specific effects, which may be very important. In contrast, the EC/US approach is a bottom-up, full fuel cycle approach, with site-specific primary data.

Another well-known study is by Pace University (1990). They used a bottom-up approach to: (i) estimate the emissions and their amounts, (ii) estimate the dispersal of the pollutants, (iii) determine the population, flora, and fauna exposed to the pollutants, (iv) estimate the impacts on the population from exposure to the pollutants, and (v) compute the monetary cost of that impact. Pace relied on numerical estimates from previous studies, and used those values to compute damages. One of the studies studied by Pace was by ECO Northwest (1987). The Bonneville Power Administration (1991), in providing estimates of environmental damages, also relied on ECO Northwest's studies. Pearce et al. (1992) recently completed a study similar in spirit to the Pace study, and drew quite heavily on the latter. They addressed more impacts than Pace, and used a literature review similar to the approach used by Pace to estimate some of the social costs associated with fuel cycles. In neither case were data collected at the primary level, nor was the full fuel cycle analyzed. The results are presented in terms of costs per kg or ton of emissions. Thus they are not site specific and do not take account of differences in external costs that arise from differences in topography or concentrations of population.

The EC/US study goes beyond the earlier approaches in terms of: (a) a more thorough characterization of the energy technologies and their discharges into the environment on a site specific basis, (b) considering all major stages of a fuel cycle rather than just electric power generation, (c) modeling the dispersion and transformation of pollutants rather than relying on previous estimates, (d) engaging in a more extensive, critical review and use of the ecology, health sciences, and economics literatures than previous studies, (e) estimating externalities by accounting for existing market, regulatory, insurance, and other conditions that internalize some damages so that they are not externalities, and (f) using models and analysis, rather than direct use of numbers from the literature.

Unlike the other studies mentioned above, and as part of a different approach, Bernow et al. (1990) of the Tellus Institute point out that it is difficult to estimate social costs based on damages. They suggest that abatement costs may be a reasonable surrogate for damages. In this approach, existing and proposed environmental regulations are analyzed to estimate the value that society implicitly places on different environmental impacts. According to Tellus, the marginal cost of abating emissions, when they are at the limit imposed by regulation, reflects the preference of regulators to require that particular level of abatement and the corresponding incremental cost, rather than allow emissions to exceed that limit and subsequently to have adverse impact on the public. Tellus reasoned that since these regulators represent the public, their views represent the costs placed on those emissions by the public.

We view such reasoning as flawed. The premise that marginal control costs represent the costs of air emissions to society implies that regulators know what individual environmental damages are, and always decide on the optimal policy where the marginal costs of control equal the marginal damages. In fact it is abundantly clear that they do not know these costs, and the political processes by which policy decisions are made do not generally have the property that they equate social damages to costs of abatement.

Another major ongoing study is funded by the state of New York. Their methodological approach is very similar to ours. The major differences between the studies are that the New York study: (a) gives less attention to technology characterization than does the EC/US study, (b) concentrates more exclusively on the electric power generation stage than the EC/US's full fuel cycle approach, and (c) unlike the EC/US study, assumes that externalities are zero for all of the renewable energy technologies. On the other hand, the New York study considers more impacts in the generation stage and is developing a computer model to do many of the calculations. A similar computer model is being developed by the EC for Europe but there is currently no plan in the EC/US study to develop such a system for the U.S.

2. The Impact-Pathway Approach

2.1 Main Assumptions Underlying the Impact Pathway Approach
The Impact Pathway Approach is illustrated in Figures 1-4 below, taken from the coal fuel cycle. Figure 1 shows the different stages of the coal fuel cycle: coal

mining, transportation, generation, and transmission. The different generation of waste is included in the analysis, as is the construction of the additional facilities (to the extent they are needed) for mining, generation, and transportation. The figure identifies the main burdens generated by such activities in terms of water, solid waste, and air. Figure 2 shows the impact pathway, tracing the burdens through the different environmental media into ambient concentrations. These concentrations will have impacts that depend on the receptors' responses—i.e., the ability to absorb wastes which affect human and ecological functions. The impacts are described in terms of health, ecological, and 'other' categories in Figure 3. These impacts in turn are valued as shown in Figure 4, which includes not only the environmental externalities, but also the non-environmental externalities, such as security of supply, administrative costs of regulation, and employment effects of changes in the energy mix.

The fundamental approach, therefore, is to track the pathway from activities to emissions to ambient concentrations to impacts to costs. This is done for a specific site (either real or hypothetical), with actual data on emissions, affected populations, etc. The time frame over which the assessment is carried out is wide enough to cover the full effects of a fuel cycle. Therefore, for example, the construction of a generating plant is included, as is the decommissioning. In order to associate these costs with generation of electricity (per kWh), it is necessary to take a lifetime view of the amount of electricity generated by the plant. This has been the case in the EC-US fuel cycle study.

Another issue that emerged at the outset was that of the discount rate. The pollutants generated from a fuel cycle may exist in the atmosphere for a considerable period of time. For the nuclear fuel cycle, potential effects have to be assessed over thousands and millions of years. Thus the discount rate is of great importance. The debate on what rate should be used has not been resolved and is unlikely ever to be fully resolved. Hence, it was decided that three different rates spanning a reasonable range should be used to estimate the external costs: a zero rate, a rate of 3%, which reflects the best guess at the social discount rate, and a rate of 7%, which reflects the kind of rate typically used for appraising large scale investments in the EC and US. Both the EC and US report for a range of discount rates insofar as it was feasible. However this was not always possible, because in some cases estimates of damages transferred from other studies embody a different discount rate and it was impossible to disentangle the basic data.

The valuation of the impacts has relied almost entirely on the use of dose response functions to evaluate physical damages, and the valuation of those damages using a variety of techniques. These include market values (adjusted for market imperfections where appropriate), estimates based on contingent valuation studies, and estimates based on travel cost studies. Hedonic studies, which value damages by estimating the impact on property values and that largely bypass the use of dose response functions, were used sparingly. The one exception was the valuation of noise costs, where such studies have produced credible estimates of damages.

In selecting the dose response functions, the literature on emissions-concentrations-impacts was searched extensively. Issues arising with the use of different estimation techniques are discussed further below, together with the main lessons emerging from the studies

2.2 What is Not Covered and What Simplifying Assumptions Have Been Made

The above discussion indicates in general terms what the EC-US fuel cycle covered. There are, of course certain simplifying assumptions that underlie the research and there are certain areas that were not covered. First, it is assumed that the additional generating capacity that is being evaluated will leave the dispatch of the utility system unaffected. Accounting for system-wide effects would require running the entire system and looking at the changes in impacts arising from the operation of the modified system. Second, although the impacts of the additional construction facilities for the plant are included in the analysis, the examples we use assume no new roads will have to be built or new mines opened to allow for the impacts of the additional plant. Third, we do not look at the indirect impacts of the fuel cycle, except in the case of energy conservation fuel cycles. Thus we do not assess the environmental costs of producing the cement to construct a generating plant[1]. In general, these second round effects are dominated by the direct effects. The exception is when the secondary processes and outputs would not have been produced were it not for the fuel cycle (as in the case of the production of certain chemicals for the production of solar cells), or when the second order effects are the dominant external effects (as in the case of energy conservation).

3. Main Lessons from the Study So Far

The study is ongoing, with the coal reports in both the EC and US ready for publication. The EC nuclear report should be available by the end of the year and the US in early March 1994. At that time the US will also have the oil, natural gas, hydro, and biomass reports ready. The EC should follow with those in the first half of 1994. The conservation fuel cycles and the solar and wind reports (in which the EC has taken the lead) will be part of the work program for 1994 and should be available by 1995.

---

[1] The US Study reports some of the indirect emissions but does not estimate the impacts of those emissions.

A number of important general points, several of which have been referred to by Krupnick (1993), have already emerged from the studies.

3.1. The Areas Where Monetary Valuation of Externalities Can Be Carried Out by Transfer from Other Studies are Limited

Damage estimates arise in a number of areas. These have been identified as health, ecological, and non-ecological. For health impacts, the transferability from other epidemiological and valuation studies is the easiest, although even here it clearly faces many problems. The valuation of mortality and morbidity effects has social characteristics that vary across nations. Many of the relevant studies have been carried out in the US (especially on the morbidity valuations). Their relevance to other OECD countries is questionable, although one may be able to develop rules of thumb to assist with this (Markandya, 1993). The mortality valuation studies are based on situations where individuals have voluntarily accepted an increased risk of death or sacrificed something to reduce that risk. Issues such as the way the death occurs, the fact that it is not voluntary, the latency period, and the age groups affected are not allowed for in the Value of a Statistical Life (VOSL), which is used in most of the work carried out so far.

For damages to market related products such as crops, valuing small changes in ambient concentrations is relatively easy. Dose response functions that are transferable exist and local price data can be applied. However, where the changes in concentrations are large, it will be necessary to model supply and demand for the different products and estimate the changes in consumer and producer surpluses. This will require an extensive local model that cannot be transferred from outside.

Damages to building materials have been a source of disagreement between the EC and US teams. The EC have argued that reasonable dose response functions and building materials inventories exist for the sites they have looked at. The US team has argued that materials inventories do not exist for most sites and that "no major modeling efforts for valuing the complex behavioral linkages necessary for a defensible materials benefit estimates have been undertaken for many years" (Krupnick, 1993). Damages due to noise are generally transferable, given the consistent relationship between property values and noise levels.

Both teams agree that transferring damage estimates for damages to environmental assets is the most difficult. Damages to recreational sites must apparently be site specific, as rules for 'benefit transfer' have yet to be established. Such rules have been established for the transfer of visibility benefits, where the level of the damage has been assessed as a function of site specific sites (Rowe, Chestnut, and Skumanich, 1990). However transfer from one country to another remains very problematic, as the EC study has found that visibility is not an issue of any importance. Indeed the only study that obtained any estimates of damage costs could not be certain that it was not picking up health effects (the embedding problem).

Finally, studies of non-use values are almost completely non-transferable. Most of them are for non-marginal changes in unique environments, whereas power generation will, at the margin, generally have only small effects on any species or environment.

To conclude, apart from health, crops, and noise, where transferability can be defended if it is carried out with care, other areas will generally require site specific studies to obtain proper estimates of the environmental costs.

3.2 Site-specific Differences Are Important and Differences With Previous Studies Are Emerging

While it is not possible to discuss quantitative results at this stage, it does appear that the estimates emerging from this study are very different from those of earlier studies. In particular, for modern coal and nuclear plants, the damages appear to be much lower than those estimated by PACE and others. Within the small collection of sites investigated (two for Europe and two for the US), differences between sites can be quite large. The implication is that taking average values, as other studies do, may seriously mislead the policy-maker in decisions about different sites and about spatial allocation of energy resources.

3.3 Global Warming Damages and Damages Arising from Long Distance Deposition Remain Highly Uncertain

Both the EC and US teams have felt that the ability to provide quantitative estimates for damages from global warming and acid rain is still very weak—much weaker than for the other areas discussed. The damages from global warming range widely, from almost no effect to catastrophe. The forest damage studies associated with acid rain are total damage studies associated with air pollution, from which one can compute some estimate of average damages for certain ecosystems. On the global warming damages it was decided that they would be reported but not added to other estimates of damages. For acid rain damages, the EC study has (reluctantly) included them, but the US study has reported no figures.

3.4 The Issue of Uncertainty is of Prime Importance and Must Be Raised at Many Levels

There are so many areas where uncertainties arise in the impact pathway that, by the time the final figure has been reported, it contains many cumulated possible errors. It is important to convey the extent of the uncertainty to policy-makers. There are several ways of doing this. One would be to present an uncertainty message system that has been developed in connection with the study, and that offers a structured qualitative appraisal of each study of data entry used to generate damages. A second would be to identify confidence intervals at as many stages of the analysis as possible. A third would be to carry uncertainties at each through to the final stage by developing a low, mid, and high estimate for each stage. There are other methods such as using subjective ratings to indicate the degree of confidence in an estimate.

In this study we have focused on the first and third methods identified above. The NUSAP system, based on work by Funtowicz and Ravetz (1990), is being used for the DOE Fuel Cycle and may be a useful tool for the documentation of choices of studies, especially the uncertainties felt by the analyst in making benefit transfers. NUSAP stands for Numerical Entry, Units, Spread of Values, Assessment of Values, and Pedigree, the categories of evaluation of a piece of information. The second approach, taking ranges of confidence at each point of the pathway and using them in the calculations, results in a distribution of values. It requires a Monte Carlo type simulation of the possible combinations of parameter values to come up with a final distribution. So far in the analysis, we have not carried out such a full simulation, although we have tried to report ranges of low, mid, and high values based on uncertainties at each stage of the pathway.

3.5   The Treatment of Risk and Risk Perception Must Be Developed
An important category of environmental damage from energy sources is that from accidents. The traditional estimation of such damages has been based on the expected damages approach, where potential losses if the accident occurs are multiplied by the expert estimates of the probability of such an accident. However, this results in values associated with the accident that do not properly reflect risk aversion in the population or the lay assessment of that risk. On the first count, estimates of damages based on an *ex ante* assessment of the impacts reveal that accounting for risk aversion could result in much higher values of loss, especially if the probability of loss is small, the size is large, and the degree of risk aversion is high. On the question of lay versus expert estimates of risks, there is interesting and emerging literature which shows that lay assessments are often much higher and, more importantly, that expert assessments are not always as securely founded on data as they would seem to be. Insofar as the basis of our valuation is individual willingness to pay, we should pay more attention to lay assessments than we do currently. Krupnick, Markandya and Nickell (1993) addressed these issues in a paper emerging from the EC-US study. Although it cannot offer definitive adjustments to conventional values of accident damages, it does provide some order of magnitude differences from conventional estimates.

3.6   Consideration of Non-environmental Externalities is Important
The preliminary research, especially, from the US side, shows that as important as the environmental externalities are, the non-environmental externalities associated with power generation are also important. These arise from market failures such as employment and occupational health concerns not being fully internalized through imperfect labor markets, improperly priced public goods such

as roads, energy security issues, and government intervention in the economy, such as liability limits, tax policy, and even public utility regulation itself.

Four types of effects have received some attention so far in the externalities literature: damages to roads from fuel transport, employment benefits from a new generation plant, occupational health and safety, and energy security. Some of these, such a energy security, are being discussed in other papers at this conference. Road damages are in excess of taxes paid by road users, in the case of large trucks such as transporters (Small, Winston, and Evans, 1989). The employment benefits are estimated on the basis of presence of involuntary unemployment. The US study has operationalized this by modeling the sources from which labor would be drawn in such an unemployed pool, and valuing the benefits at the difference between the value of the output produced by such a worker and his reservation wage. Preliminary results indicate that the benefits could be substantial. Whether these benefits are in fact external, however, is still questionable. The EC has not yet undertaken any analysis of such effects. Finally, in the area of occupational health, three reasons the market wages would not reflect the increased risk of disease and thereby internalize the costs are worker immobility, difficulties in the perception of risk, and monopsony influences in the industry. These may all be questioned, but the teams in the EC-US study calculate the environmental cost in the event that the costs are not internalized. It finds that they are significant in almost all cases, compared to the costs to the general public, though not necessarily greater.

4. Relevance of Externalities for Policy

The relevance of external costs for policy arises in many contexts. A rational use of energy demands that the price paid by the final consumer be equal to the full social cost. This requires an estimate of the environmental costs as well as other non-environmental costs. The key question we must answer is how much of the external cost has already been internalized through existing regulation, and how much such internalization matters. In the U.S., these issues are probably the most visible in the regulation of investor-owned utilities by State Public Utility Commissions (PUCs). Over recent years, an increasing number of States are requiring that utilities explicitly consider externalities in their resource planning. Under these requirements, utilities must consider externalities in their cost comparisons, just like any other cost. This debate is ongoing, with some economists arguing that only those externalities that have not been internalized should be included and others arguing that for a PUC, which must take environmental policy as given, the marginal damage as estimated here is the relevant externality for planning purposes. (Joscow, 1992; Freeman et al, 1992). Whatever position one takes, the calculation of marginal damages, as carried out in this study, will be important.

In the EC the debate is not so much on the addition of external costs for private generators, as on the appropriate basis for environmental taxes, strategic planning of energy development within the EC, subsidies to renewable energy sources and the like. In this context the estimated external costs will be of paramount importance. As an example, in the UK there has been a heated debate on the justification of coal and nuclear subsidies to encourage private generators to use that fuel, although gas is a cheaper alternative at present price configurations. The external environmental costs, security of supply issues, and external costs associated with increased unemployment are the critical parameters around which decisions will be made.

In the planning contexts there is an ongoing EC project to incorporate results of the EC/US study in a computable general equilibrium model of the energy-economy.

The dissemination of the results of this study in the EC will be conducted by national implementation teams who will adapt the methodologies to their national needs and apply them to advise their governments on the appropriate energy policies, in an EC context. This process has already begun, with considerable interest both from the research institutes and policy making bodies in the member countries.

In the U.S., the new Administration is currently developing its policy on externalities. An important aspect of the process in developing its policy is the active solicitation of the opinions of the various stakeholders, such as the Public Utility Commissions. Developing computer software so that more users can utilize the results of the U.S. portions of the study should be a major short-term priority—as it is on the EC side as well as in the New York State study.

5.   Problem Areas and Future Research Developments

This study must be seen as providing a methodology, with some estimates of key parameters, rather than as a full set of numbers that can be used in all circumstances and for all fuel cycles. There are main difficulties yet to be resolved.

(a)   The comparisons across fuel cycles are difficult. The comparability across the cycles could not be maintained. The plants cannot all be sited at the same place when it makes no sense to have, say, a hydro plant in the location chosen for a coal plant. Hence the reference environments vary and the analysis becomes even more site specific.

(b)   The choice of a single plant for the unit of analysis has some limitations. First, the scale of the effect may be too small to pick up aggregate effects of several plants (e.g., thresholds may dictate that the effect of the single plant case is zero, but this is not true when many plants are operating). To the extent that dose response functions are linear and go through the origin, this is not a problem. But not all functions are of that type. Second, the impacts of new generation investment on other plants is not taken into account. Thus dispatch decisions, or imports or exports of power, are not modeled in this framework.

(c)   Further work must be done on the extent of the spatial area that is analyzed. So far, work has focused on areas over which dispersion models yield reliable results. But these areas are expanding and there are reasons to believe that, even for discharges other than greenhouse gases, a plant has some effects over a wide geographical area. This issue needs to be resolved and worked on.

(d)   The compliance status of different plants needs to be assessed. Largely we have assumed that plants comply with existing regulations. The extent of compliance, however, varies across fuel cycles. This is one source of complaint from the nuclear lobby, for example, which claims that its level of compliance is higher than some other cycles. This issue needs to be addressed.

(e)   While many externalities can be quantified, our scientific and economic knowledge do not allow us to quantify reliably some potentially very important ones. The most significant of these are the effects associated with emissions of CO_2 and other greenhouse gases, the impacts of acid rain damage from site specific emissions, and the regional effects of discharges on ecosystems.

(f)   Also, as emphasized elsewhere (Lee, 1992), considerable uncertainty about the magnitude of externalities plagues each step of the methodology. More information is needed on the anticipated designs and discharges from advanced technologies so that they can be analyzed. Improved air transport models are needed which are accurate but not so data-intensive. More studies are needed on health impact-pathways and thresholds, in some cases to confirm previous estimates of exposure-response rates. Much more research is also needed to develop general ecological exposure-response functions which are not site- or study-specific. Many more economic studies are needed to estimate the values of many health and environmental endpoints, as well as to confirm current estimates.

6.   Conclusions

The EC-US study has made a contribution to the estimation of external costs and to the use of these costs in energy planning and regulation. It has developed a coherent methodology for plant by plant analysis of the costs and has demonstrated the use of that methodology with real examples. The work is ongoing and the results should be of wide interest.

The major product of the EC/US study is certainly *not* a set of numbers that are estimates of externalities. Rather, it is a methodological framework, with the various analytical methods, models, exposure-response functions, and economic valuation functions which have been compiled and adapted for use within that framework.

The EC/US study uses numerical examples to demonstrate the use of the methodological approach. One clear conclusion from those examples is that externalities are generally *project- and site-specific*: they depend on the design and location of the plant. Both the magnitude and the relative importance of different externalities vary as a result.

The emerging results appear somewhat different from other more derivative and aggregated studies. This will be confirmed in the near future and is an important finding. However, much remains to be done, as has been pointed out in this brief paper.

References

Bernow, S., B. Biewald,. and D. Marron (1990) *Environmental Externalities Measurement: Quantification, Valuation and Monetization*, Tellus Institute, Boston, Massachusetts.

Burtraw, D., W. Harrington, A. Myrick Freeman III, and A.J. Krupnick. (1992). *The Analytics of Social Costing in a Regulated Industry*, Quality of the Environment Division Discussion Paper QE93-01 (Washington, D.C., Resources for the Future), October.

Bonneville Power Administration (1991) *Environmental Costs and Benefits: Documentation and Supplementary Information*, Portland, Oregon.

Chestnut, L.G., and R.D. Rowe. (1988). *Ambient Particulate Matter and Ozone Benefit Analysis for Denver*. Draft report prepared for U.S. Environmental Protection Agency, Denver, Colo. (January).

ECO Northwest (1987) *Generic Coal Study: Quantification and Valuation of Environmental Impacts*, prepared for the Bonneville Power Administration, Portland Oregon.

Freeman A.M. III, D. Burtraw, W. Harrington and A.J. Krupnick. (1992). *Externalities- How to Do it Right*, The Electricity Journal, Vol. 5, no. 7, pp. 1-25.

Funtowicz, S. and J. Ravetz. (1990). *Uncertainty and Quality in Science for Policy*. Dordrecht, The Netherlands, Kluwer Academic Publishers.

Hohmeyer, O. (1988) *Social Costs of Energy Consumption: External Effects of Electricity Generation in the Federal Republic of Germany*, Springer-Verlag, New York.

Joscow, P. (1992). *Dealing with Environmental Externalities- Lets Do It Right*, Edison Electric Institute, US.

Krupnick. A.J., A. Markandya, A., E. Nickell. (1993). *External Costs of Nuclear Power: Ex ante Damages and Lay Risks*, Resources for the Future Discussion Paper, Washington DC.

Krupnick. A.J. (1993). *The Social Costs of Fuel Cycles: Lessons Learned.*, Resources for the Future Discussion Paper, Washington DC.

Lee, R. (1992). *Estimating the Impacts, Damages and Benefits of Fuel Cycles: Insights from an Ongoing Study*, to be published in a book edited by H. Herz, O. Hohmeyer, and R. L. Ottinger, based on the proceedings of a conference on the Incorporation of Social Costs of Energy in Resource Acquisition Decisions, Racine, Wisconsin, September 8-11, 1992.

Markandya, A. (1993). *Air Pollution and Energy Policies: The Role of Environmental Damage Estimation*, Development Discussion Paper, No. 451, Harvard Institute for International Development, Cambridge, Ma.

Mitchell, R.C. and R.T. Carson. (1986). *Valuing Drinking Water Risk Reductions Using the Contingent Valuation Method: A Methodological Study of Risks from THM and Giardia*, Report prepared for the U.S. Environmental Protection Agency, Washington, D.C.

National Acid Precipitation Assessment Program [NAPAP]. (1989). *Acid Deposition: State of Science and Technology*, Report 27 in *Methods of Valuing Acidic Deposition and Air Pollution Effects*, Washington, D.C. (September).

Pace University Center for Environmental Legal Studies. (1990). *Environmental Costs of Electricity*, Oceana Publications, New York.

Pearce, D., C. Bann, and S. Georgiou. (1992). *The Social Cost of Fuel Cycles*, Centre for Social and Economic Research on the Global Environment, University College London, London, United Kingdom.

Rowe, R.D., L.G. Chestnut, and M. Skumanich. (1990). *Controlling Wintertime Visibility Impacts at the Grand Canyon National Park: Social and Economic Benefit Analysis*, Prepared by RCG/Halger, Bailly, Inc., for the U.S. Environmental Protection Agency, Boulder, Colo.

Viscusi, W. K. (1992). *Pricing Environmental Risks*, Policy Study Number 112 (St. Louis, Mo. Washington University Center for the Study of American Business), June.

The Evaluation of External Costs from Energy Sources:
The EC-US Fuel Cycle Study
Figure 1

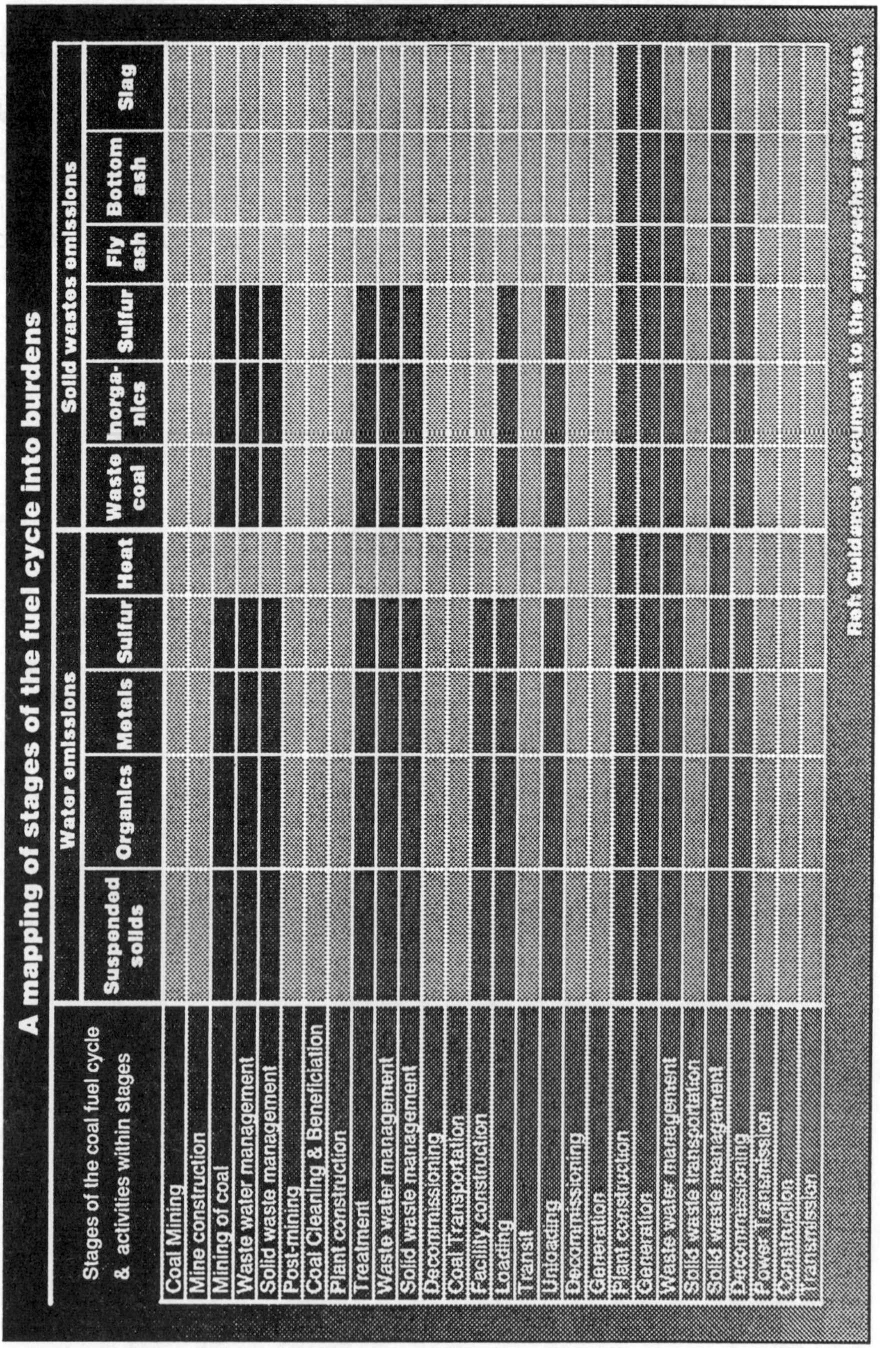

A mapping of stages of the fuel cycle into burdens

A mapping of stages of the fuel cycle into burdens

Figure 2

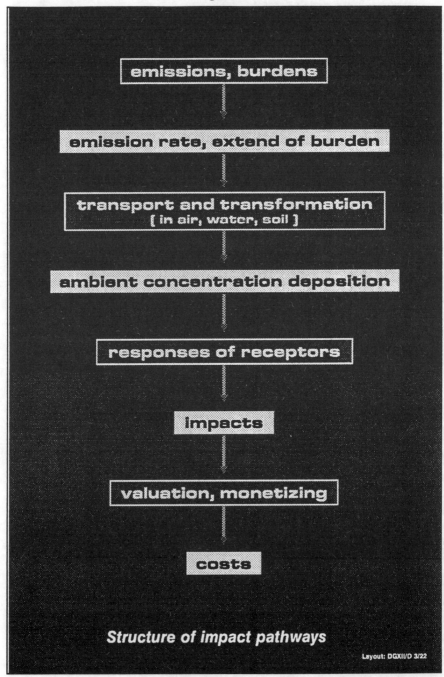

Structure of impact pathways

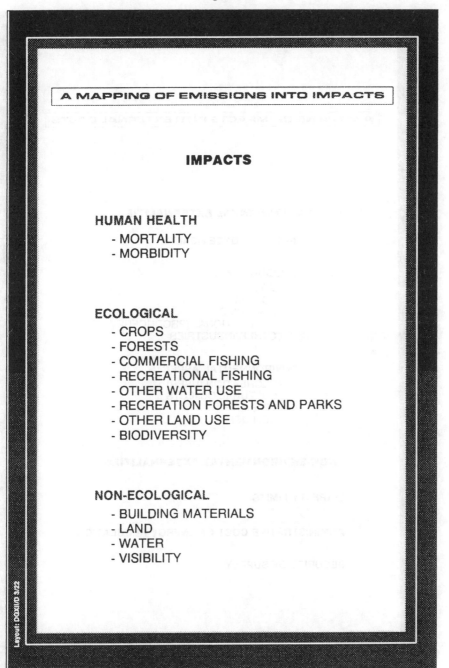

A MAPPING OF EMISSIONS INTO IMPACTS

IMPACTS

HUMAN HEALTH
- MORTALITY
- MORBIDITY

ECOLOGICAL
- CROPS
- FORESTS
- COMMERCIAL FISHING
- RECREATIONAL FISHING
- OTHER WATER USE
- RECREATION FORESTS AND PARKS
- OTHER LAND USE
- BIODIVERSITY

NON-ECOLOGICAL
- BUILDING MATERIALS
- LAND
- WATER
- VISIBILITY

Layout: DGXII/D 3/22

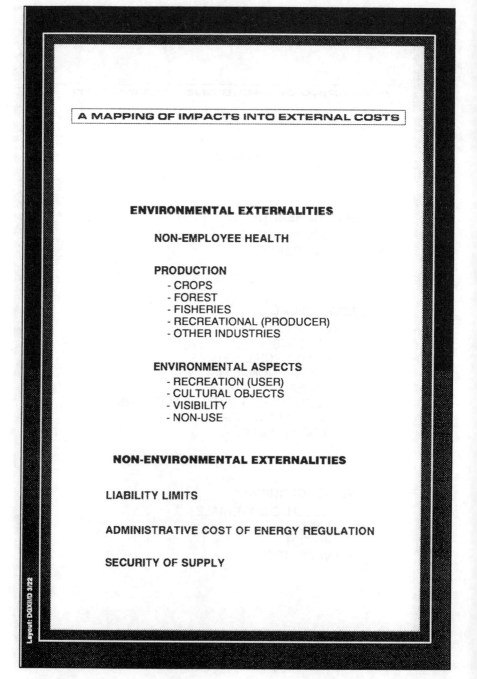

A MAPPING OF IMPACTS INTO EXTERNAL COSTS

ENVIRONMENTAL EXTERNALITIES

NON-EMPLOYEE HEALTH

PRODUCTION
- CROPS
- FOREST
- FISHERIES
- RECREATIONAL (PRODUCER)
- OTHER INDUSTRIES

ENVIRONMENTAL ASPECTS
- RECREATION (USER)
- CULTURAL OBJECTS
- VISIBILITY
- NON-USE

NON-ENVIRONMENTAL EXTERNALITIES

LIABILITY LIMITS

ADMINISTRATIVE COST OF ENERGY REGULATION

SECURITY OF SUPPLY

Layout: DGXII/D 3/22

PERSPECTIVE ON ENERGY SECURITY AND OTHER NON ENVIRONMENTAL EXTERNALITIES IN ELECTRICITY GENERATION

Douglas R. Bohi[1]

Abstract

Applications of the term externality to nonenvironmental matters are often controversial and ambiguous. This paper argues that these externalities are also rarer or less important than sometimes alleged. The paper examines various potential energy security externalities and concludes that none of them are relevant to decisions regarding electric generation. Externalities may exist with regard to effects on local employment and the local infrastructure, although their importance is location specific and their measurement is highly subjective. In short, the consideration of this subset of externalities may confuse policymakers more than it helps them.

[1]Senior Fellow and Director, Energy and Natural Resources Division, Resources for the Future, Washington, D.C. Prepared for presentation to an international symposium on "Power Generation Choices: An International Perspective on Costs, Risks and Externalities", Washington D.C., September 24, 1993. This paper draws from Bohi and Toman [1993], and has benefited from the comments of Dallas Burtraw and Michael Toman, although neither necessarily agrees with the outcome.

POINT DE VUE SUR LA SÉCURITÉ ÉNERGETIQUE ET LES AUTRES EXTERNALITÉS NON ENVIRONNEMENTALES DE LA PRODUCTION D'ÉLECTRICITÉ

Résumé

L'utilisation du terme "externalité" pour désigner des facteurs non environnementaux est souvent controversée et ambiguë. L'auteur de la présente communication montre que ces externalités sont moins fréquentes ou importantes qu'on ne le prétend parfois. Il passe en revue les différentes externalités potentielles de la sécurité énergétique et conclut qu'aucune d'entre-elles n'intervient dans les décisions concernant la production d'électricité. Des externalités peuvent être liées aux effets sur l'emploi et les infrastructures au niveau local, mais leur importance varie selon les cas et leur mesure reste très subjective. En bref, la prise en compte de ce sous-ensemble d'externalités risque de constituer davantage une gêne qu'une aide pour les décideurs.

PROSPECTS FOR INTERNALISATION OF EXTERNALITIES

WHERE WE STAND - WHAT IS AHEAD ?

Peter M.S. Jones

Independent Consultant, Newbury, England; formerly Chief Economic Adviser, AEA Technology, and Chairman, Nuclear Energy Agency's, Nuclear Development Committee.

Abstract

The paper examines the ways in which technical, economic and political problems affect the internalisation of externalities, using gaseous emissions form fossil fuel burning as the principal example. It examines the current position on internalisation in OECD countries and concludes that even though market mechanisms may be used to adjust the relative attractiveness of different fuels, the choice of "adders" will have to retain a considerable degree of arbitrariness and be based on technically informed political judgements into the foreseeable future. This should not however discourage efforts to improve the state of knowledge.

PERSPECTIVES D'INTERNALISATION DES EXTERNALITÉS

BILAN ET PRÉVISIONS

Résumé

La présente communication analyse comment les problèmes techniques, économiques et politiques affectent l'internalisation des externalités, en se fondant principalement sur l'exemple des émissions gazeuses issues de la combustion des combustibles fossiles. Elle présente la position actuelle des pays de l'OCDE concernant l'internalisation et conclut que, même s'il est possible de jouer sur les mécanismes de marché pour équilibrer les avantages relatifs des différents combustibles, le choix des "paramètres complémentaires" restera dans une très large mesure arbitraire et se fondera sur des jugements politiques techniquement bien étayés dans un avenir prévisible. Cela ne doit pas toutefois décourager les efforts déployés pour faire progresser l'état des connaissances.

INTRODUCTION

There are two separate and distinct facets to the question to be addressed in this paper. The first relates to our knowledge of and ability to evaluate the externalities associated with energy production and use, and the second to the likelihood that governments will do anything that will bring about a closer relationship between the overall societal costs and prices in the market-place.

During the course of this conference a great deal will have been said about the present position in the OECD countries on both of these aspects. Since this paper is being written before we have had the benefit of having heard the other contributions, some of the views to be expressed may need modification. Nevertheless, I believe the facts are clear and unlikely to be changed significantly in the short term.

Indeed, relatively little has changed in the past 26 years since I first entered the field of economic assessment and its application to decision processes in industry and government. Even at that time we were trying to determine in quantitative economic terms the effects on society and the economy as a whole of air pollution[1], the incidence of fires[2], marine oil pollution[3], and a wide range of other technological innovations[4]. Our purpose was to establish whether research or specific development programmes were justified by the prospective benefits, taking due account of eventual markets and the estimated prospects of success[5], or to identify technological areas of opportunity where innovation could be expected to beneficial. After many years of such analysis it was possible to identify very clearly the limitations to the application of quantitative criteria in practical decision making in our fields of interest[6].

At that time some governments were sympathetically disposed to the concept of introducing social welfare, environmental, and other wider impacts into their decision processes in a quantitative manner. The US Congress' Office of Technology Assessment was formed[7], a Futures Group reported directly to the Swedish Prime Minister[8], whilst in the United Kingdom the Programmes Analysis Unit covered the areas within the then Ministry of technology's extensive remit[4].

Some of the thinking prevalent in the late 1960s and the 1970s has been institutionalised in such forms as the environmental impact assessments statutorily required in the USA and the European Economic Community. Economic Cost-Benefit analysis was also built into the International Commission for Radiological Protection's definition of the ALARA (as-low-as-reasonably-achievable) criterion[9], and to its application in countries like the United Kingdom[10].

However, progress has been patchy, and the application of quantitative criteria, even when they have been clearly defined, is often considerably modified by the superimposition of judgemental, administrative, or political considerations.

This paper will look at some of the reasons of this, starting with the problems of evaluation and progressing to the political aspects. It will then take stock of where we stand, with particular reference to the nuclear industry, and look to the prospects for the future, and how these might influence public and political perceptions of nuclear power.

The views expressed are wholly those of the author and they do not necessarily correspond with those of his previous employers or the members of the expert groups he had the privilege of leading in a number of studies for the Nuclear and International Energy Agencies during the 1980s and early 1990s.

PROBLEMS OF EVALUATION

For simplicity and ease of understanding the following paragraphs look first at the problems of evaluation in terms of emissions associated with fossil fuel use. However, the problems, whilst different in kind, also exist for other forms of benefit or detriment, and some of these will be looked at later.

There a number of distinct stages that have to be gone through in order to evaluate in monetary terms those effects of energy production and use that are not reflected in the market place. Fortunately there does now seem to be a general recognition that these effects need to be evaluated on a cradle to grave basis, covering the exploration for, and extraction, processing and transportation of fuels and materials of construction, as well as the effects of their use and the disposal or dispersion of the effluents and waste products arising at all stages of the process[11]. This is particularly important where the renewable sources and energy conservation measures are concerned[12] because, although many of them have no direct emissions, they generally require relatively large quantities of steel and cement, or other materials, the production of which does lead to significant emissions of acid gases and greenhouse gases.

The first requirement is to identify qualitatively and comprehensively the external costs or benefits that arise in one way or another from fuel use. This is no light task. Even knowing which emissions, wastes or activities may contribute to external costs may be difficult. It was only in the early 1970s that the importance of nitrogen oxide emissions from vehicles came to be appreciated[1], despite the known toxic nature of the gas. The UK's Central Electricity Generation Board argued until well into the 1980s that its policies of dispersing acid gases from coal and oil combustion via high stacks ensured that they caused no environment damage. There is still no consensus on the full significance of the past and continuing use of lead additives in fuels for internal combustion engines.

Outside the fuel field, the importance of chlorofluorocarbon emissions in terms of their effects on the ozone layer has only been known for few years. Many other examples could be quoted in the fields of pharmaceuticals, pesticides, etc. Even when the facts have been known, action has often only been initiated when public consciousness has been aroused[13].

Even when the potential for adverse consequences should be recognisable in advance it may go unremarked until after the event. The spread of bilharzia following hydroelectric schemes in Africa, the instability of coal mine spoil heaps like that at Aberfan, and the tragic fire at Chernobyl in the Ukraine are examples.

The potential effects of sulphur dioxide through its contribution to acid rain appear to be well known, although surprises still arise. It is only recently that the possible role or atmospheric sulphate particles in reducing the earth's surface temperature has been considered. The obvious effects of acid emissions on masonry and their contribution to the rusting of outdoor metalwork by dissolving protective coatings, have long been understood[1]. The effects of such gases on other paintwork, fabrics, paper, crops, trees are less well established. the effects on human and animal life, both directly and indirectly, through their effects on the habitat, are only gradually being recognised; except in the case of the consequences of their acute effects in high concentrations.

It is certain that however careful we are in our analyses we will be likely to cover only a part of the overall impacts of any pollutant; usually the part that is most obvious and most easily measured. Is it a large part of the whole or merely the tip of the iceberg?

Having identified the means by which external effects could occur we are still faced with the problem of quantifying the consequences in meaningful terms and attributing them to the specific agents and their sources. It is not sufficient to know that masonry can be eroded or zinc galvanising dissolved; crop yields reduced or land acidified; or human pulmonary function impaired. We need to know by how much and to be able to scale this to obtain a measure of the overall effect within the geographical boundaries laid down for the analysis[1].

If we can obtain a general measure of the ambient pollution level and from it deduce its likely consequences, based on an understanding of the fundamental mechanisms at work; or if we can relate the consequences statistically to the pollution level, as in conventional epidemiological studies, there is still a need to attribute the ambient pollution to the correct sources.

There are many pitfalls awaiting the unwary. The problem of synergy can be important. The health effects of acid gases like sulphur dioxide are not determined solely by their atmospheric concentration, but are also influenced by the presence or absence of particulate matter[1] and, one would imagine, its particle size. The effects of nitrogen oxides are strongly influenced by the presence of hydrocarbon gases and sunlight. The contributions to ambient levels and damage effects in

centres of population are likely to be far higher from emissions at low elevations from vehicle exhausts and domestic chimneys, than from the emissions at high elevations from power stations or factories with tall smoke stacks[14].

Some effects are local (visual intrusion, solid waste tips, noise and dust); others are regional (acid gases, water pollution, smoke and micro-climatic change). In both cases their effects will be strongly dependent on the location of the source in relation to centres of population, to the topography and the local climate. (This is not the case for carbon dioxide where the scale or the effect is global rather than regional.) One consequence of this is that there is no reason why the external costs appropriate to the USA, for example, should be the same as those appropriate for Western Europe or Japan, Geography, climate, population density, individual wealth, and revealed preferences all differ significantly across OECD and even more when a wider range of countries is considered.

Leaving these questions aside for the moment, there remains what is probably the most difficult and contentious task of all, the monetary evaluation of the effects that have been identified. Some effects are in principle tangible and measurable and have a clear market value. Reduced crop yields, restoration and repair of buildings are examples. However, the equally tangible and measurable (in principle) damage to unique architecture, to works of art or paper records is far harder to deal with. The Acropolis in Athens, the Sphinx in Egypt, the medieval cathedrals of Europe are examples where the blame can be unequivocally attributed to general pollution levels from fuel combustion. It is also a matter of record that the damage done during this century far outweighs the accumulated damage incurred over the many centuries or even millennia that have elapsed since these and similar monuments were constructed.

Equally tangible are effects on morbidity and mortality which figure prominently in the external costs of energy sources as they are usually presented[1][15]. But what is the value to be attached to illness or premature death? The most prominent economists in the field currently favour the use of revealed preferences for such evaluations[15][16]. That is, the value attached to premature death is deduced from the responses of samples of the population to questions on how much they would be prepared to pay to reduce the levels of risk of early demise. In many countries[17] this approach has replaced that used previously, which attributed a value related to the average contributions individuals make over their lifetime to gross domestic product, i, e, the average salary[1][18].

The present author, who may well be in a minority of one at this point in time, believes the latter to be better founded when one is concerned with decisions on public welfare. It properly reflects an upper bound to what society is willing and able to pay to individual over his lifetime, within his community, to enable him to enjoy his life to the fullest extent possible. I can see no reason why society should attach a higher valuation (albeit his) should he die prematurely. The distinction may appear somewhat academic, but it can have a dramatic effect on the external costs associated with particular energy sources.

I have used the term upper bound advisedly, because some economists also favour the introduction of risk aversion premia into their valuations[15][19]. The arguments for such a course are again based on revealed preferences. People (or at least the media) are more concerned about accidents involving larger numbers of deaths or injuries, and are said to be disproportionately more averse to higher levels of risk, although the evidence to support the latter claim is far from convincing. Even if it were true, then consideration of the full range of risks to which individuals are exposed in their lifetime, leads me to the conclusion[20] that society would not be prepared to accept the sacrifice that would be involved in meeting the expenditures implied to be necessary to reduce the risks to negligible levels.

This is very evident in countries with publicly funded health and welfare services. The expenditures, though large, are constrained as an act of policy to levels well below those that would be justified by the very high values some economists would suggest to be appropriate on the basic of the revealed valuations of individuals.

If all this seems difficult, what about the intangible consequences? The means devised by economists for measuring the value of quality of life type effects are ingenious if nothing else. Such things as the preservation of habitats for wildlife or even the preservation of species, or the value of a view or area of countryside, a clean as opposed to a dirty environment, have again been approached from the perspective of revealed preferences. Proxies such has property values (after attempting to eliminate all the other factors that influence them)[21], the distances people are prepared to travel and the numbers travelling to visit leisure sites[22], and the ubiquitous questionnaire have been employed. They may or may not give guidance in relation to carefully defined situations that have been the subject of specific studies, but how can they be generalised, and would the results be the same if a wide range of "attractions" were under threat at one the same time?

The greenhouse effect adds yet another dimension. Whilst there is no question but that the effect is real, its predicted consequences are essentially hypothetical, based on large mathematical models that may or may not capture all the significant parameters, and may or may not contain the appropriate equations to represent the behaviour of the real world. There is a considerable degree of consensus[23] about the likely global temperature rises given specified increases in the concentration of carbon dioxide in the atmosphere, but are there complex positive or negative feedback mechanisms that have yet to be recognised? The possible release of trapped methane from permafrost regions as they thaw out has been suggested as a major positive feedback. Possible biological mechanisms exist that could provide stabilising negative feedback.

However, even if the average global temperature rises could be predicted accurately, there remains the question of regional variations of temperature and rainfall, and importantly the frequency and severity of climatic extremes. There appear to be a growing number of such extreme events which, whilst within the range of long term historic experience, seem to be occurring too frequently and in too many places to be due to chance alone; yet one can not say that they are attributable to increasing levels of carbon dioxide in the atmosphere. The UK has seen two violent and damaging hurricanes, a run of unusually hot dry summers (pre-Mount Pinaturbo), and several years of sever drought in southern countries, mixed with torrential rains and severe flooding. The USA has had droughts, storms and the recent catastrophic flooding along the valleys of the Mississippi and Missouri rivers, freak snowfalls and frosts. Africa has suffered prolonged drought and famine. Europe and Asia seem to have had more frequent and severe floods, droughts and storms in recent years.

It may be that these are transient phenomena, or they may reflect a real trend that has nothing to do with the release of greenhouse gases from fossil fuel combustion. When will we know?

In view of the uncertainties in the model predictions it is no surprise to find that the physical, ecological and human consequences of the greenhouse effect are subject to even greater uncertainty. If there is a climate change and if it is sudden there will be little time for mankind and nature to adapt and the costs could be high. A slow change allows time for adaptation and species migration. Both winners and losers are to be expected, but who and to what extend is problematic.

Yet another dimension to the problem is raised by the difficulty of attributing relative values to the gains and losses of the present as opposed to future generations. Over short periods economists deal with the problem using discounting which weights near term costs and benefits more heavily than those in the future[24]. However, if standard discount rates are used, the interests of future generations are sacrificed for short term gain[25].

This can be justified if one believes, as everyone did in the 1950s and 1960s, that scientific, technical and economic progress will leave them richer and better equipped to cope with the problems we bequeath them[26]. Now, perhaps, we are not so sure. Nevertheless, I would argue that it is still wrong to adjust one's assessments, by measures such as altering the discount rate, to give greater weight to the future. The fact is that such an approach is entirely arbitrary with no sound economic basis for the choice of weights. Far better to take a view on the value that future society will attach to the matters covered by the analysis [27]. This is not easier nor is it necessarily more accurate, but it does focus attention on the real issue of changing resource (tangible and intangible) scarcity and the value standards that are being employed, including those being attributed to the future.

The use of discount rates based on real and _realistic_ rates of return on investment makes allowance for the fact that society can increase its real wealth by investment now, and this provides the basis for the belief that future generations will be better off.

The fact that intergenerational equity poses problems is a characteristic of modern industrial society and the capability man has developed for radically altering his environment for better or worse. Recognition of this has led to the call for the development and use of sustainable technologies, paralleling management and development in agriculture[28].

In essence many of the problems that should now be faced up to by governments on behalf of mankind are ethical rather than economic. Attempts to force them into a pseudo-economic framework does no one any good, least of all those charged with making decisions on our behalf.

OTHER EXTERNALITIES

This paper has concentrated so far on environmental externalities, with particular reference to emissions. There are many other environmental impacts that also have to be considered. Visual intrusion from plants and mining operations, subsidence, noise, smell, disruptions to society during construction and operation stages, solid and liquid wastes. Most are local to the mines, transport routes and plants. Some have wider regional implications[1].

There are also several other impacts that can arise from the choice of specific energy options that are not currently reflected in the costs to the consumer, although some are taken into account when governments as opposed to markets determine the choices made[30]. Examples include the consequences of choices for employment, for gross domestic products, for the balance of payments and future terms of trade.

Security of supply, contributions to a country's technological base and spin-off effects are also often cited[30].

Selection of the option with lowest costs will normally be best for a country's GDP and employment prospects in the long run, although there can be transient adverse consequences as the economy adapts to change[30]. Use of the lower cost option will release resources which can be redeployed elsewhere in the economy, with a lag, to increase the level of output or services available. Preserving an expensive option precludes this possibility and maintains the economy in a less competitive position, with needlessly high prices, which can affect the terms of trade and lead to depreciation of a nation's currency, making everyone worse off.

Balance of payments effects can have similar consequences. Use of indigenous resources rather than imports, or favouring an option with lower import costs, or one that releases resources for export, will have positive consequences for a country's balance of trade. This should benefit the exchange rate if currency is floating or permit the use of lower interest rates if currency parities are fixed. The extend of the adjustment depends on the price elasticities of imports (-0.5) and exports (-1.5)[31]. For the UK economy it has been estimated that domestic fuels should be favoured provided their price is no more than about 66 percent greater than that of the available imports[32]. This differential is not reflected in costs paid by the generator and it is as much an externality as environmental damage.

Security of supply has been given great weight in government decisions in the past[33]: not surprisingly in the light of events during the 1970s. It can be enhanced by reliance on indigenous resources, by diversity in types of fuel and their sources, and by stockpiling fuels. Insofar as these strategies diverge from commercial preferences of fuel users like generation companies (for the least cost fuel with minimum stock holding), they add to the overall costs of energy supply. The extra costs are relatively easy to assess, but the benefits of the enhanced security can only be established ex-post or on the basis of assumed disruption scenarios.

Even ex-post analysis is difficult because there are almost invariably a range of factors involved. Examples include the contribution of nuclear power to the post-1984 decline in world oil prices and to the defeat of the year long miners' strike in the UK in 1984[34].

The value of any benefit arising from diversity is therefore highly speculative. At best one can pose the question of whether the potential benefits arising from specific means of achieving diversity on a plausible range of disruption or price escalation scenarios are small or large compared with the costs. This may give guidance on the value or otherwise of the insurance offered by the chosen strategies[35]. An alternative approach was explored by Ulph[36], who attempted to estimate the value attached to reduced risks of fuel price fluctuations, but this too had to rely on postulated scenarios.

WHY INTERNALIZE?

It is assumed in this text that the object of internalisation of external costs is to give the right price signals to those making decisions. These decisions can however take a number of forms. They may be by potential consumers solely concerned with energy use. They may be by private sector companies considering investment in energy extraction, production or supply. They may be by public sector companies concerned with production or distribution. They may be by governments interested in regulation, research and development or general energy policy. And they may be by international bodies.

Some of these are concerned with the supply and consumption of energy from the plants that exist; others with the types of plants that should be developed or built in the future. The externalities associated with the different time-frames are different.

Ideally use of existing plants should take account of marginal direct and social costs, omitting sunk costs and future unavoidable costs arising from past activities which are not affected by the current use or non-use of the plants. For nuclear plants the unavoidable costs include those of management of existing stocks of spent fuel and plant decommissioning[37]. This approach has been endorsed for nuclear plants in the UK by the Government[29] and the Trade and Industry Select Committee[38]. External costs should also be treated on the same marginal basis. There is a question however of whether the marginal costs should be solely those appropriate to the current stock of plants, i.e. short run marginal costs (SRMC), or those needed to give consumers forewarning of the longer term implications of their consumption, i.e. the long run marginal costs (LRMC).

The social costs associated with plant construction, visual intrusion and eventual plant dismantling should be omitted from SRMCs but would be included in LRMCs. The use of LRMC pricing has been widely advocated as a means of helping consumers make optimal decisions on their fuel choices and appliance purchases, and to provide the right signals on capacity requirements to suppliers [39]. when there is surplus capacity, as is the case in many countries at present, the use of SRMCs is more appropriate.

For new plant investment it is appropriate to consider the full direct and external costs that will be incurred on a cradle to grave basis. This would include environmental and macroeconomic externalities, like balance of payments and security of supply benefits over the plants projected life. The same is true for LRMCs which, for electricity supply, are generally equal to the average cost of generation from the future new plant with the lowest costs[23].

For R and D decisions by government it may be appropriate to consider spin-off and strategic benefits to the economy of different development options. The use of adders (or subtracters) and internalisation are not the most efficient way of dealing with these aspects in non-market situations like public sector decision making.

Where external costs are variable and dependant on aspects of a plants design and use, then their internalisation would encourage an operator to optimise his plant to achieve minimum total costs. The resultant minimum cost is that which provides the right signals to consumers provided the social costs associated with the optimised level of emissions are included.

THE POLITICAL DIMENSION

The previous sections have served to illustrate that there are major problems in evaluating external costs. Assuming that these could be overcome, wholly or in part, would governments take advantage of the new-found knowledge?

The situation in the United Kingdom offers us an ideal opportunity to test the political will[40]. The recent government White Paper, *The Prospects for Coal* [29], laid out a number of key elements of government policy. One of these was to ensure that energy is provided to customers in a commercial environment in which the consumers pay the full costs of the energy resources they consume (my emphasis).

Another is to have full regard to the impact of the energy sector on the environment, including measures to meet the Government's international commitments. These commitments relate to the targets agreed within the European Economic Community for reductions in the emissions of acid gases and carbon dioxide that arise from the combustion of fossil fuels.

The former Secretary of State for Energy, Mr. Wakeham, when he announced in September 1990 the Government's intention to review the economics of nuclear power in the UK in 1994 (now brought forward to 1993), said that future PWR stations would need to be assessed to be economic over their life as a whole, and that the benefits of fuel diversity and the environmental advantages of non-fossil fuel generation would need to be taken into account in the assessment[41].

Most OECD countries have formally recognised the fact that energy production, conversion and use, which are essential to the existence and maintenance of any modern industrial society, bring with them less desirable side effects that need to be controlled. The recognition is neither recent nor revolutionary. What is new is the acceptance by governments, both within and outside OECD, that the environment has a global dimension. There are also new approaches, both technical and economic, to the issue of environmental protection.

Recent UK Governments have been more committed than most to the view that free markets provide the most efficient means of allocating national resources, including those that have generally been regarded as part of the national infrastructure and which had been assumed to carry a degree of natural monopoly. It is also the Government's intent that the newly created privatised energy companies should operate with the very minimum of government interference[29].

A logical inference is that the different energy options available in the UK should reflect their full costs to society in the prices charged to the consumer[40]. In this way the consumer could strike a balance between his use of energy and his investment in its conservation; in his use of electricity as opposed to other energy forms where substitution was possible; and, given the choice, between nuclear electricity, electricity from renewable sources and fossil-fuelled electricity, and the varied range of options that is available within each of these categories.

However, whilst fine in theory, there are severe practical difficulties in the way of such a free market solution :

(a)   Setting energy prices equal to their full social cost will not lead to an overall optimum use of energy if all non-energy prices in the economy remain on the current basis that excludes social costs.

(b)   The typical consumer has no market mechanism for expressing preferences amongst the alternative means of power generation. Electricity is an undifferentiated product with a single price, delivered through a single channel, and it will remain so for the foreseeable future.

(c)   The scope for the introduction of true competition in supply is limited by the high costs of entry into the supply industry and the relatively high costs of transporting and distributing energy, particularly to the smaller user.

The present structure of the UK electricity industry does afford some large scale users not covered by the franchise markets a potential measure of choice. They can choose to generate their own power by whatever means they like, including the permission; and they can choose between the major generators for direct supply of power. In practice Nuclear Electric has been barred from competing, but were this restriction removed the user could choose between an essentially nuclear source and a mix of fossil based supplies as operated by the National Power and Power Gen.

Even in the UK therefore there remain major obstacles to the use of markets and pricing as the sole route to achieving socially optimal energy choices. Whether this matters or not depends on the extent to which the costs not reflected in the market place are significant relative to the direct financial costs to the generators of electricity production.

The final terms of reference for the UK's nuclear review are not known at the time of writing this paper, although they should be known before the conference itself. However, taken at its face value, and assuming no change in Government policy, it would appear that explicit recognition will be given to the need to incorporate external costs into energy prices in order that the market can respond appropriately, or, at the very least, to ensure that they are reflected in the decision processes concerning investment in new capacity and the preference given to sources available to supply power to the Power Pool. The latter is the UK's means

of providing a short term spot market for electricity, based on bidding by suppliers 24 hours in advance to supply in specified 30 minute periods[42].

Whether or not this will happen is a matter of conjecture. Although a number of European governments have introduced specific taxes related to the carbon content of fuels[43], the United States has been considering a broad brush energy tax based on the energy content of fuels, and the European Economic Community an energy/carbon tax based on both energy content and carbon content, but selectively excluding renewable sources and with exemptions for large energy users. According to the press the UK government is opposed to the EEC tax, but a decision to apply value added tax to fuels taken announced in this year's budget has been referred to as an environmental measure in the *Prospects for Coal White Paper*[29]. This tax, applied at the point of sale, like the proposed US tax, does not differentiate between sources and does nothing to internalise external costs. Their only effect is to favour conservation relative to energy consumption.

Indeed, none of proposed or actual fiscal measures are based on the estimated levels of external costs. Where they exist they are essentially politically determined taxes based on what is judged to be acceptable and is considered likely to assist governments in achieving, wholly or partially, their targets for emission reduction. The sole exception to this is the attempted introduction of adders into decision making in the USA. These are based on estimates of external costs, although the values used very greatly from State to State[44].

A major problem for governments is the claim by industry that the unilateral imposition of taxes will change the terms of trade internationaly, hamper exports, and act as a disincentive to the inward investment that most countries welcome. In the present difficult economic climate, with high levels of unemployment in many OECD countries, the risks of adverse economic impacts from any measures directed at tighter environmental controls are a deterrent to action.

There are, of course, other ways of internalising costs. It can be done by regulation that puts a limit to emissions, forcing companies to introduce appropriate control technology and adding to their direct costs. Alternatively the use of tradeable emission permits can set upper bounds to the level of emissions and create a market in which emissions have a value that is set by the demand for their products and the costs of control technologies. Both approaches have the drawback that they impose costs that are arbitrary in economic terms, based solely on a judgemental view of what levels of emissions can be regarded as tolerable. These judgements may be well informed and soundly based on scientific evidence or may be based on political expediency.

What is not generally recognised is that the adoption of statutory limits to emissions, permitted levels of radiation exposure, etc., is fundamentally contrary to the principle of economic optimisation. As noted earlier, the optimisation principle allows a trade-off between control costs and supply cost, including social cost externalities arising from supply, so that overall costs to society are minimised. Introduction of limits that are either above or below the economic optimum imposes extra costs on society. This may appear unimportant to environmentalists but one consequence of setting standards that are too restrictive is a reduction in the level of funds available for other socially desirable purposes[20]. Clearly there is no direct trade-off since the financial consequences are diffused through the market. However, in principle, it can be argued that to spend too much on, say, nuclear safety, may save statistical lives but only at the expense of a lost opportunity to save a larger number of statistical lives in other sectors of the economy. For this reason the tendency for nuclear organisations to err on the side of extreme caution in plant design and operation is not necessarily in the best interests of the industry or society as a a whole[20], although the organisations concerned may benefit from an improved public image. Will governments discourage the practice?

A further political complication arises from the fact that the detrimental effects of emissions and other external costs are functions of location and the social and economic condition of the country in which they occur. This is true for both the physical consequences and the effects on morbidity and mortality. A World Bank economist pointed this out a few years ago, and had the temerity to draw the logical conclusion that it might be rational to locate polluting activities in countries where the revealed costs would be lower, rather than pay the price required to ameliorate the problem in an advance industrial nation with higher revealed costs. He became the immediate object of widespread and undeserved criticism.

Insistence on the application of Western standards of environmental quality in developing countries can only retard their economic development. Yet national pride can stand in the way of logic, and companies that have taken advantage of the less demanding requirements in some developing countries have been subject to media criticism. The main area where it may be appropriate to strive for the highest standards are developments having global implications, like exploitation of the rain forests, but assistance with the costs may be needed to compensate for the nationaly sub-optimal policies that have to be adopted.

WHERE DO WE STAND?

It will be evident from the foregoing text that there are a great many problems, both economic and political, that have to be solved before there can be widespread internalisation of social costs of energy sources.

Past work on the evaluation of externalities has been of very variable quality. Economists do not always appreciate the limitations of scientific and technical knowledge and understanding that surround the suggested causes and the

observed or postulated effects. Non-economists are unlikely to understand the limitations of the economic hypotheses and assumptions that underpin the valuations attached to environmental detriment. No-one can be confident that all the effects have been captured or of the relationship between those that have been covered and those that are known to be omitted or, worse, have yet to be recognised at all.

Bearing in mind the difficulty of identifying the effects, the difficulty in quantifying those that are known to occur, the difficulty of associating known effects with their specific causes and the difficulty of attaching meaningful value to many of the more important effects; remembering the problems of intergenerational equity and international divergence; it would be a brave economist who laid claim to knowing the external cost associated with any source or pollutant to much better than an order of magnitude; even in some cases several orders of magnitude. That this is the case is revealed by the wide spread of values to be found in the literature[15][45][46].

This may be of no importance in situations where the external costs are demonstrably very small relative to the overall direct financial costs of energy production. It is however a major barrier to internalisation in circumstances where the external costs are potentially similar in magnitude to the direct costs.

The former appears to be the case for new nuclear plants built to the standards demanded in OECD countries[30][46], whereas the latter appears to be the case for fossil fuelled combustion when the consequences of greenhouse warming are taken into account[15][30][45][46]. The costs could be small or equal to or greater than the direct costs of coal-fired, oil-fired or gas-fired generation. With such uncertainly the choice of a so called adder is little more than a gamble.

In this author's opinion it is far better in the present state of knowledge to recognise the limitations of the economic analyses and to present those responsible for taking energy decisions with disaggregated data expressed in physical and temporal terms, with due acknowledgement of technical uncertainty. The reduction of all consequences to a single present worth value conceals many ethical and value judgements. It is surely more appropriate for accountable elected representatives to make these value judgements explicitly, than for officials or economic technicians to bury them implicitly as a series of hidden assumptions, no matter how well meant they are.

Politicians looking for a "respectable" basis for energy taxes may be lulled into a false sense of security if they are provided with numerical costs, particularly if they seem to reflect o degree of international consensus.

This is not to say that the efforts devoted to attempting to identify and evaluate external impacts are wasted. It is extremely important that the magnitudes be established and progress towards a more comprehensive understanding maintained, particularly in areas such as the greenhouse effect and low probability high consequence accidents. Without some effort to put the costs or potential costs into perspective politicians will be vulnerable to the pressures from the many vociferous lobbies, and may be driven to imposing needlessly costly and potentially damaging controls or taxes.

THE FUTURE

Nuclear power can only benefit from a more detailed and clearer understanding of the external costs arising from energy production and use. For nuclear generation itself it should help to dispel some of the widespread misconceptions about back-end costs, and provide those who are prepared to listen with a better appreciation of the significance of nuclear risks relative to those of more familiar energy sources. If, as I believe, the residual external costs associated with the construction and operation of modern nuclear plants in OECD are no more than about 1 percent of the direct financial costs, nuclear has nothing to lose. In the current regulatory frame-work its environmental costs are internalised already. Indeed the non-environmental externalities I have described could make it appropriate to apply subtracters to nuclear electricity costs.

We are long way from knowing the true external costs of fossil fuel use. They are undoubtedly greater than those of nuclear power and renewable sources, but it could take several decades of intensive work to reach the point where even the physical effects can be established with confidence. This should not stop us from introducing allowances for the effects we do know of, but we need to be very careful not to apply inappropriate data merely because they are there. There is not a unique cost for any pollutant regardless of how and where it is released.

It may be possible to make some progress on economic valuation, but in my judgement this will arise more from the recognition of governments that they have to take explicit and consistent decisions, than from the development of ever more esoteric economic methodology. In particular, governments, on behalf of society, have to decide what countries can afford to spend on avoidance of ill-health and premature death, and how the balance should be made between the interests of the present and future generations.

The external costs of any one energy source differ from region to region and country to country, and it is quite proper for these differences to be taken into account by governments and markets. This may however lead to be taken into account by governments and markets. This may however lead to delays in reaching international agreement on tax or regulatory regimes, because of fears over the macroeconomic consequences in countries that feel themselves to be disadvantaged.

INTRODUCTION

Most people have a pretty good idea of what is meant by the term "externality" when it is applied to the environmental effects of electric generation, but the term becomes vague and is used to refer to many different things when it is applied to nonenvironmental effects. Nonenvironmental externalities are also controversial because of the way they are used in policy discussions. They constitute a way of introducing positive (or "good") externalities into the debate that can serve to balance against the negative environmental effects of electric generation. For example, new generating units are alleged to have positive spillover effects on local employment that help offset the negative environmental effects of living near a plant. Nonenvironmental externalities are also used to favor one kind of fuel over another. For example, energy security externalities are often cited as a reason for favoring domestic energy sources over imported sources, which in many countries means favoring coal or natural gas over oil. This argument in favor of coal is also used to offset the environmental argument against coal.

Attempts to identify and measure nonenvironmental externalities soon encounter an inherent ambiguity in the use of the term. An externality is well-defined only in a highly theoretical context that provides little useful guidance for measurement and policy analysis. More practical standards are required for measuring an externality that rely on subjective judgements about the feasibility of the economy to achieve higher levels of efficiency and the feasibility of government policy to realize the improvement. In some cases the government may have no practical option for addressing an externality, while in other cases the cost-effectiveness of available options may be such as to rule against taking any action.

A related consideration in judging the feasibility of government policy concerns the level of aggregation at which the externality and the policy apply. Externalities may occur at the aggregate macroeconomic level or at the microeconomic level of the individual firm and consumer. Policies that affect only a few firms or individuals are of no use in addressing macroeconomic externalities and, conversely, macroeconomic policies are seldom suitable for addressing microeconomic externalities.

My purpose here today is to explain these issues in more detail and to give a preliminary assessment of the employment and energy security externalities mentioned above. My conclusion is that the only valid arguments for including nonenvironmental externalities in electricity planning arise at the local level, and are specific to the technology of the plant and to characteristics of its location. A warning is in order even in this limited dimension, however. It is easy to overstate any positive effects and ignore any negative effects (or vice versa) of localized externalities, making these arguments a convenient refuge for scoundrels seeking to advance their special interests. Skeptics will understandably doubt whether these externalities have any meaning other than as indicators of the value judgments (or vested interests) of the persons measuring them.

THE PROBLEM OF DEFINING A NONENVIRONMENTAL EXTERNALITY

When applied to environmental effects, the term externality is used to refer to the spillover costs or benefits of one person's activities on another person's welfare.[2] Moreover, to establish that these spillover effects are not taken into account in private decisions, "The essential feature of an external effect is that the effect is not a deliberate creation but an unintended or incidental by-product of some otherwise legitimate activity".[3] When these spillover effects are ignored, social costs and benefits will deviate from private costs and benefits. When the externality refers to costs of production, for example, the condition that private costs are less than social costs means that too much output will be produced; when it refers to benefits from production, too little output will be produced. Either condition indicates that the economy is not operating as efficiently as it could in the sense that everyone could benefit from a reallocation of resources.

One source of confusion that bothered the economics profession for many years, and some policy analysts even today, is that not every spillover is an externality. If I take a particular job or win a specific contract, then someone else can not get that job or win that contract and may be worse off because of it. The outcome will affect the distribution of income but not the total level of income; hence, there is no efficiency loss for the economy and no externality. The term "pecuniary externality" is used to refer to cases where spillovers affect only the distribution of income. This paper will ignore pecuniary externalities that arise in the domestic economy, because national income is unaffected, but will consider those that alter the international distribution of income because national income could be affected.

For nonenvironmental externalities, a definition that focuses on spillover effects is not broad enough. Efficiency losses can occur when there is no direct spillover from one party's action to another's welfare. For example, market failures such as natural monopolies, and market imperfections such as rigid wages and prices, lower economic efficiency even though spillover effects are not the cause of the problem. Since the scope of nonenvironmental effects is open-ended, the definition of externality should be broadened to include all market imperfections that reduce efficiency.[4]

---

[2]See Baumol and Oates [1988], Ch. 2 and Mishan [1971].

[3]Mishan [1971], p. 2.

[4]This broader definition was originally advanced by Bator [1958]. As noted by Baumol and Oates [1988], the narrower interpretation of externalities captures most environmental externalities and avoids the ambiguities that come with the broader definition.

When the definition of an externality refers to any economic condition that causes the economy to operate less efficiently than it otherwise could, a critical measurement issue becomes obvious: what is the benchmark that defines how well the economy can work? Unfortunately, the only unambiguous conceptual benchmark is a perfectly competitive economy where markets are complete and work perfectly, all participants are equally well informed, and there are no transactions costs. Such a benchmark is obviously meaningless for measuring externalities for the purpose of making policy recommendations because it cannot possibly be attained. The benchmark should be a realistic, attainable optimum.

Here lies the rub. The realistic, attainable optimum must necessarily include market imperfections that constrain the economy from reaching an even higher level of efficiency. In other words, the choice of the benchmark will simultaneously specify those externalities that are candidates to be internalized and those externalities that are not. The choice of which externalities to measure is necessarily subjective because the only positive statement that can be made about what is attainable is what actually exists. All other potentially attainable scenarios are hypothetical and subjective.

The definition of what is attainable should recognize that we live in an imperfect world, with imperfect institutions and imperfect knowledge about how to improve them. There is no merit in worrying about externalities that can not be addressed. A related issue concerns those externalities for which no cost-effective policy options are available. When a policy action yields benefits that are smaller than its costs, the policy should not be adopted and the externality that gives rise to the policy need not be measured. All such externalities can be ignored for purposes of designing a particular public policy, for the same reason that all externalities that do not affect efficiency can be ignored.[5]

Finally, the level of aggregation at which electricity policy is implemented must be consistent with the level of aggregation of the externality to be addressed. Electricity policy may be administered at the national level, where it is applied to every generating unit built in the country, or at the local level where it is applied to just one or a few units. Similarly, energy security externalities may occur at the national level in relation to total energy imports and the market price of energy, or at the local level in relation to local employment conditions and other factors. Clearly, national energy security externalities cannot be addressed with local electricity policies because of the limited influence of the policies on energy markets. Hence, these national externalities are irrelevant to local policy decisions. On the other hand, a national electricity policy could be designed to address local externalities, though local policies may be better crafted to deal with the specifics of local externalities.

---

[5]Optimum efficiency does not require that all externalities be eliminated, only that their existence not effect the optimum amounts of other goods. See Mishan [1971], p. 15.

The foregoing considerations help explain why opinions may differ about the existence or significance of any given externality. These issues will surface again as we proceed with the assessment of energy security and employment externalities related to electricity generation.

SOURCES OF POSSIBLE NONENVIRONMENTAL EXTERNALITIES

As noted already, all externalities result from some form of market imperfection that reduces economic efficiency. These imperfections manifest themselves in at least four different ways. The first way is through the spillover effects that characterize environmental externalities, though broadened here to include all unintended economic effects of one person's actions on another's welfare.

A second source of externalities is the condition of increasing returns and declining marginal costs of production that characterize natural monopolies, where the lowest cost mode of production is a single producer. Since monopolies tend to produce less and charge more than firms in a competitive market, an unrestrained market will operate less efficiently than possible.[6] To obtain the benefit of lower average costs provided by a single firm, and at the same time act to prevent the firm from earning excess profits, the government sometimes intervenes to create regulated monopolies such as electric utilities. Regulation is not perfect, however, so that there is a trade-off between imperfect competition and imperfect regulation.

Economies of scale also arise in relation to the size and scope of the infrastructure in which the firm operates, so that changes in the infrastructure could alter production costs. The construction of an electric generating plant may not qualify as an addition to the infrastructure, but its location could bring with it some improvements in the local infrastructure as well as increased demands on the existing infrastructure.

A third source of externalities is institutional rigidities that prevent prices and wages from performing their usual informational and market clearing functions, and prevent resources from flowing to their highest return. This category recognizes that institutions are often created to make economic life easier and more manageable (i.e., more efficient), but at the same time they can interfere with the functions of an efficient market. For example, labor contracting practices have developed over time in response to social and economic needs that have led to wage contracts, annual salary adjustments, and other terms and conditions. These practices tend to reduce the ability of wages to respond to changes in economic conditions. In particular, a reduction in labor demand could lead to

---

[6]A perfect price-discriminating monopolist may produce as much as a perfectly competitive firm, but perfect price discrimination is impossible in practice.

unemployment rather than a reduction in the wage rate, even though both workers and employers might prefer a lower wage to unemployment. Price regulation of electric utilities creates another kind of rigidity that lowers economic efficiency. This institution would constitute the source of an externality worth considering if alternative, more efficient forms of regulation were available.

The fourth source of externalities arises from an assortment of market imperfections such as imperfect information and incomplete markets.[7] When private agents have incomplete information about prices, product quality, profit opportunities, and so on, they will not be able to make optimal production and consumption decisions and the economy will not operate as efficiently as it could. Similarly, the lack of complete markets for insurance and capital means that opportunities for investment will be foregone because risks and transactions costs are too high.

In each of these categories situations can arise that make it difficult to establish a practical benchmark for describing the level of efficiency that the economy is capable of achieving. For example, the results of existing labor contracting institutions may be the best that society is capable of achieving, or at least not worth the cost of changing, and the same might be said about the net benefits of providing some kinds of additional information or some extensions of existing insurance and capital markets. Also, as noted, public utility regulation is imperfect, yet no practical model exists that is capable of attaining all of the potential net benefits from regulating natural monopolies.

POTENTIAL NATIONAL ENERGY SECURITY EXTERNALITIES

Energy security, as the term is used here, will refer to the potential loss of aggregate domestic output that would result from a change in the price or availability of energy.[8] An energy security externality would be present if the private sector of the economy failed to properly take into account the possibility of such losses. To the extent that the costs associated with a disruption in energy markets are taken into account in private plans and decisions, they should not be included in any assessment of the value of an externality. Only those costs that are external to private decisions need to be considered. Thus, the presence of an externality means that the private sector is underestimating the full social cost of using energy, which is inefficient because it leads to excessive consumption of energy, insufficient domestic production of energy and, as a consequence, excessive imports of energy.

---

[7]Distributional inequities are another form of market failure, although they are often ignored (as here) when the focus is on efficiency rather than equity.

[8]Changes in availability may not be captured in changes in the price because of price controls that are common in energy sectors such as electricity and natural gas.

Our interest in the issue has to do with the reasons why the private sector undertakes insufficient actions in response to energy insecurity, and the policy options that might be used to compensate for the inadequacies. These issues and policies are essentially economic in character. In contrast, a disruption in energy supply or price may be the result of terrorism, war, or some other noneconomic event that is inherently external to the economy, and which might warrant military, diplomatic, or some other noneconomic policy action (e.g., guarding nuclear power stations) to reduce the security risks. Analysis of these noneconomic policies is beyond the scope of this paper,[9] except to say that the cost of the action should not exceed the expected value of the economic dislocation costs (plus any noneconomic costs).[10]

The discussion of potential externalities below focuses on oil as the primary source of energy insecurity. This is because the oil market is most susceptible to disruption and because oil is the only fuel capable of transmitting a disruption to the world economy.[11]

All of the issues considered in this section apply only to policies implemented at a level capable of affecting aggregate fuel demands and market prices for fuels. This means that the policy must affect many electric generation plants.

OPEC Market Power

If the members of OPEC operate as a cartel in which output is controlled in order to influence the market price of oil, then world oil output may be too low and the price of oil may be too high compared to that which would occur in a competitive market.[12] Moreover, importing countries would be subject to the risk of fluctuations in the scarcity of oil as exporters flex their muscle. Consequently, importing countries may be justified in implementing policies that would reduce the demand for imports to recapture some of the monopoly rents extracted by OPEC and to reduce OPEC's market power.

---

[9]Note that deliberate actions taken by one party to inflict damage upon another do not qualify as externalities in the narrow definition of the term (see footnote 3), since the expected results of a deliberate act are internalized in the decision to act.

[10]It is sometimes argued that military expenditures, such as those connected to the recent Gulf War, represent an external cost of oil insecurity that should be added to the price of oil. Actually, such expenditures should be considered as the cost of mitigating insecurity rather than as the cost of insecurity itself. The distinction is important because military actions represent a policy response, and the cost of the policy response should be gauged according to the magnitude of the externality, not the other way around.

[11]Disruptions in natural gas supply or price may exert similar, though weaker, effects as oil disruptions in countries dependent on international gas trade for a significant portion of their energy consumption, such as in Western Europe.

[12]There is some ambiguity about how long an output-constraining cartel will maintain the price above a competitive level. For a depletable resource, slowing the rate of output will conserve production for tomorrow, so that the rate of output under a cartel will eventually exceed that under competition.

With regard to electricity generation, one such policy might be to discourage the use of oil, and possibly the use of close substitutes for oil (e.g., gas), for electric generation. Obviously, such a policy makes little sense if it is implemented at a level that affects only one or two generating units. The question is whether there is a credible externality on which to base a national policy for electric generation.

While there is no doubt that the members of OPEC control a large enough share of total world oil production to influence the price of oil, there is some question as to whether they are willing and able to coordinate their efforts as required. Studies of past behavior are mixed regarding the extent of OPEC influence over the price. A minority of studies argue that OPEC has exerted no demonstrable influence over the price, and that the price shocks of the 1970s reflect political events and demand-side responses rather than OPEC actions, while most researchers conclude that OPEC has exercised albeit imperfect control over the price.[13]

These studies focus on the period between 1973 and 1986--that is, between the date of the first price shock and the date of the collapse of the world price, when it returned to a level in real terms only slightly higher than where it started in 1973. Whatever one may think about the extent to which OPEC controlled price behavior during these years, the fact that the price still remains near the 1986 level indicates that the exercise of market power was temporary at best. It seems reasonable to conclude, therefore, that OPEC market power is not a fixed characteristic of the marketplace, but is intermittent at most.

It is difficult to imagine a sensible policy on electric generation that could be applied on an intermittent basis. For example, once a plant is built the flexibility to respond to changes in fuel preferences is limited. Built-in fuel substitution capability is expensive and begs the question whether a fixed cost borne indefinitely is worth the benefits gained from combating market power that is uncertain and intermittent.

Importer Market Power

The United States currently consumes about 28 percent of total world oil production, and the members of the OECD consume about 63 percent. The size of these numbers suggests that the U.S. or the OECD could, if desired, restrict oil demand and drive down the world price. In effect, importers could use their monopsony market power to reduce the income transfers to oil exporters. Note that the change in import demand must be large enough to influence the world

---

[13]See the discussion of the literature in Toman [1993]; in particular, compare MacAvoy [1982] and Jones [1992] with Adelman [1980] and Griffin [1985].

market price. While there is some question about the magnitude of the change in demand that is required to influence the price, there is no doubt that the magnitude represented by the fuel consumption of one or even a few individual electric generating plants would not be sufficient. As above, this factor is relevant only to policies capable of affecting aggregate fuel consumption.

There is reason to question whether the externality is worth measuring. To begin with, while it may be true that an importer can gain by restricting imports, there will be no increase in economic efficiency as a result, only a shift in the distribution of income between importers and exporters of oil.[14] This means that import controls will increase domestic income only at the expense of foreign income. Even when the policy objective is narrowly focused on raising domestic income, the history of international trade indicates that such beggar-thy-neighbor policies will likely result in successive rounds of retaliation that will reduce the volume of trade and make all parties worse off. Moreover, since there are many commodities other than oil where importers may possess monopsony power, there is nothing to prevent a policy directed at oil from spilling over to other traded commodities.

These considerations suggest that oil must be sufficiently different from other commodities to warrant constraints on the flow of trade. The argument that oil exporters already exercise market power is a potential justification, although we have seen that this argument is weak at best. Ironically, if OPEC is not already operating as a cartel, constraints imposed by importing countries are more likely to cause retaliation than if OPEC is functioning as a cartel, simply because OPEC has more to lose from retaliation if it is already acting as a cartel.

We conclude that the possibility of retaliation means that any attempt to capture a trade-related externality is uncertain and temporary at best.

Spillover Effects on Inflation and Exchange Rates

A rise in oil prices is said to have a negative indirect effect on the economy because it leads to a boost in inflation or a reduction in international currency exchange rates.[15] Again, the arguments are moot with respect to policies that affect only one or two electric generation plants and, as argued below, too weak to influence more comprehensive policies.

The argument that higher oil prices will lead to currency depreciation for oil importing countries (and hence to higher costs for all other imported goods) focuses on the initial change in trade balances and ignores subsequent related

---

[14]See Bohi and Montgomery [1982], Chapter 2, for a demonstration of the argument.
[15]Broadman and Hogan [1988] make this argument.

changes in trade and investment. Currencies of oil importing countries will initially depreciate relative to the currencies of the oil exporting countries because oil demand is price inelastic. However, the ultimate effect on the exchange rates of oil importing countries will depend on how much the oil exporting countries respend and invest in specific oil importing countries, as well as any related adjustments in trade and investment flows among the oil importing countries that might be caused by the higher cost of oil imports. The final result is ambiguous for any specific country, and what is true for one time period may change for the next.

Regardless of what the results may be, it is likely that any attempt to secure the gains by restricting trade would be met with retaliation, just as in the case of oil. Moreover, in this case the issue involves trade relations among oil importing countries, not just between oil importing and oil exporting countries, so the potential justification for restricting trade based on OPEC market power does not apply. Thus, there is even less reason to factor this issue into energy policy.

The connection between energy prices and inflation also suggests a spillover effect from the price of energy to the cost of producing other goods and services. To assess this issue one must make a distinction between inflation and a rise in prices, and between cause and effect. A one-time rise in oil prices will cause a one-time rise in other prices (one-time even though some prices adjust faster than others), not an increase in inflation (which refers to a *rate* of increase in prices). Moreover, inflation means a devaluation of the currency relative to commodities, which can happen only with an increase in the supply of money (or a reduction in money demand, which is ignored here). An increase in the money supply is the responsibility of the monetary authority, suggesting that the blame for an inflationary event is better placed at the feet of the monetary authorities than on the pressures caused by an increase in the price of any specific commodity. Again, the case for calculating an externality is ambiguous.

Unemployment due to Rigid Wages and Prices

Rapid increases in oil prices (and, hence, in all energy prices) are thought to cause increases in unemployment and reductions in output because wages and prices are unable to adjust fast enough to prevent these undesired spillover effects.[16] The adjustment costs are short term in nature, and will dissipate over time as markets respond, but are thought to cause substantial losses in output.[17]

---

[16]Output will also decline because more resources are required to pay for energy (whether domestically produced or imported), but this response is due to private adjustments to energy prices, so the costs are internalized in private decisions.

[17]Bruno and Sachs [1985]. In addition to these cyclical effects, higher energy prices will cause long term adjustments in energy-related capital investment and innovation that may have significant effects on the economy, as in Jorgenson [1988]. However, prices and wages are sufficiently flexible in the long run that a market failure argument does not apply to these adjustment costs.

That the losses are important, or even measurable, are questionable propositions, as argued below.

The most important source of adjustment cost concerns the possibility that wages will not fall to maintain employment when the price of energy rises.[18] Briefly, the argument is that higher energy prices will reduce the use of energy in production and lower the productivity of labor. Lower productivity in turn means higher labor costs, which employers will try to trim by either lowering wages or the amount of employment. If institutional constraints prevent wages from falling, employers have no option but to reduce employment. The loss in employment, and corresponding reduction in potential production in the economy, is sometimes regarded as an external cost of oil (because only disruptions in oil markets are thought to be capable of causing prices for all energy sources to rise so rapidly).

To some extent, at least, wage rigidity results from institutional arrangements in the labor market that represent society's best choice among currently available methods for structuring transactions for labor services. As a practical matter, it is not possible to substitute new institutions, even if they were known, to reduce the cost of energy price volatility. However, it is possible to choose energy sources with less volatile prices as a way of avoiding the adjustment problem, or building buffer stocks of crude oil to try to insulate the domestic market from external shocks. Hence, the issue is relevant to energy policy, but whether the policies are worth pursuing depends upon the importance of the costs being avoided.

Unfortunately, empirical studies of the macroeconomic effects of energy price shocks do not resolve the question of how important the externalities are. In fact, these studies do not try to distinguish between costs that are already internalized in private decisions and costs that are external. All we know is that the total costs are sufficiently ambiguous to question the importance of any underlying externality.

While the oil price shocks in 1973-74 and 1979-80 were both accompanied by recessions in many industrial countries, there are reasons to doubt whether oil prices were responsible. One reason for doubt is that Japan managed to avoid a recession in 1979-80 (even though it did not in 1973-74), raising the question of what Japan did differently the second time that others failed to copy. A second reason is that a worldwide economic boom failed to materialize after the collapse of oil prices in 1986, as one might expect if energy prices are an important cause of business cycles. Third, the analysis of disaggregated industry data from the four largest industrial countries fails to reveal a systematic connection between the importance of energy in production and changes in industrial employment and

---

[18]A related adjustment cost concerns the premature obsolescence of capital that becomes too expensive to operate at high energy prices. Also, commodity prices may not change with energy input costs as required to clear markets. These potential spillover effects are discussed in Bohi [1989] who finds no basis for attributing a significant market failure.

output, as one should expect if energy prices were responsible for the underlying dislocations.[19] Finally, the worldwide experience with recessions in 1973-74 and 1979-80 corresponds to the countries that had implemented tight money policies and the expansionary policy followed by Japan in 1979-80 may explain why that country did not suffer another recession.[20]

If nothing definitive can be said about the gross economic costs of energy price shocks, it follows that even less can be said about the subset of those costs that might result from any externalities. A great deal more study of this issue is required before a credible quantitative result can be derived, and before the issue can play a role in guiding energy policy.

LOCAL EMPLOYMENT AND OTHER EFFECTS

While national energy security issues do not appear to be important to electric generation policy, a similar set of issues arise at the local level that could provide a rationale for more localized policies. We will consider three possible candidates for externalities, though the list is by no means exhaustive.

First, the location of an electric generating station can have a significant effect on the level of local employment because of new jobs at the plant or because of new jobs in businesses that are created to service new employees of the plant. Second, the choice of generating technology used at the plant can subject the community to potential employment risk because of volatility in the price of fuel used by the plant. The presence of an externality in both of these cases depends upon how well the labor market works in the locality in question. Finally, the location of a generating plant will tend to add to demands on the local infrastructure and could bring with it a contribution to the infrastructure. The net spillover effect on the rest of the community could constitute an externality.

An Increase in Local Labor Demand

The location of a new generating station will increase the demand for labor and draw in idle workers, upgrade workers from lower income jobs in the community, or attract new workers from neighboring areas. Total employment and income in the community will rise, but no externality is necessarily involved. These adjustments are consistent with a well-functioning market.

A positive externality is involved if the location of the plant results in the employment of workers who would remain unemployed otherwise. That is, the

---

[19]See Bohl [1989].

[20]Bohi [1989] gives a description of the monetary policies adopted by Germany, Japan, the United Kingdom, and the United States during the 1970s.

existence of a breakdown in the local labor market is required to establish the existence of an externality, where for some reason unemployed labor will not migrate to other areas to gain employment, and will remain unemployed unless there is an increase in local job opportunities. Regional labor immobility of this kind is not exactly rare, but also not so commonplace that one can assume that any new generating station will help address a local market failure.

Identifying and measuring the existence of such an externality can become quite involved. For example, it must be established that the increase in labor demand will reduce local unemployment rather than simply result in an influx of labor from neighboring areas. It is possible that local employment will increase while local unemployment does not fall because the plant requires skills that do not match the capabilities of the unemployed population. Extending the externality argument from the local community to the neighboring areas that are the source of the new workers is not likely to be credible, however, simply because the demonstrated willingness of workers to move from these areas contradicts the market failure argument.

Disruptive Effects of Changes in Fuel Prices

A rapid increase in the cost of fuel will necessitate an increase in electricity rates and lead to a corresponding reduction in the quantity of electricity demanded. The result could be the shutdown or temporary mothballing of some plants, or a reduction in the rank order of plants most affected by the increase in fuel costs. The closure or reduction in use of a plant could create spillover costs for the community in which it is located, in effect reversing some of the positive effects identified above (to the extent that they exist). The risk that such a disruption might occur would vary, of course, with the type of fuel used by the plant. Oil would have a higher risk than coal, for example.

An indirect effect of fuel price shocks may occur if sharp increases in electricity rates force other businesses to scale back or shutdown their operations. This outcome may occur with firms that use electric-intensive production processes such that production costs become uncompetitive at high electricity prices. Another possibility mirrors the rigid wage argument described above: a reduction in the use of electricity by local businesses may cause a reduction in labor productivity and a corresponding increase in the cost of labor. If employers do not have the option of lowering wages (because of institutional constraints), they may have to reduce employment and production.

No externality is associated with these cutbacks as long as local labor and commodity markets work properly; that is, unless the location of the generation plant results in positive spillover effects in the first place. The risk of disruption means that the original externality benefits may have to be adjusted to cover the

possibility that the benefits may disappear. Of course, fuels that are more price-stable will result in a higher net benefit than fuels with greater price volatility.

Improvements in the Local Infrastructure

Another kind of externality arises if the location of an electric generation plant alters the local infrastructure in such a way as to change the cost of supplying other goods and services produced in the community. For example, if the location of the plant brings with it improvements in the local transportation system or in telecommunications, these improvements will have positive spillover effects on the rest of community. In general, the growth of the local business community stimulated by a new power plant may produce scale economies that have beneficial effects on production costs throughout the community. Alternatively, an increase in demand for local services could, in the absence of any scale economies, result in higher supply costs.

There are many possible candidates in the infrastructure category that could be considered as externalities, simply because there are so many aspects to the infrastructure that could be affected. Included are items as diverse as the local road system, police and fire protection, and the educational system. Moreover, the net effect on each aspect of the infrastructure could be positive or negative. The local tax base may expand because of siting a new plant that will provide for more education and police protection, for example, but the increase in demands for services from the increase in population could outstrip the increase in supply. Similarly, the new plant may bring with it new roads, but also an increase in road damage from truck traffic that could exceed the increase in maintenance revenues.

Calculating the net effects on each aspect of the infrastructure could be an arduous task, raising the question of whether the information gained is worth the cost. This kind of information could be useful in determining the location of a plant, but it is unlikely to affect decisions about the number of plants or their technology. If the payoff from the information is limited, so should the investment in its creation.

CONCLUSION

We conclude that there is no need to include energy security externalities that may arise at the national or international level in decisions regarding electric generation capacity. Externalities present at the local level are more relevant to electric generation decisions, though it is arguable whether an effort should be made to quantify the externalities in a formal way so that they can be combined with other externalities in making powerplant decisions.

The list of potential local externalities is open-ended, their values are location-specific, and their assessment is highly subjective. Consequently, there is considerable room for differences of opinion about the items to be considered and the values to be attached. Since the study of externalities is intended to guide policy making rather than confuse it, it is reasonable to ask whether the effort required to put nonenvironmental externalities on an analytical par with environmental externalities is ultimately worth the trouble. A no answer does not mean that the underlying issues will be completely ignored in electric generation decisions. For example, issues regarding the impact of a new power plant on the local community will be raised by the community when it reacts to a siting proposal.

REFERENCES

Adelman, M.A. 1980. "The Clumsy Cartel," *Energy Journal,* vol. 1, no. 1, pp. 43-52.

Adelman, M.A. 1986. "Scarcity and World Oil Prices," *Review of Economics Statistics,* vol. 68, no.3, pp. 387-397.

Bator, F.M. 1958. "The Anatomy of Market Failure," *Quarterly Journal of Economics,* vol. 72, pp. 351-79.

Baumol, W. J. and W.E. Oates. 1988. *The Theory of Environmental Policy,* Cambridge: Cambridge University Press.

Bohi, D. R. and W. David Montgomery. 1982. *Oil Prices, Energy Security, and Import Policy,* Washington: Resources for the Future

Bohi, D. R. 1984. *Energy Price Shocks and Macroeconomic Performance,* Washington: Resources for the Future.

Bohi, D. R. and M. A. Toman. 1993. "Energy Security: Externalities and Policies," *Energy Policy,* forthcoming.

Broadman, H.G. and W. W. Hogan. 1988. "Is an Oil Tariff Justified? The Numbers Say Yes," *Energy Journal,* vol. 9, no. 3, pp. 7-30.

Bruno, M. and J. Sachs. 1985. *Economics of Worldwide Stagflation,* Cambridge: Harvard University Press.

Griffin, J.M. 1953. "OPEC Behavior: A Test of Alternative Hypothesis," *American Economic Review,* vol. 25, no. 5, pp. 954-963.

Jones, C.T. 1990. "OPEC Behavior Under Falling Prices: Implications for Cartel Stolidity," *Energy Journal,* vol. 11, no. 3, pp. 117-130.

Jorgenson, D.W. 1988. "Productivity and Postwar U.S. Economic Growth," *Journal of Economic Perspectives,* vol. 2, no. 4, pp. 23-41.

MacAvoy, P. 1982. *Crude Oil Prices as Determined by OPEC and Market Fundamentals,* Cambridge: Ballinger.

Mishan, E.J. 1971. "The Postwar Literature on Externalities: An Interpretative Essay," *Journal of Economic Literature,* vol. 9, no. 1, pp. 1-28.

Toman, M.A. 1993. "The Economics of Energy Security: Theory, Evidence, Policy," in A.V. Kneese and J.L. Sweeney, eds., *Handbook of National Resource and Energy Economics, vol. 3: Economics of Energy and Minerals,* Amsterdam: North Holland.

It seems inevitable that moves towards cost internalisation will be largely dictated by political judgement based on scientific advice for the foreseeable future. If the very imperfect and incomplete adders suggested by the sorts of analyses we have heard about at this conference are consistent with these judgements, their use will add e veneer of professional respectability to what are, and will remain for some time to come, inherently political decisions.

REFERENCES

1. Jones P M S, Taylor K, Clifton M, Storey D J, *A technical and Economic Assessment of Air Pollution in the United Kingdom*, PAU-M20 HMSO, London, 1972.

2. Jones P M S, Llewelyn G I W, Guppy C B, Whitty L, *Fire Research Station*, Programmes Analysis Unit, 1968; *Annual Report of the Fire Research Station Steering Committee*, HMSO, London, 1968.

3. Jones P M S, Taylor K, Storey D J, Marsden P S S F, *The Environmental and Financial Consequences of Oil Pollution from Ships*, PAU-M24, International Maritime Consultative Organisation, London, 1973.

4. *The Programmes Analysis Unit; 1967-1971 and The Programmes Analysis Unit; The Second Five Years*, HMSO, London, 1972 and 1977.

5. Objectives set out in documents of ref. 4 above.

6. Jones P M S, *Project Appraisal*, in *Collected Essays by PAU Authors*, PAU-M25, HMSO, London, 1974.

7. US Congress, Senate Committee on Rules and Administration, *Office of Technology Assessment for the Congress*, US Government Printing Office, Washington, 1972.

8. Private communication.

9. International Commission for Radiological Protection, *Quantitative Basis for developing an Index of Harm*, Annals of the ICRP, Pergamon Press, Oxford, 1985.

10. National Radiological Protection Board, *Cost Benefit Analysis in the Optimisation of Radiological Protection*, NRPB, Chilton 1982.

11. Jones P M S, *Nuclear Power and the Greenhouse Effect*, Surrey Energy Economics Centre, SEEDS 51, Surrey University, 1990.

12. Inhaber H, *Is Nuclear Riskier then Other Energy Forms*, in Nuclear Power Policy and Prospects, Wileys, Chichester, 1987.

13. Carson R, *Silent Spring*, for other examples see Jones P M S, *Discontinuities in Social Attitudes*, in Social Forecasting for Company Planning, Macmillan, London, 1982.

14. Scorer R S, J. Inst. of Fuel, 1957, p. 110.

15. Pearce D, Bannard C, Giorgiou S, *The Social Cost of Fuel Cycles*, HMSO, London, 1992.

16. Jones-Lee W M, *The Economics of Safety and Physical Risk*, Blackwells, Oxford, 1989.

17. Persson, V, *The Value of Risk Reduction*, Institute of Swedish Health Economics, Stockholm, 1989.

18. Dawson, R F, *Cost of Road Accidents*, Road Research Laboratory, Garston, LR79, 1967.

19. National Radiological Protection Board, *Cost Benefit Analysis in the Optimisation of Radiological Protection*, NRPB, Chilton, ASP9, 1986.

20. Brown M L, Blackman T W, Jones P M S, Mc Keague R, *One Organisation's Experience*, in ALARA Principles and Practice, Adam Hilger, London, 1987.

21. Ridker RG, *The Economic Costs of Air Pollution, Praeger*, London, 1966.

22. Burton TL, Fulcher M N, *Measurement of Recreation Benefits*, J. Econ. Studies, 1968, 3, 35-48.

23. Intergovernmental Panel on Climate Change, *Climate Change: The IPCC Scientific Assessment*, Cambridge University Press, 1990.

24. Nuclear Energy Agency/International Energy Agency, *Comparative Costs of Generating Electricity (Annex 4)*, OECD, Paris, 1989.

25. Mishan E T, *Cost Benefit Analysis*, Allen & Unwin, London, 1975.

26. Kahn H, Brown W, Martel L, *The Next 200 Years*, Morrow, New York, 1976.

27. Jones P M S, *Intergenerational Equity*, Futures, 1978, 10, 68.

28. Pearce D, Markandya A, Barbier B B, *Blueprint for a Green Economy*, Earthscan, London, 1989.

29. Department of Trade and Industry, *Prospects for Coal*, HMSO, London, 1993.

30. Nuclear Energy Agency, *Broad Economic Impacts of Nuclear Power*, OECD, Paris, 1992.

31. Williamson, National Institute Economic Review, 1984.

32. Glyn, Memorandum to Trade and Industry Committee Coal Review, H237 (iv), HMSO, London, 1993.

33. Jones P M S, *Nuclear Power: Policy and Prospects*, Wiley, Chichester, 1987.

34. Jones P M S, *A World Without Nuclear Power*, NEA Newsletter, 1989, 7(1), 12.

35. Barnes M, *Hinkley Point Inquiry*, Vol II, HMSO, London, 1990, p.354.

36. Ulph, *Method of Quantification of the Benefits of Diversity in relation to Fossil Fuel Prices*, S4165, Hinkley Inquiry Paper, Department of energy, London.

37. Jones P M S, *Nuclear Power and the Coal Review*, in Fuels for Electricity Supply, Surrey Energy Economics Centre, SEEDS 69, Surrey University, Guildford, 1993.

38. Trade and Industry Select Committee, *British Energy Policy and the Market for Coal*, HC237, HMSO, 1993.

39. Turvey R, *Optimal Pricing and Investment in Electricity Supply*, Allen and Unwin, London, 1968.

40. Jones P M S, *Paying the Full Cost of Energy Consumption*, Atom, 1993, 428, 41.

41. Wakeham J, Press Release, UK Department of Energy, 1990.

42. Jones P M S, *Nuclear Economics, Externalities and Privatisation*, Evidence to Environmental Assessment Board Hearings, Ontario, 1993.

43. See Table 10.3 in ref. 24.

44. *Current US Utility Sector Market Strategies to Incorporate Environmental Externalities*, Paper to OECD Group on Environment and Development, 1991.

45. *Energy and the Environment*, Senior Symposium on Energy and the Environment, Helsinki, IAEA, Vienna, 1991.

Session 4 - Discussion

W. Munn

I have a comment and a question for you and many of our others who are concerned with this question of risk. Several decades ago, here in the United States, an activist organisation which called it itself Critical Mass had as a keynote speaker, one of the foremost sociologist in the world, who told her audience at that time that the American people were not afraid of dying, they were afraid of dying slowly. They should never let them forget that.

There have been, probably, a minimum of several dozens of spin-off organisations from that meeting since that time and it has been my observation, working in the area of public information, that they have done an excellent job of never letting the American people forget that they are afraid of dying slowly, which is what they believe will occur if they are exposed to even one additional millirem of radiation. That being the case, as you have pointed out, risks and benefits are very difficult to evaluate. Not only are they difficult to evaluate, but my numerous conversations with psychologists in respect to the fears that people have about energies have led me to believe that facts don't matter. It does not make any difference. We can win the argument hands down when it comes to evaluation of risks for various technologies. What we cannot win is the battle in addressing their fears because we do not seem to have found the formula for doing that. In our struggle to try to identify the cost of externalities, I would submit that a major cost is our failure, our own personal failures, to become adept at how to address fears rather than facts, simply because it seems to be the fears that are the overriding events in the long run.

I would like to have your comment on whether you think it is possible for us to perhaps incorporate in our concerns for externalities and their costs, the costs that perhaps we have already suffered as a result of our inability to address those fears. Whether they might not be a valid incorporation in our view of externalities in the future?

J. Grawe

This is a question that is not easy to answer and I think it is a crucial question for the electricity industry. I can completely agree with you that risk evaluation is quite different from risk perception by the public. This has hardly anything to do with each other, if I understood all the studies of psychologists and sociologists that I have read. So that means there is a big gap between objective risks and subjective risks. It doesn't mean that subjective is something better or worse than objective. It is just another way of regarding things, and it is a consequence of the gap between acceptability of a risk or a technology and the acceptance in reality.

To answer the question, "Can we do anything about that?" would need another talk and I am certainly not the right man to give this talk. My personal view is that we should never be tired of telling the facts, but to tell them in a way that people can understand them. When someone tells me that he is afraid of dying slowly, I would answer him at first that it seems very strange to me that everyone lives very much longer than his parents and in particular his or her grandparents. If we die slowly that means that we live longer, but it would be too long to talk about the best way to discuss risks and benefits with people taking up their specific problems. We cannot help them, we will not have any effect if we just tell them what the results of our studies are. We have to go to meet them at their place and meet their fears and their hopes.

R. Shelton

I think that this particular question maybe of interest to more than one of the panellists. It certainly is an issue that has come up almost in every session in some form. We may like to return to that particular issue. I know that one of the panellists here has thought on this issue at some length as well. Is there another question on the last presentation?

Unidentified

A question for Dr. Grawe. In the dismantling scenarios you estimated about 6 per cent of the plant material as contaminated and activated. I am just curious to find out your presumptions and basis concerning the type of plant, its size, what are the assumptions. If it's a PWR, did you assume any leak from primary to secondary? Is it a relatively clean plant or a dirty plant, and so forth?.

J. Grawe

I could send you the figures, I don't have them here, if you leave me your card. But anyway, it is a fully contaminated plant and at present we have three nuclear power plants which are being dismantled now. Different types, different sizes, different grade of contamination, but these are the figures for a complete PWR, fully used, large nuclear power plant.

M. Thibau

Mr. Chairman, this is not properly a question, it is more of a remark. It seems that I am the only participant in this convention whose country does not belong to the OECD. So I would like to convey to you a view of someone that does not belong to the First World, and see it through the optics of the Third World. I can assure you to start with, that although our economies are Third World, our ecological activists are definitely First World, as good as any you can have here. So what happens to us is that we have to face the realities of all those externalities that we are talking about here during the discussion for approval of our projects. But the overwhelming externality for us has not been mentioned up to now, and I think we are at the end of our work, and I am not going to hear anybody mention the externality that is really conditioning: it is money, or more precisely lack of money.

The big difference is that, for the OECD Members, when you have a project that after all is approved and is financed if it proves its rentability, it gets the finance, no question, it is taken for granted. For us, quite the contrary, we have to fight for money. So after we get a project, we have to get the money to finance it, and not being able get the total amount of money, we have to frame what we have to do, that is our externality, which in my opinion counts. We have to put a frame to the little available money we are able to get. So if it is difficult for you, it is much more difficult for us.

L. Williams

I think that your point is very valid. There is a financial constraint, a capital constraint on investment and this means that priorities have to be allocated somewhat differently. There is a literature that does look at the relationship between environmental externalities and discounting, which tries to see how much we should try and adjust the discount rate or adjust the way in which we appraise the estimates because of the possibility of future environmental damage. But I don't think that was the nature of your question. Your remark was that simply funds are not available and therefore that is an important constraint.

R. Shelton

I might add that we recognise in the formulation of the programme, that there was a large missing element, and I think a couple of presentations like John Foster's presentation recognised that missing part as very important in the developing countries. I think the decision of the programme committee was that, given the limitations that we had in terms of time, that we would focus on OECD countries. But a similar conference dealing with developing countries issues, environmental externalities and other issues is certainly appropriate.

Last night at dinner an engineer commented to me that energy issues used to be so simple when people like engineers and physicists dealt with the issue. Now we have turned loose the economists and none of us understand it anymore.

I have known Doug for some time and I have always found it interesting, as many of you know, that he is a world recognised expert in energy security and he is one

391

of the few experts in the world that I know, who has gone around the world, telling the world, that what he is an expert in is not important. Are there questions for Doug?

T. Takesita

May I ask a question on the coverage of the externalities and on the social costing externality. Social costing exists not only in the output side of the facility development, especially environmental effects, but also may exist on the input side, for example, research and development or subsidies for the facility introduction. For example, there are subsidies to photovoltaic development and also their introduction to the societies. In the case of Japan, R&D for new energy systems is covered extensively by special taxes and as explained by Mr. Yoda of CRIEPI yesterday, this is considered to be part of the security by diversification. How will this kind of thing be considered in the United States or in Europe?

Also, the subsidy to energy development is increasing, but their commercial application, real commercial introduction, will take quite a long time ahead. That means the product introduction may take about ten years. So they become the external cost, or must they be reflected later on these real costs or not?

D. Bohi

You are quite right. There are externalities associated with R&D. I talked only about energy security and local employment. Certainly the United States, I think, takes a similar view about the role of the government in subsidising research and development because there are market failures there that make it such that the private sector, if left to its own devices, will underinvest in research and development. A simple argument is that the benefits of R&D can't be appropriated fully by the entity that is making the investment. That makes the conditions such that the private sector will underinvest. It can't get all the benefits out. Therefore the government should step in and do it and certainly that goes on in this country.

I don't know about Europe. Your question also was how is this done in Europe. I suppose I should turn to you to answer that question.

J. Grawe

In principle I agree with you that R&D financed by the government is an externality. The question is that you have to regard future or present and future R&D or even those expenditures in the past. I think it depends on what you are willing to show. By the way, this is one of the fields where there are disputes in Germany between Hohmeyer (in first study there were a lot of deficiencies) but he sticks to that idea that the complete R&D expenditures for the various technologies, even if they go back thirty years ago, have to be put into the total picture. On the other hand, the team at Stuttgart University disagree with that and say that we can only regard present and future of R&D expenditures by public authorities. In principle, you are right. I think it will be regarded everywhere in the same way.

D. Bohi

Well it depends on the purpose of the exercise too. If you calculate an externality to determine a policy action, all of that sunk cost associated with past subsidies and investment is irrelevant because it can't be changed. It is the future costs and benefits that should be taken into account for making you policy determination.

Unidentified

One remark to that (underinvestment by the industry). You assume that government is an institution that is so perfectly wise that it knows what will really be the future needs of society. Therefore if private industry does not invest in a certain technology, then government will either force it, fight it or give incentives, pretending, that way, some of the externalities are internalised!

D. Bohi

That is a very good point because I have been backed into that corner before of trying to justify the actions of the bureaucrats as being the best of all possible actions. You are right. It is another part of the subjective nature of evaluating these externalities.

Unidentified

Dr. Bohi, with regard to the issue of possible relationship or not between increases in world oil prices and recessions, I wonder if there might possibly be some argument having to do with velocity of recycling of money having a relationship on recessions. In other words, greater outflows of money from one country to another country. It takes a while for it to come back and therefore, perhaps, in the absence of a consciously stimulative money supply, if the money supply simply doesn't change, is there another possibility that you could have a recession caused by a large increase in outflows for oil?

D. Bohi

As a matter of fact, back in the mid-1970s, the United States Council of Economic Advisors had estimated that the first oil shock of 1973-74 had reduced aggregate demand for that very reason you alluded to. Enough to cause the recession in that period. (However), even if that is correct, notice that the policy action that should be taken to remove the problem is not an energy policy action, but it is an expansionary policy that should be undertaken by the monetary authorities. That will restore the aggregate demand, you would prevent deflation and so on, but it is not an energy policy.

Unidentified

But recognising that the action does have to be taken by these other monetary authorities, nonetheless, if they have fears about inflation, if they expand the monetary supply so much, is it not true you would have to attribute some of the recession to the oil price and not lay it all on the monetary authorities?

D. Bohi

It is certainly true that they have more to worry about than just the oil price increase and its effect. But I think that really what you are trying to get at here, is whether the monetary authorities are smart enough to do the right thing at the right time, or whether we have to engage in an energy policy which calculates an externality to cover the possibility that the monetary authorities are going to do the wrong thing. That does not seem to me to be a very wise way to go about calculating energy externality.

Unidentified

Except that I am not certain that you or I or any group of people could determine as to what exactly the right thing is for the monetary authorities.

D. Bohi

Japan presents a very fine counter-example of what the United States, France, Britain and Italy did during the second recession in 1979-1980. During the first recession in 73-74, all of them experienced a severe downturn in their GDP and total employment. They all were uniform in throwing on very tight monetary policies too, so that the central banks were essentially driving the economies into the recession further than it would have gone or maybe even totally. In 1979-80, Japan was the only one of those countries that refused to do the same thing with its money policy. It was the only one of them that avoided a recession the second time.

R. Shelton

I am sorry, having had a three-hour conversation with you on the telephone (on this subject) this could go on for a while.

I am now going to introduce the most difficult talk, I think, at this symposium. It is the concluding remarks, and Pierre Girouard is going to tell us what we have learned in this very interesting two-day period.

CLOSING REMARKS

P. Girouard
OECD/NEA

Electricity, you have heard, is growing at 4% per year. It might even grow faster if electricity were to be more used in the transportation sector; and this might well be the case in order to limit pollution from that area of the economy. Given the electricity demand, there will be a need to employ all resources to meet the growing need. But no one resource, including conservation techniques, is problem free.

Therefore difficult choices will need to be made in the selection of the "best" energy technology. You have heard the different situations occurring in different countries. Avoiding specific cases, one thing seems certain - all types of technology have their place, and a strategic view needs to be adopted.

As has been said many times, all energy production and use has certain drawbacks. The need to choose the lesser evil is inevitable. But no convincing or internationally accepted method of comparison exists, making the task of decision makers all the more difficult. Inaction is not an acceptable option. The basic question then becomes: "is there a soundly-based approach to making the decisions?" Notice that I will say "soundly-based approach", I will not say right-decision. I do not believe economists are in a position to determine that.

For new power plants there are generally only two types of fuels that can compete based on costs for base-load operation: nuclear and fossil fuels, the latter essentially being coal and gas. Renewable fuels are not yet making a large impact and few people believe they will in the near future. This is borne out by numerous studies including the tri-yearly study conducted by the NEA/IEA/UNIPEDE/IAEA on electricity generating costs.

The following slide shows a comparison of costs between nuclear power plants and coal plants. This is similar to results obtained a few years ago. The item that has the most influence is the discount rates. Two rates are shown: 5 and 10 per cent. Essentially what is below 1 (I have put a five per cent band on each side of that number) is favourable to nuclear. What is above will be favourable to coal. We have been doing this every three years for almost a decade. This is the best data on an international basis that you can accumulate.

But if one were to follow the impression such a graph provides, one would think that nuclear should make a much more substantial impact than it has recently. But one of things that is not included is risk and externalities. That is one of the problems and the need to fill this gap was a motivation for us to pursue it to this symposium.

Comparisons are to date based on certain standard approaches, such as levelised costs in this case. Costing methods for comparisons are becoming more sophisticated in response to the need for integration of the issues mentioned above. Governments are asking consumers to pay the full costs of the energy consumed. Attempts are being made to internalise costs related to externalities. Methods to incorporate risk factors and to determine optimum system flexibility are also being developed.

The first step, and this is only a first step, in developing a more complete costing method is to put as many factors as possible into a common form, i.e. to monetize factors not yet included in present costs. Risks (financial risks) are usually more amenable to such treatment than other factors. The NEA has studied the impact of incorporating the economic impact of accident risks by calculating the cost of insurance. The results do not change the cost of nuclear power by more than 1%, well within the margin of uncertainty. But there is another "externality" associated with accidents; and that is the public perception which can imply a "cost" higher than the actual economic risk.

The importance of the public and social aspects is known to all. Their impact has changed over the years. For example in Germany commercial considerations previously dominated. Political restrictions, while they added a financial burden, did not have a lasting effect on power generation choices. The situation has changed. Political factors and social acceptability may determine the choices for the future.

Changes in public acceptability are part of the risks. In the case of nuclear power this has been translated into complicated and long licensing procedures resulting in increased financial risks. The public is also forcing, in certain cases, new nuclear safety criteria such as the no-exclusion-zone requirement. Whether the response from designers increases public acceptability and reduces financial risk is questionable.

High availability, competitive costs, a complete programme for handling nuclear waste are not sufficient to guarantee public acceptability or the success of a nuclear programme. This was demonstrated in Mr. Nyquist's paper. In many countries, society is implicitly putting an adder into the costs and saying nuclear power is too "expensive" to keep. But how is this adder included in other technologies? Are the costs of all the alternatives considered too expensive if one were to include the externalities, including global warming? From recent NEA/IEA studies Greenhouse Gases are the single most important externality. One can speculate how the CO_2/energy tax, as part of an attempt to internalise externalities, will affect the future energy mix. Given the uncertainties surrounding the quantification of global warming, I will avoid pursuing the subject here.

From D. Bohi's presentation, environmental externalities would seem much more important, well defined, and pertinent to policy decisions. This is also borne out by the NEA's study on the "Broad Economic Impact of Nuclear Power". Non-environmental externalities such as employment or security of supply do seem to confuse the issue. That is not to say that these externalities are not important at a national or local level; they certainly influence political decisions. But dealing only with the economic issues one should concentrate more on the environmental externalities as these tend to dominate.

Professor Grawe's paper presented these environmental externalities so well that I will not dwell on the issue. Note that if conservation is to be considered part of the planning process, externalities related to conservation must also be part of that process.

In the last paper of this symposium, the point is made that the state of the art, in explicitly incorporating externalities in the planning process, is progressing but in final analysis is limited. We cannot aggregate the diverse data into a single number. Many ethical and value judgements affect the monetization of the impacts of energy use. It may be unsatisfactory to state that science cannot provide a mechanistic mean of making "intelligent" choices. But that is not its purpose. Economists cannot dictate the "right" choices. Externalities must reflect the value society puts on these. Man must choose for himself -not only for himself but for all future generations- science can only help in pointing out the consequences. In this, the analysts have a major duty. Proper decision making should be transparent and for that, it needs a methodology for comparison. The state of the art needs to move forward. I am pleased to see the collaborative work between the CEC and ORNL on this issue.

The presentation by L. Draper illustrates the changing scene in the US, i.e. the breakup of vertically integrated utilities. L. J. Williams' paper echoes the problem of a changing situation. This may be a harbinger for many countries. This new situation complicates the process of integrating the issues into a coherent policy. To note but two problems related to time and to space:

— Looking at time -- the impact of environmental legislation and the uncertainties related to nuclear power seem to be dictating a short term view. Short term approaches hinder integration of policy questions which have to deal with problems which rebound across generations. Japan, by contrast, is taking a long term view on energy.

As evaluations of the impacts of the technologies are becoming even more sophisticated, the approach and choices must be taken at the national level. This seems to be conflicting with the breakup of national utilities and the decentralisation of the control over production even though decentralised power resources have many merits.

Governments will have to give guidance to researchers and work hand in hand with the researchers to determine the best tools to achieve well defined goals.

In these days of search for harmony between countries, and that is the raison d'être for the OECD, it would be useful if a common approach could be taken. Internalisation issues have at least in one case become a trade issue. I will end this part of my remarks with two observations:

— At least one speaker noted that if an integrated approach was applied consistently, the power generation choices would be quite different from what may be seen to emerge from the present situation.
— Some utilities have avoided some of the problems associated with risks and externalities by transferring some of these problems to other utilities, i.e. by importing electricity. Though this probably may result in an optimisation of total costs if the basic approaches in the different countries are similar, it begs the question whether an international approach is needed. Again one could think of the discussion on the abatement of greenhouse gases.

One of the original purpose of this symposium was to bring together experts in the field with decision makers and to improve understanding on each countries' situation. In this we have been successful. I hope that the dialogue will bring us closer to an accepted approach for dealing with externalities.

If I may conclude by thanking all those involved -- starting with the speakers who presented papers of such a high calibre that any summary by myself would be a step down in quality.

If the organisation of the symposium went so smoothly it is in no small measure due to the chairman of the programme committee, John Gray, who managed to bring together such a distinguished group, and to Robert Shelton and his team who did all the local organising. I wish to thank the members of the organising committee and all those who helped bring about this symposium.

Before we close there is one other group that I wish to thank - and that is you, the audience, who by your participation and interest made this a success.

Thank you.

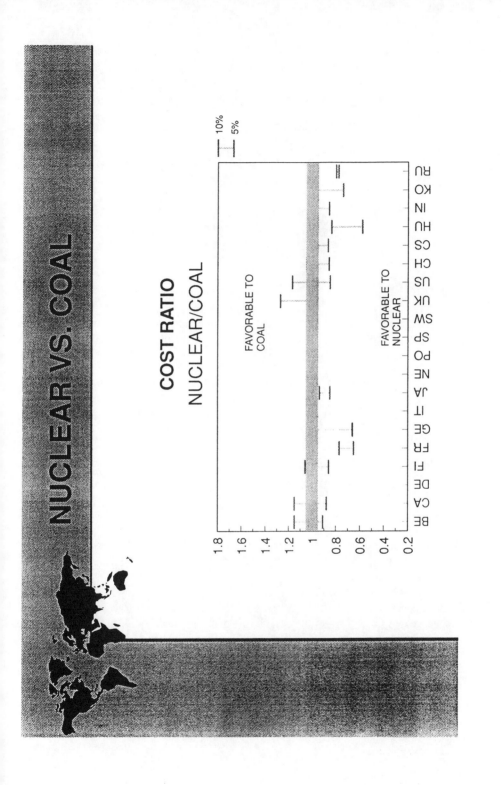

NUCLEAR VS. COAL

COST RATIO
NUCLEAR/COAL

FAVORABLE TO COAL

FAVORABLE TO NUCLEAR

BE CA DE FI FR GE IT JA NE PO SP SW UK US CH CS HU IN KO RU

10%
5%

LIST OF PARTICIPANTS
LISTE DES PARTICIPANTS

AIROZO D., Nucleonics Week,1200 G St NW, Suite 1100, Washington, DC 20005, United States

AKIYAMA Y., President & Director, The Kansai Electric Power Co., Inc., 3-3-22 Nakanoshima, Kita-ku, Osaka, 530-70, Japan

ALLEN S., Energy Research Group, Inc., 400 Fifth Avenue, Waltham, Massachusetts 02154, United States

ALLEGEIER H.J., Director, Commission of the European Communities, Directorate-General Science, Research and Development, rue de la Loi 200, 1049 Brussels, Belgium

ANDERSON T.D., Energy Division, Oak Ridge National Laboratory, Post Office Box 2008, Oak Ridge, Tennessee 37831-6187, United States

ASAI, N., The Kansai Electric Power Co., Inc., 375 Park Avenue, Suite 2607, New York, New York 10152, United States

ASCROFT-HUTTON W.W., HM Nuclear Installations Inspectorate, St. Peter's House, Ballion Road, Bootle, Merseyside L20 3LZ, United Kingdom

BAINS N.S., Niagara Mohawk Power Corporation, 301 Plainfield Road, Syracuse, New York 13217, United States

BALLESTER N.M., University of Maryland University College, Office of Special Programs, Presidential Building, 6525 Belcrest Rd., Room 340, Hyattsville, Maryland 20782, United States

BARABASCHI S., Director for Research, ANSALDO Group, Pza Carignano 2, I-16128 Genoa, Italy

BARNES A., Japan Electric Power Information Center, 1120 Connecticut Avenue, NW, Suite 1070, Washington, DC 20036, United States

BARTH M.S., Bechtel Power Corporation, 9801 Washingtonian Blvd. 2A-2 (R), Gaithersburg, Maryland 20878-5356, United States

BAUER D.C., Martin Marietta Energy Systems, Inc., Oak Ridge National Laboratory, 600 Maryland Avenue S.W. - Suite 306, Washington, DC 20024, United States

BAYNE P., U.S. Council for Energy Awareness, 1776 I Street N.W., Suite 400, Washington, DC 20006, United States

BERMANIS H.L., Raytheon Engineers & Constructors, 1215 Jefferson Davis Highway, Suite 1500, Arlington, Virginia 22202, United States

BERTEL E., Planning & Economic Studies Section, International Atomic Energy Agency, P. O. Box 200, A-1400, Vienna, Austria

BOHI D.R., Director, Energy & Natural Resources Division, Resources for the Future, 1616 P. Street, N.W., Washington, DC 20036, United States

BOUWSMA A., BNA, Washington, DC, United States

CANNON J.B., Energy Division, Oak Ridge National Laboratory, Post Office Box 2008, Oak Ridge, Tennessee 37831-6189, United States

CARLE R., Directeur Général Adjoint, Electricité de France, 32 rue de Monceau, 75384 Paris Cedex 08, France

CARLIN J., Department of Energy, EI-531, 518/H St, Washington, DC 20585, United States

CARRET O., Electricité de France International, 1730 Rhode Island Avenue, NW, Suite 509, Washington, DC 20036, United States

CASTRO, R.A., International Access Corporation, 1901 Pennsylvania Ave., NW, Suite #300, Washington, DC 20006, United States

COPE D.R., Executive Director, UK Centre for Economic & Environmental Development, 3E King's Parade, Cambridge CB2 1SJ, United Kingdom

CURLEE T. R., Energy Division, Oak Ridge National Laboratory, Post Office Box 2008, Oak Ridge, Tennessee 37831-6205, United States

DECAMPS F., ONDRAF-NIRAS, Place Madou 1, B 25, Brussels B-1030, Belgium

DELVOYE J., Tractebel, Avenue Ariane 7, B-1200 Brussels, Belgium

DENNY F.I., Edison Electric Institute, 701 Pennsylvania Ave., NW, Washington, DC 20004-2696, United States

DEVELL L., Studsvik AB, S-61182 Kykoping, Sweden

DISAPIA R., ENEA, Viale R. Margherita 125, Rome 00152, Italy

DRAPER, E.L. Jr., Chairman, President, and Chief Executive Officer, American
Electric Power Company, Inc., 1 Riverside Plaza, Columbus,
Ohio 43216-6631, United States

DURANTE R.W., AECL Technologies, 9210 Corporate Boulevard, Suite 410,
Rockville, Maryland 20850, United States

EVANS P.C., 2633 Garfield Street NW, Washington, DC 20008, United States

EYRE B.L., Chief Executive, AEA Technology, Corporate Headquarters, Harwell,
Didcot, Oxfordshire OX11 ORA, United Kingdom

EYSYMONTT G.P., Autex, Inc., 7213 Chestnut Street, Chevy Chase,
Maryland 20815, United States

FELD S., U.S. Nuclear Regulatory Commission, Mailstop NL/S 129,
Washington, DC 20555, United States

FOSTER J.S., President, World Energy Council, 10 Thornbury Crescent,
Islington Cres., Ontario M9A 2M2, Canada

FOX E.C., Engineering Technology Division, Oak Ridge National Laboratory,
Post Office Box 2009, Oak Ridge, Tennessee 37831-8063, United States

FREDERICKS J, Saskatchewan Energy and Mines, 1914 Hamilton Street, Regina,
Saskatchewan, Canada

FRI R.W., President, Resources for the Future, 1616 P Street N.W.,
Washington, DC 20036, United States

FULLER E.D., Senior Vice President, TENERA, 1340 Saratoga-Sunnyvale Road,
Suite 206, San Jose, California 95129, United States

GALE H., TRW Environmental Safety Systems, 2650 Park Tower Drive, Suite 800,
Vienna, Virginia 22180, United States

GASSMANN J., NOK, Parkstrasse 23, P.O. Box 5401, Baden, Switzerland

GEBERT L., Department of Energy, Energy Efficiency & Renewable Energy,
EE-71, 1000 Independence Avenue, Washington, DC 20585, United States

GIROUARD P., NEA/OECD, Le Seine St. Germain, 12 Boulevard des Iles, 92130 Issy-les-Moulineaux, France

GRAHME T., Department of Energy, 1000 Independence Avenue, Washington, DC 20585, United States

GRAWE J., Vereinigung Deutscher, Elektrizitätswerke VDEW eV, Postfach 70 11 51, Stresemannallee 23, D-6000 Frankfurt/Main 70, Germany

GRAY J.E., 508 Queen Street, Alexandria, Virginia 22314, United States

GUINDON S., Department of Natural Resources, 580 Booth Street, Ottawa, Ontario, Canada K1A OE4

GUNTER G., Director of Nuclear Development, New Brunswick Electric Power, Commission, 515 King Street, Fredericton, New Brunswick E3B 4X1, Canada

HARA T., Mitsubishi Corporation, 1-5 Dojima-Homa 1-Chome, Osaka, Japan

HAYASHI T., The Kansai Electric Power Co., Inc., 3-22 Nakanoshima 3-chome, Kitaku, Osaka 530, Japan

HEDVALL P., Asea Brown Boveri AB, Dept. DU, Västerås S-721 83, Sweden

HORIE Y., Central Research Institute of Electric Power Industry, 1-6-1 Ohtemachi, Chiyoda-ku, Tokyo 100, Japan

HORII H., Tohoku Electric Power Co., Inc., New York Office, 65 East 55th Street, Suite 2304, New York, New York 10022, United States

IKEDA M., Electric Power Development Company, 1825 K Street, Suite 1205, Washington, DC 20006, United States

IMANAGA T., Japan Electric Power Information Center, 1120 Connecticut Avenue, NW, Suite 1070, Washington, DC 20036, United States

IMHOFF C.H., Battelle Pacific Northwest Laboratory, P.O. Box 999, K8-54, Richland, Washington 99352, United States

ISHIOKA O., Japan Electric Power Information Center, 1120 Connecticut Avenue, NW, Suite 1070, Washington, DC 20036, United States

KAMMERER K.J., San Diego Gas & Electric, P.O. Box 1831, San Diego, California 92112-4150, United States

KING J., TRW Environmental Safety Systems, 2650 Park Tower Drive, Suite 800, Vienna, Virginia 22180, United States

KIYOSE R., Department of Nuclear Engineering, Tokai University, c/o Nuclear Safety Research Association, Meikoh Bld., 1-18-2 Shimbashi Minato-ku, Tokyo 105, Japan

KREWITT W., University of Stuttgart, Institut für Energiewirtschaft, Hessbrühlstr. 49a, 70565 Stuttgart, Germany

KÜFFER K., Director, Nordostschweizerische Kraftwerke AG, Parkstr. 23, CH-5400 Baden, Switzerland

LADA W., National Atomic Energy Agency, Warsaw, u1 Krueza 36, Poland

LAFLEUR J.D., Jr., 14701 Poplar Hill Road, Germantown, Maryland 20874, United States

DEVEZEAUX DE LAVERGNE J.G., COGEMA, 2 Rue Paul Dautier, BP 4, 78141 Velizy-Villacoublay, France

LEE R.M., Energy Division, Oak Ridge National Laboratory, Post Office Box 2008, Oak Ridge, Tennessee 37831-6205, United States

LELAND M., Voice of America, Washington, DC, United States

LEPSCKY C., ENEA, P. O. Box 2538, 001100 Rome, Italy

MacGREGOR P.R., General Electric Company, 1 River Road, Bldg. #2, Room 639, Schenectady, New York 12345, United States

MAGWOOD W.D. Magwood, IV, Edison Electric Institute, 701 Pennsylvania Avenue, NW, Washington, DC 20004, United States

MARKANDYA A., Harvard Institute for International Development, 1 Elliott Street, Cambridge, Massachusetts 02138, United States

MARRIAGE A., Director of Power System Planning, Ontario Hydro, 700 University Ave., H-1, Toronto, Ontario M5G H6, Canada

MASSON L., Vice President System Planning, Hydro Québec, 75 Blvd. René-Lévesque Blvd. Ouest, Montréal, Québec H2Z 1A4, Canada

McCLOUD D.E., Technology Advancements, Tennessee Valley Authority, CST 17A-C, 1101 Market Street, Chattanooga, Tennessee 37402, United States

MICHAELS G.E., Chemical Technology Division, Oak Ridge National Laboratory, Post Office Box 2008, Oak Ridge, Tennessee 37831-6495, United States

MINAMIYAMA H., Hokkaido Electric Power Company, Higashi 1, Ohdori, Chuoku, Sapporo 060, Japan

MIZOBUCHI M., Sikoku Electoric Power Co., Inc., Nuclear Policy Office, 2-5, Marunouchi, Takamatsu 760-91, Japan

MUKAI J., The Japan Atomic Power Company, No. 6-1 1-Chome Ohtemachi, Chiyoda-ku, Tokyo 100, Japan

MUNN W., American Nuclear Society, Westinghouse Hanford Company H0-39, P.O. Box 1970, Richland, Washington 99352, United States

MURABE Y., Japan Electric Power Information Center, 1120 Connecticut Avenue, NW, Suite 1070, Washington, DC 20036, United States

MURPHY P.W., American Nuclear Society, 10th Floor, 370 L'Enfant Promenade, SW, Washington, DC 20024, United States

MURRAY A., Australian Nuclear Science & Technology Organization (ANSTO), Australian Embassy/ANSTO, 1601 Massachusetts Ave., NW, Washington, DC 20036, United States

NAGANO K., Central Research Institute of Electric Power Industry, Otemachi Bldg. 1-6-1 Otemachi, Chiyoda-ku, Tokyo 100, Japan

NAUDET M., Commissariat a l'Energie Atomique, Service des Etudes Economiques, 33, rue de la Fédération, 75752 Paris, France

NISHINOIRI K., Hitachi, Ltd., Energy and Environmental Systems Div., Hitachi Ltd., 6 Kanda-Surugadai 4-chome, Chiyoda-ku, Tokyo 101, Japan

NORBERT L., CEA/DSE, 29-33 rue de la Fédération, 75752 Paris Cédex 15, France

NORRIS J.F., Jr., Duke Engineering & Services, Inc., P. O. Box 1004, Charlotte, North Carolina 28201-1004, United States

NYQUIST C.E., President, Vattenfall, S-162 87 Vallingby, Jämtlandsgatan 99, Sweden

OCHI H., Chubu Electric Power Co., Inc., 900 17th Street NW, Suite 1220, Washington, DC 20006, United States

OVERHOLT P.N., Department of Energy, M/S E-477, NE-44, 1000 Independence Ave. SW, Washington, DC 20585, United States

PONCELET J.P., ONDRAF-NIRAS, Place Madou 1, B 25, Brussels B-1030, Belgium

PUGH C.E., Oak Ridge National Laboratory, Post Office Box 2009, Bldg. 9201-3, MS 8063, Oak Ridge, Tennessee 37831-8063, United States

RESÉNDIZ D., Comisión Federal de Electricidad (México), Ródano 14, piso 6, Col. Cuauhtémoc, México City, México

ROACH M.W., Tennessee Valley Authority, 1101 Market Street, LP 4J, Chattanooga, Tennessee 37402, United States

ROHM H., Department of Energy, Office of Nuclear Energy, Washington, DC 20585, United States

RUPERT T., PRC Engineering Systems, Inc., 1500 PRC Drive, McLean, Virginia 22102, United States

SATO K., Central Research Institute of Electric Power Industry, 1-6-1 Ohtemachi, Chiyoda-ku, Tokyo 100, Japan

SATO S., The Central Electric Power Council, (Keidanren Bldg.) 9-4, Ohtemachi, 1-chome Chiyoda-ku, Tokyo, Japan

SHAPAR H., Shaw, Pittman, 3200 N. S.K., NW, Washington, DC 20037, United States

SHAVE D.F., Stone & Webster Engineering Corporation, 245 Summer Street, Boston, Massachusetts 02210, United States

SHELTON R.B., Director, Energy Division, Oak Ridge National Laboratory, Building 4500N, MS-6187, Post Office Box 2008, Oak Ridge, Tennessee 37831-6187, United States

SHIGEMITSU T., Nuclear Fuel Industries, Ltd., 3-7, Tosabori, 1-Chome, Nishi-ku, Osaka, Japan

SHIMOYAMA S., The Japan Atomic Power Company, 1-6-1 Ohtemachi, Chiyoda-ku, Tokyo, Japan

SIMON W.A., General Atomics, P.O. Box 85608, San Diego, California 92186-9784, United States

SISK R.B., Westinghouse Electric Corporation, P.O. Box 855, Pittsburgh, Pennsylvania 15230-0855, United States

SMITH H., Oil and Natural Gas Policy Office, U.S. Department of Energy, 1000 Independence Avenue, SW, Washington DC 20585, United States

STRAUSS L., Director, Member of the Board of Directors, Bayernwerk AG, Postfach 20 03 40, 8000 Muchen 2, Germany

SUNDT N.A., Energy, Economics and Climate Change, 1347 Massachusetts Avenue, SE, Washington, DC 20003, United States

SUTO Y., Tohoku Electric Power Co., Inc., 7-1, Ichibancho 3-chome, Aoba-ku, Sendai, Miyagi 980, Japan

SUZUKI T., Massachusetts Institute of Technology, E40-393, Energy Laboratory, Cambridge, Massachusetts 02139, United States

TAKESHITA T., Technova Inc., 2-2-2 Uchisaiwai-cho Chiyoda-ku, Tokyo 100, Japan

TAKEUCHI E., Chubu Electric Power Co., Inc., 5561 Sakura, Hamaoka-cho, Ogasa-gun, Shizuoka Pref., Japan

TANIGUCHI T., Institute of Applied Energy, Shinbashi Sy Bldg. 1-14-2 Nishi-Shinbashi, Minato-ku, Tokyo 105, Japan

TAYLOR M., The Uranium Institute, 12th Floor, Bowater House, 68 Knightsbridge, London SWIX 7LT, United Kingdom

THIBAU M., Brazilian National Committee-World Energy Council, Rua Sambaiba 157/201 Leblon, Rio de Janeiro, Brazil 22450-140

TIERNEY S.F., Assistant Secretary of Energy, Office of Policy, Planning and Program Evaluation, Department of Energy, 1000 Independence Avenue, SW, Room 7C-016, Washington, DC 20585, United States

TSUTSUMI M., Power Reactor and Nuclear Fuel Development Corporation, 2600 Virginia Avenue, NW, Suite 715, Washington, DC 20037, United States

UEMATSU K., Director General, OECD Nuclear Energy Agency, Le Seine St. Germain, 12, Bd. des Iles, 92130 Issy-les-Moulineaux, France

VERNON A., McKinsey, Blvd. Manuel Avila Comacho #1, 13th Floor, Mexico, D.F. 11560, Mexico

WALTER J.F., Ogden Environmental & Energy Services, P. O. Box 10130, Fairfax, Virginia 22030, United States

WAMSTED D., The Energy Daily, 627 National press Building, Washington, DC 20045, United States

WARD D.P., Sargent & Lundy, 55 East Monroe Street, Chicago, Illinois 60603, United States

WILLBY C., HM Nuclear Installations Inspectorate, Baynards House, 1 Chepston Place, London W241F, United Kingdom

WILLIAMS K., Oak Ridge National Laboratory, Post Office Box 2009, MS 8038, Bldg. 9102-1, Oak Ridge, Tennessee 37831-8038, United States

WILLIAMS L.J., Integrated Energy Systems Division, Electric Power Research Institute, 3412 Hillview Avenue, Post Office Box 10412, Palo Alto, California 94303, United States

YAEGASHI T., Tohoku Electric Power Co. Inc., 7-1 Ichibancho 3-chome, Aoba-ku Sendai, Miyagi 980, Japan

YAJIMA A., Vice President in Charge of Planning Division, Central Research Institute of Electric Power Industry, 1-6-1 Otemachi, Chiyoda-ku, Tokyo 100, Japan

YAMAJI K., Central Research Institute of Electric Power Industry, Otemachi Bldg., 1-6-1 Otemachi, Chiyoda-ku, Tokyo 100, Japan

YAMATANI Y., The Kansai Electric Power Co., Inc., 3-22 Nakanoshima 3-chome, Kitaku, Osaka 530, Japan

YASINSKY J.B., Group President, Westinghouse Electric Corp., 11 Stanwix Street, Pittsburg, Pennsylvania 15222-1384, United States

YASUI H., Tokyo Electric Power Company, 1901 L Street NW, Suite 720, Washington, DC 20036, United States

YODA S., President, CRIEPI, 1-6-1 Ohtemachi, Chiyoda-ku, Tokyo 100, Japan

YOKOYAMA H., Central Research Institute of Electric Power Industry, 1-121-702 Naitoh-cho, Shinjuku-ku, Tokyo 160, Japan

MAIN SALES OUTLETS OF OECD PUBLICATIONS
PRINCIPAUX POINTS DE VENTE DES PUBLICATIONS DE L'OCDE

ARGENTINA – ARGENTINE
Carlos Hirsch S.R.L.
Galería Güemes, Florida 165, 4° Piso
1333 Buenos Aires   Tel. (1) 331.1787 y 331.2391
Telefax: (1) 331.1787

AUSTRALIA – AUSTRALIE
D.A. Information Services
648 Whitehorse Road, P.O.B 163
Mitcham, Victoria 3132   Tel. (03) 873.4411
Telefax: (03) 873.5679

AUSTRIA – AUTRICHE
Gerold & Co.
Graben 31
Wien I   Tel. (0222) 533.50.14

BELGIUM – BELGIQUE
Jean De Lannoy
Avenue du Roi 202
B-1060 Bruxelles   Tel. (02) 538.51.69/538.08.41
Telefax: (02) 538.08.41

CANADA
Renouf Publishing Company Ltd.
1294 Algoma Road
Ottawa, ON K1B 3W8   Tel. (613) 741.4333
Telefax: (613) 741.5439
Stores:
61 Sparks Street
Ottawa, ON K1P 5R1   Tel. (613) 238.8985
211 Yonge Street
Toronto, ON M5B 1M4   Tel. (416) 363.3171
Telefax: (416)363.59.63

Les Éditions La Liberté Inc.
3020 Chemin Sainte-Foy
Sainte-Foy, PQ G1X 3V6   Tel. (418) 658.3763
Telefax: (418) 658.3763

Federal Publications Inc.
165 University Avenue, Suite 701
Toronto, ON M5H 3B8   Tel. (416) 860.1611
Telefax: (416) 860.1608

Les Publications Fédérales
1185 Université
Montréal, QC H3B 3A7   Tel. (514) 954.1633
Telefax : (514) 954.1635

CHINA – CHINE
China National Publications Import
Export Corporation (CNPIEC)
16 Gongti E. Road, Chaoyang District
P.O. Box 88 or 50
Beijing 100704 PR   Tel. (01) 506.6688
Telefax: (01) 506.3101

DENMARK – DANEMARK
Munksgaard Book and Subscription Service
35, Nørre Søgade, P.O. Box 2148
DK-1016 København K   Tel. (33) 12.85.70
Telefax: (33) 12.93.87

FINLAND – FINLANDE
Akateeminen Kirjakauppa
Keskuskatu 1, P.O. Box 128
00100 Helsinki

Subscription Services/Agence d'abonnements :
P.O. Box 23
00371 Helsinki   Tel. (358 0) 12141
Telefax: (358 0) 121.4450

FRANCE
OECD/OCDE
Mail Orders/Commandes par correspondance:
2, rue André-Pascal
75775 Paris Cedex 16   Tel. (33-1) 45.24.82.00
Telefax: (33-1) 49.10.42.76
Telex: 640048 OCDE

OECD Bookshop/Librairie de l'OCDE :
33, rue Octave-Feuillet
75016 Paris   Tel. (33-1) 45.24.81.67
(33-1) 45.24.81.81
Documentation Française
29, quai Voltaire
75007 Paris   Tel. 40.15.70.00
Gibert Jeune (Droit-Économie)
6, place Saint-Michel
75006 Paris   Tel. 43.25.91.19
Librairie du Commerce International
10, avenue d'Iéna
75016 Paris   Tel. 40.73.34.60
Librairie Dunod
Université Paris-Dauphine
Place du Maréchal de Lattre de Tassigny
75016 Paris   Tel. (1) 44.05.40.13
Librairie Lavoisier
11, rue Lavoisier
75008 Paris   Tel. 42.65.39.95
Librairie L.G.D.J. - Montchrestien
20, rue Soufflot
75005 Paris   Tel. 46.33.89.85
Librairie des Sciences Politiques
30, rue Saint-Guillaume
75007 Paris   Tel. 45.48.36.02
P.U.F.
49, boulevard Saint-Michel
75005 Paris   Tel. 43.25.83.40
Librairie de l'Université
12a, rue Nazareth
13100 Aix-en-Provence   Tel. (16) 42.26.18.08
Documentation Française
165, rue Garibaldi
69003 Lyon   Tel. (16) 78.63.32.23
Librairie Decitre
29, place Bellecour
69002 Lyon   Tel. (16) 72.40.54.54

GERMANY – ALLEMAGNE
OECD Publications and Information Centre
August-Bebel-Allee 6
D-53175 Bonn   Tel. (0228) 959.120
Telefax: (0228) 959.12.17

GREECE – GRÈCE
Librairie Kauffmann
Mavrokordatou 9
106 78 Athens   Tel. (01) 32.55.321
Telefax: (01) 36.33.967

HONG-KONG
Swindon Book Co. Ltd.
13–15 Lock Road
Kowloon, Hong Kong   Tel. 366.80.31
Telefax: 739.49.75

HUNGARY – HONGRIE
Euro Info Service
Margitsziget, Európa Ház
1138 Budapest   Tel. (1) 111.62.16
Telefax : (1) 111.60.61

ICELAND – ISLANDE
Mál Mog Menning
Laugavegi 18, Pósthólf 392
121 Reykjavik   Tel. 162.35.23

INDIA – INDE
Oxford Book and Stationery Co.
Scindia House
New Delhi 110001   Tel.(11) 331.5896/5308
Telefax: (11) 332.5993

17 Park Street
Calcutta 700016   Tel. 240832

INDONESIA – INDONÉSIE
Pdii-Lipi
P.O. Box 269/JKSMG/88
Jakarta 12790   Tel. 583467
Telex: 62 875

IRELAND – IRLANDE
TDC Publishers – Library Suppliers
12 North Frederick Street
Dublin 1   Tel. (01) 874.48.35
Telefax: (01) 874.84.16

ISRAEL
Praedicta
5 Shatner Street
P.O. Box 34030
Jerusalem 91430   Tel. (2) 52.84.90/1/2
Telefax: (2) 52.84.93

ITALY – ITALIE
Libreria Commissionaria Sansoni
Via Duca di Calabria 1/1
50125 Firenze   Tel. (055) 64.54.15
Telefax: (055) 64.12.57
Via Bartolini 29
20155 Milano   Tel. (02) 36.50.83
Editrice e Libreria Herder
Piazza Montecitorio 120
00186 Roma   Tel. 679.46.28
Telefax: 678.47.51
Libreria Hoepli
Via Hoepli 5
20121 Milano   Tel. (02) 86.54.46
Telefax: (02) 805.28.86
Libreria Scientifica
Dott. Lucio de Biasio 'Aeiou'
Via Coronelli, 6
20146 Milano   Tel. (02) 48.95.45.52
Telefax: (02) 48.95.45.48

JAPAN – JAPON
OECD Publications and Information Centre
Landic Akasaka Building
2-3-4 Akasaka, Minato-ku
Tokyo 107   Tel. (81.3) 3586.2016
Telefax: (81.3) 3584.7929

KOREA – CORÉE
Kyobo Book Centre Co. Ltd.
P.O. Box 1658, Kwang Hwa Moon
Seoul   Tel. 730.78.91
Telefax: 735.00.30

MALAYSIA – MALAISIE
Co-operative Bookshop Ltd.
University of Malaya
P.O. Box 1127, Jalan Pantai Baru
59700 Kuala Lumpur
Malaysia   Tel. 756.5000/756.5425
Telefax: 757.3661

MEXICO – MEXIQUE
Revistas y Periodicos Internacionales S.A. de C.V.
Florencia 57 - 1004
Mexico, D.F. 06600   Tel. 207.81.00
Telefax : 208.39.79

NETHERLANDS – PAYS-BAS
SDU Uitgeverij Plantijnstraat
Externe Fondsen
Postbus 20014
2500 EA's-Gravenhage   Tel. (070) 37.89.880
Voor bestellingen: Telefax: (070) 34.75.778

NEW ZEALAND
NOUVELLE-ZÉLANDE
Legislation Services
P.O. Box 12418
Thorndon, Wellington   Tel. (04) 496.5652
Telefax: (04) 496.5698

NORWAY – NORVÈGE
Narvesen Info Center – NIC
Bertrand Narvesens vei 2
P.O. Box 6125 Etterstad
0602 Oslo 6   Tel. (022) 57.33.00
                                    Telefax: (022) 68.19.01

PAKISTAN
Mirza Book Agency
65 Shahrah Quaid-E-Azam
Lahore 54000   Tel. (42) 353.601
                                    Telefax: (42) 231.730

PHILIPPINE – PHILIPPINES
International Book Center
5th Floor, Filipinas Life Bldg.
Ayala Avenue
Metro Manila   Tel. 81.96.76
                               Telex 23312 RHP PH

PORTUGAL
Livraria Portugal
Rua do Carmo 70-74
Apart. 2681
1200 Lisboa   Tel.: (01) 347.49.82/5
                               Telefax: (01) 347.02.64

SINGAPORE – SINGAPOUR
Gower Asia Pacific Pte Ltd.
Golden Wheel Building
41, Kallang Pudding Road, No. 04-03
Singapore 1334   Tel. 741.5166
                                    Telefax: 742.9356

SPAIN – ESPAGNE
Mundi-Prensa Libros S.A.
Castelló 37, Apartado 1223
Madrid 28001   Tel. (91) 431.33.99
                                    Telefax: (91) 575.39.98

Libreria Internacional AEDOS
Consejo de Ciento 391
08009 – Barcelona   Tel. (93) 488.30.09
                                    Telefax: (93) 487.76.59
Llibreria de la Generalitat
Palau Moja
Rambla dels Estudis, 118
08002 – Barcelona
               (Subscripcions) Tel. (93) 318.80.12
               (Publicacions) Tel. (93) 302.67.23
                               Telefax: (93) 412.18.54

SRI LANKA
Centre for Policy Research
c/o Colombo Agencies Ltd.
No. 300-304, Galle Road
Colombo 3   Tel. (1) 574240, 573551-2
                 Telefax: (1) 575394, 510711

SWEDEN – SUÈDE
Fritzes Information Center
Box 16356
Regeringsgatan 12
106 47 Stockholm   Tel. (08) 690.90.90
                                    Telefax: (08) 20.50.21

Subscription Agency/Agence d'abonnements :
Wennergren-Williams Info AB
P.O. Box 1305
171 25 Solna   Tel. (08) 705.97.50
                                    Téléfax : (08) 27.00.71

SWITZERLAND – SUISSE
Maditec S.A. (Books and Periodicals - Livres
et périodiques)
Chemin des Palettes 4
Case postale 266
1020 Renens   Tel. (021) 635.08.65
                                    Telefax: (021) 635.07.80

Librairie Payot S.A.
4, place Pépinet
CP 3212
1002 Lausanne   Tel. (021) 341.33.48
                                    Telefax: (021) 341.33.45

Librairie Unilivres
6, rue de Candolle
1205 Genève   Tel. (022) 320.26.23
                                    Telefax: (022) 329.73.18

Subscription Agency/Agence d'abonnements :
Dynapresse Marketing S.A.
38 avenue Vibert
1227 Carouge   Tel.: (022) 308.07.89
                                    Telefax : (022) 308.07.99

See also – Voir aussi :
OECD Publications and Information Centre
August-Bebel-Allee 6
D-53175 Bonn (Germany)   Tel. (0228) 959.120
                         Telefax: (0228) 959.12.17

TAIWAN – FORMOSE
Good Faith Worldwide Int'l. Co. Ltd.
9th Floor, No. 118, Sec. 2
Chung Hsiao E. Road
Taipei   Tel. (02) 391.7396/391.7397
                     Telefax: (02) 394.9176

THAILAND – THAÏLANDE
Suksit Siam Co. Ltd.
113, 115 Fuang Nakhon Rd.
Opp. Wat Rajbopith
Bangkok 10200   Tel. (662) 225.9531/2
                                    Telefax: (662) 222.5188

TURKEY – TURQUIE
Kültür Yayinlari Is-Türk Ltd. Sti.
Atatürk Bulvari No. 191/Kat 13
Kavaklidere/Ankara   Tel. 428.11.40 Ext. 2458
Dolmabahce Cad. No. 29
Besiktas/Istanbul   Tel. 260.71.88
                                    Telex: 43482B

UNITED KINGDOM – ROYAUME-UNI
HMSO
Gen. enquiries   Tel. (071) 873 0011
Postal orders only:
P.O. Box 276, London SW8 5DT
Personal Callers HMSO Bookshop
49 High Holborn, London WC1V 6HB
                               Telefax: (071) 873 8200
Branches at: Belfast, Birmingham, Bristol, Edin-
burgh, Manchester

UNITED STATES – ÉTATS-UNIS
OECD Publications and Information Centre
2001 L Street N.W., Suite 700
Washington, D.C. 20036-4910 Tel. (202) 785.6323
                            Telefax: (202) 785.0350

VENEZUELA
Libreria del Este
Avda F. Miranda 52, Aptdo. 60337
Edificio Galipán
Caracas 106   Tel. 951.1705/951.2307/951.1297
               Telegram: Libreste Caracas

Subscription to OECD periodicals may also be
placed through main subscription agencies.

Les abonnements aux publications périodiques de
l'OCDE peuvent être souscrits auprès des
principales agences d'abonnement.

Orders and inquiries from countries where Distributors have not yet been appointed should be sent to:
OECD Publications Service, 2 rue André-Pascal,
75775 Paris Cedex 16, France.

Les commandes provenant de pays où l'OCDE n'a
pas encore désigné de distributeur devraient être
adressées à : OCDE, Service des Publications,
2, rue André-Pascal, 75775 Paris Cedex 16, France.

6-1994